OFFSHORE ENGINEERING

AN INTRODUCTION

First Published 1995

Second Edition 2000

© Angus Mather 1995, 2000

ISBN 1 85609 186 4

WITHERBY

PUBLISHERS

All rights reserved

Printed in Great Britain by
Witherby & Co Ltd
32–36 Ayleshury Street
London EC1R 0ET, England

British Library Cataloguing in Publication Data

MATHER, ANGUS
Offshore Engineering – An Introduction, 2nd Edition
1. Title

ISBN 1 85609 186 4

All rights reserved. No part of this publication may be reproduced, stored in a retrieval system, or transmitted in any form or by any means, electronic, mechanical, photocopying, recording or otherwise, without the prior permission of the publisher and copyright owner.

While the principles discussed and the details given in this book are the product of careful study, the author and the publisher cannot in any way guarantee the suitability of recommendations made in this book for individual problems, and they shall not be under any legal liability of any kind in respect of or arising out of the form of contents of this book or any error therein, or the reliance of any person thereon.

Cover – The Britannia installation.
Picture reproduced with kind permission of Conoco (U.K.) Limited and Chevron U.K. Limited.

Offshore Engineering

An Introduction

Second Edition

By

Angus Mather

Published by
Witherby & Company Limited
32–36 Aylesbury Street, London EC1R 0ET
Tel No. 0207-251 5341   Fax No. 0207-251 1296
International Tel No. +44 207 251 5341   Fax No. +44 207 251 1296
E-mail: books@witherbys.co.uk
www: witherbys.com

*Tension Leg Platform (TLP), the Conoco Hutton
(now owned by Oryx).*

FOREWORD

I certainly felt honoured on being asked to write a foreword for such an impressive and comprehensive manual, covering, as it does, every aspect of todays offshore exploration and production — a section of the industry which hardly existed in the early 50's when my oilfield career commenced. Without a doubt this will be the essential reference source in the future.

Regarding earlier days, I seem to recall that a well was drilled approximately 1 mile off the coast of Borneo in 1952, believed to be the first major submarine operation in what, in those days, was known as the British Commonwealth. This particular well had a rather unique distinction: the sole means of transport to and from the shore, for personnel and equipment, was via an aerial ropeway! Some things have changed.

Today the offshore engineer is required to carry out an enormous range of tasks and a full appreciation of the industry is an essential prerequisite to success. This book should be on the bookshelf, if not a permanent part of every engineers baggage. Virtually every element of equipment used and each phase of the procedures, various specifications and relevant regulations etc. are covered in understandable terms. It will be as valuable to the new recruit to the industry as to the experienced engineer embarking on a new field of work.

Brendan G. McKeown O.B.E.
Norfolk
September 1999

Former Amoco Drilling Superintendent, Brendan McKeown remembers September 16th 1969 with great pride and considers himself to be privileged to be present at the beginning of one of the great British success stories of the 20th century, the discovery of North Sea Oil. How did he view it at the time? "Its difficult. You're an oil man, its your life's business and all of a sudden you are in on a first. Its like winning a cup winner's medal. It was marvellous."

The event Brendan is referring to is the discovery of the first commercial oil discovery in the North Sea, well 28/18-1, now known as the Arbroath field, located some 130 miles off the coast of Scotland.

Born and educated in Belfast, Northern Ireland, Brendan McKeown graduated from Queen's University in 1946 with a B.Sc. in Civil Engineering. Initially, Brendan was employed in Local Government at Stormont as a civil engineer before joining the Kuwait Oil Company. In 1956 Brendan undertook a post graduate diploma in Oil Technology at the Imperial College, and then spent the next 10 years *drilling* in Kuwait and Iran. From 1967 until his retirement, Brendan worked for Amoco in the role of Drilling Superintendent, Production Superintendent, Operations Manager, and finally as Production Manager. On retiring, Brendan was honoured by Her Majesty the Queen, receiving an OBE for *Services to the Oil Industry*.

THE AUTHOR

The author, Angus Mather commenced his career with the Ministry of Defence at Her Majesty's Dockyard, Devonport, in 1970 as a Technician Apprentice. This was followed by 10 years service as an Engineer Officer in the British Merchant Navy prior to entering the oil and gas industry. The next 15 years were spent employed both in the contracting industry as Engineering Manager of Kvaerner Oil and Gas Services, and as an Engineer Surveyor for Lloyds Register of Shipping, involved with oil and gas related projects, both on and offshore, in the UK, the Middle East and the Gulf of Mexico.

PREFACE

Offshore engineering encompasses a considerable number of very specialized and often completely unrelated disciplines. They can be categorised into three core activities, namely Construction, Production and Reservoir Engineering and this book has been written, not as a definitive manual but to provide the reader with a basic explanation of these various activities.

It is hoped that the material contained within this publication will provide the new recruit to the industry with a basic appreciation of what is a relatively complex subject, whilst at the same time providing the more experienced individual with a fuller appreciation of activities outside of their own particular speciality. The decision as to what topics should be included and the depth to which they should be discussed are based largely on the authors personal experience of what information is required to create an overall picture.

Whilst frequent references are made to the oil and gas industries of the North Sea the bulk of the text is of a more general nature and thus applicable to offshore engineering on a world-wide basis. Units of measurement are quoted in both imperial and metric with preference being given to the unit most frequently associated with a particular discipline.

Angus C. Mather
Suffolk
March 2000

ACKNOWLEDGEMENTS

The author acknowledges with thanks the following Companies for their contributions and cooperation in the preparation of this book.

ABB Offshore Technology
Amerada Hess Limited
Amoco (U.K.) Exploration Company
ARCO British Limited
Baker Service Tools
Bettis Actuators and Controls
British Gas plc
BK Kendle Engineering Limited
BP Exploration
Camco Products and Services
Chevron U.K. Limited
Claxton Engineering Services Limited
Conoco (U.K.) Limited
Cooper Oil Tool
Dockwise N.V.
Framo Engineering AS
Halliburton Energy Services
Hydril Inc.
Hydrovision Limited
Institute of Marine Engineers
Institute of Petroleum
Kongsberg Offshore a.s.
Kvaerner Surveys
Lloyd's Register of Shipping
Measurement Technology Limited
Oilfield Publications Limited
RGB Stainless Limited
Schlumberger Oilfield Services
Shaffer Inc.
Shell (U.K.) Exploration and Production
Slingsby Engineering Limited
SLP Engineering Limited
SSR (International) Limited
Statoil
Stena Offshore Limited
Sun Oil Britain Limited
Thomas Telford Limited (Offshore Engineer)
Varco International Inc
Weir Pumps Limited

Photographic acknowledgements:
Charles Hodge Photography, Lowestoft
Tony Wilks
David Gold
Richard Loughery
John Price

CONTENTS

	Page
Foreword	v
The Author	vi
Preface	vii
Acknowledgements	viii
Abbreviations	xv
Glossary	xxi
Standards, Guidance Notes and Codes of Practise	xxvii
Table of Colour Plates	xxxi

Chapter One – Offshore Structures and Support Vessels — 1

Part 1 – Offshore Structures — 3

Fixed Steel Structures – Concrete Gravity Base – Tension Leg Platform – Semi-submersible Vessel – Floating Production Systems – Shuttle Tanker – Self-elevating Jack-up – Single Point Mooring

Part 2 – Support Vessels — 19

Offshore Supply Boat – Stand-by Vessels

Part 3 – Offshore Installations –Description — 23

Accommodation – Wellheads – Spider Deck – Conductors – Caisson etc

Part 4 – Fixed Steel Structures – Installation — 26

Introduction – Sub-sea Template – Jacket – Topside Installation – Abandonment

Chapter Two – The North Sea History and Legislation — 33

Section One – The North Sea

Part 1 – The Story so Far — 35

History – Gas fields – Oil fields – World Statistics – Field Developments – The Future

Part 2 – Oil and Gas Distribution — 43

Oil Distribution – Gas Distribution

Section Two – Legislation

Part 1 – History 45

Introduction – Initial Legislation – Mineral Workings Act – Health and Safety at Work Act – Certification – Annual Surveys – Major Surveys – Certification – Certifying Authority – Certificate of Fitness – The Cullen Inquiry

Part 2 – New Legislation 50

Summary of Applicable Legislation – The Safety Case – Goal Setting – Verification – Summary

Part 3 – The Safety Case 53

The Safety Case Regulations – Safety Management System – Risk Assessments – Verification Schemes – The Verification Process – Examination Process – Independent and Competent Persons – Records and Review of the Scheme – Responsibility

Part 4 – Statutory Instruments 63

Safety Case Regulations – Amendments to The Safety Case – Prevention of Fire, and Explosion and Emergency Response – Design and Construction – Management and Administration – Pipeline Safety Regulations – Provision and Use of Work Equipment Regulations

Part 5 – Associated Information 93

The Health and Safety Executive – UKOOA

Chapter Three – Safety Systems 97

Part 1 – The Safety Case 99

Statutory Operations Manual – The Safety Case

Part 2 – Fire-fighting Equipment 103

History-UK Continental Shelf – Examination and Testing – Active Fire Protection – Passive Fire Protection

Part 3 – Life-Saving Appliances 119

History-UK Continental Shelf – General Requirements – Life-saving Equipment

Part 4 – Navigational Aids 126

Visual Navigational Aids – Audible Navigational Aids – Navigation and Failure – Heli-deck Illumination – Identification Panels

Part 5 – Hazardous Areas 130

Non-Hazardous Areas – Hazardous Areas – Enclosed Compartments – Hazardous Area Equipment – Commissioning, Inspection, Maintenance, Testing and Repair – Explosion Proof Electrical Equipment

Part 6 – Emergency Systems 141

Emergency Support Systems – Emergency Shutdown Systems – Fire Loop – Emergency Pipeline Valves

Chapter Four — Piping Systems and Process Pressure Vessels   147

Part 1 — Piping Systems   149

Hydrocarbon Process — Process Gas — Condensate — Corrosion Inhibitor — Hydrate Inhibitor — Fuel Gas — Vent — Flare — Utility — Diesel Oil — Compressed Air — Fire Water — Drains — Sodium Hypochlorite

Part 2 — Process Pressure Vessels   155

Production Separators — Slug Catchers — Knock Out Drums — Oil/Water Separators — Skimmers — Coalescers — Hydrocylones — Holding Tanks

Part 3 — Piping and Pressure Vessel Design   161

Piping Systems Design — Pressure Vessels — Installation and Layout — Specifications

Part 4 — Piping Systems — Construction   165

Pipe Schedules — Fittings — Valve Types — Class Ratings — Specifications

Chapter Five — Production   175

Part 1 — Gas Production   179

Gas Process — Wellheads — Slug Catchers — Production Header — Gas Compression — Liquid Process — Condensate System — Produced Water System — Wet Gas

Part 2 — Oil Production   187

Oil Processing — Associated Gas — The Gas Process — The Dehydration Process — Natural Gas Liquids — Stabilisation — Sweetening

Part 3 — Enhanced Oil Recovery and the Oil Drive Mechanism   199

Oil Drive Mechanisms — Enhanced Oil Recovery — Water Injection — Gas Injection — Gas Lift — Deepwell/Submergible Pumps

Chapter Six — Underwater Engineering   209

Part 1 — Diving   211

Air Diving — Saturation Diving — Equipment — Diving Support Vessels — Remotely Operated Vessels

Part 2 — Underwater Surveys   217

Splash Zone Examination — Swimaround Survey — Non-Destructive Examination — Flooded Member Survey — Marine Growth Measurement — Scour Survey — Cathodic Protection Examination — Differential Settlement Survey — Air Gap Measurement

Part 3 — Sub-sea Wells — 221

Subsea Development Options — Subsea Wells — Manifolds and Templates — Sub-sea Flowbase — Pipelines and Risers — Sub-sea Wells — Christmas Trees and Wellheads — Diverless Wellhead Systems — Guideframes — Wellhead Protection Structure

Part 4 — Sub-sea Developments — 229

Sub-sea Separation — Multiphase Pumping and Metering — Sub-sea Pressure Boosting

Chapter Seven — Drilling — 237

Part 1 — Introduction — 241

The Formation of Oil and Gas — Exptoration — Seismic Surveys — Drilling Programme

Part 2 — The Well — 245

Well Construction — Intermediate Casing — Completions — Cement Job

Part 3 — Equipment — 252

Drilling Derrick — Rotary Table — Top Drive — Drill String — Drilling Mud

Part 4 — Operations — 259

Drill Crew — Making Hole — Making a Trip

Part 5 — Well Control Equipment — 263

Diverter — Blowout Preventer (BOP) — BOP Operations

Part 6 — Drill Ship Equipment — 269

Floating Drill Ship Equipment

Part 7 — Deviated Wells — 275

Directional Drilling — Horizontal Wells — Geosteering — Measurement While Drilling — Mud Pulse Telemetry — Multi-Lateral Wells

Chapter Eight — The Well Component Parts — 285

Part 1 — Christmas Tree — 287

Surface Trees — Wellhead Control Panel — Gate Valves and Actuators — Horizontal Trees

Part 2 — Surface Wellhead — 291

Base Plate/Landing Ring — Starting Head — Casing Head Spools — Tubing Head Spools — Christmas Tree

Part 3 — Mudline Suspension Systems — 295

Pre-drilling of Wells — Sub-sea Wells

Part 4 — Mudline Safety Valve — (SCSSV) — 297

Sub-Surface Safety Valves — Tubing Retrievable — Wireline Retrievable

Part 5 — Production Packer — 299

Chapter Nine — Well Maintenance — 301

Part 1 — Wireline Operations — 303

Wireline Equipment — Lubricators — Wireline Tool String — Wireline Operations

Part 2 — Workover Operations — 311

Well Stimulation — Fracturing — Acidizing — Production Tubing Modifications

Appendix I — Structural Steel — 315

Primary and Secondary Steels — Material Specifications — Material Properties

Appendix II — Welding — 319

Terminology — Welding Specifications — Supervision

Appendix III — Non-Destructive Examination — 325

Surface Inspection Techniques — Volumetric Inspection Techniques — NDE Operator Qualifications — Non-Destructive Examination Specifications

Appendix IV — Units of Measurement — 328

Appendix V — Table of Line Pipe Dimensions — 329

Index — 331

ABBREVIATIONS

ABS	American Bureau of Shipping
AC	Alternating current
ACoP	Approved code of practise
ACFM	Alternating Current Field Measurement
AIT	Auto ignition temperature
ALARP	As low as reasonably practicable
ANSI	American National Standards Institute
APAU	Accident Prevention Advisory Unit (of HSE)
API	American Petroleum Institute
ARPA	Automatic Radar Plotting Aid
ASME	American Society of Mechanical Engineers
ASNT	American Society for Non-Destructive Testing
ASTM	American Society for Testing and Materials
AWS	American Welding Society
BA	Breathing apparatus
BASEEFA	British Approval Service for Electrical Equipment in Flammable Atmospheres
BHAB	British Helicopter Advisory Board
BOP	Blowout preventer
BPD	Barrels per day
BS	British Standard
BSI	British Standards Institute
BV	Bureau Veritas
CA	Certifying Authority
CAA	Civil Aviation Authority
CAP	Civil Aviation Publication
CALM	Catenary anchor leg mooring
CAPEX	Capital expenditure
CE	Carbon Equivalent
CE	Community European
CHAOS	Consequences of hazards and accidents on offshore structures
CIMAH	Control of Industrial Major Accident Hazards
CoF	Certificate of Fitness
COSHH	Control of Substances Hazardous to Health
CP	Cathodic protection
CRINE	Cost reduction in the new era
CSON	Continental Shelf Operations Notice
CSWIP	Certification scheme for weld inspection personnel
CTOD	Crack tip opening displacement
DC	Direct current
DCR	Design and Construction Regulations
DDC	Deck decompression chamber
DEG	Duoethylene glycol
DEn	Department of Energy
DIN	Deutsches Institut fur Normung
DNV	Det Norske Veritas

DoT	Department of Transport
DP	Design pressure
DP	Dynamic positioning
DSM	Diving safety memorandum
DSV	Diving support vessel
DT	Design temperature
EECS	Electrical Equipment Certification Service
EEMUA	Engineering Equipment Material Users Association
EERA	Evacuation, escape and rescue analysis
EOR	Enhanced oil recovery
EPC	Engineer, procure and construct
EPIC	Engineer, procure, install and construct
EPIRB	Emergency position indicating radio beacon
EPS	Emergency power supply
ESD	Emergency shutdown
ESDV	Emergency shutdown valve
FARSI	Functionality, availability, reliability, survivability and interaction.
FEA	Fire and explosion analysis
FES	Fire and explosion strategy
F&G	Fire and gas
FPS	Floating production system
FPSO	Floating, production, storage and offloading vessel
FPDSO	Floating, production, drilling, storage and offloading vessel
FPV	Floating production vessel
FRAMS	Floating riser and mooring system
FRA	Fire risk analysis
FRC	Fast rescue craft
FSA	Formal safety assessment
FSO	Floating storage and offloading vessel
FSU	Floating storage unit
GA	General alarm
GBS	Gravity base structure
GL	Germanischer Lloyd
HAZ	Heat affected zone
HAZAN	Hazard analysis
HAZOP	Hazard and operability study
HF	High frequency
HGT	High pressure grease tube
HIPS	High integrity protection system
HP	High pressure
HPHT	High pressure and high temperature
HSC	Health and Safety Commission
HSE	Health and Safety Executive
HSWA	Health and Safety at Work
HVAC	Heating, ventilating and air conditioning
ICP	Independent, Competent Persons
IMO	International Maritime Organization

IP	Institute of Petroleum
ISO	International Standards Organisation
JT	Joule Thompson
KO	Knock out
LAT	Lowest atmospheric tide
LCV	Level control valve
LED	Light emitting diode
LEL	Lower explosive limit
LIC	Level indicator controller
LNG	Liquefied natural gas
LOLER	Lifting Operations and Lifting Equipment Regulations
LP	Low pressure
LPG	Liquefied petroleum gas
LRS	Lloyd's Register of Shipping
LSA	Life saving appliances
MAPD	Major accident prevention document
MAR	Management and Administration Regulations
MCA	Maritime and Coastguard Agency
MEG	Monoethylene Glycol
MHAU	Major Hazards Assessment Unit (HSE)
MHSWR	Management of Health and Safety at Work Regulations
MIG	Metal inert gas, welding
MMSCF	Millions standard cubic feet
MMSCFD	Millions of standard cubic feet a day
MLSS	Mudline suspension system
MLSV	Mudline safety valve
MPI	Magnetic particle inspection
MSA	Marine Safety Agency (UK), now the MCA
MSS	Manufacturers Standardization Society of the Valve and Fittings Industry, Inc. (USA)
MWA	Mineral Workings Act
MWP	Maximum working pressure
NACE	National Association of Corrosion Engineers (USA)
NAMAS	National Material Accreditation Service
NB	Nominal bore
NDE	Non-destructive examination
NDT	Non-destructive testing
NEL	National Engineering Laboratory
NRV	Non-return valve
NUI	Normally unattended installation
OCB	Offshore Certification Bureau
OIM	Offshore Installation Manager
OPEX	Operating expenditure
OSD	Offshore Safety Division
OWS	Oily water separator
PA	Public address

PCN	Personnel certification in non-destructive testing
PCV	Pressure control valve
PED	Petroleum Engineering Division
PFEER	Prevention of Fire and Explosion, and Emergency Response Regulations
PG	Plate girder
PM	Preventative maintenance
POB	Persons on board
PPA	Petroleum Productions Act
PPE	Personal protective equipment
PPM	Planned Preventative Maintenance
PS	Performance standard
PSPA	Petroleum and Submarine Pipe-lines Act
PSV	Pressure safety valve
PTW	Permit to work
PUWER	Provision and Use of Work Equipment Regulations
PWHT	Post weld heat treatment
QA	Quality assurance
QC	Quality control
QMS	Quality management system
QRA	Quantitative risk assessment
QRS	Quantitative risk study
RGIT	Robert Gordon Institute of Technology
RIV	Rapid intervention vessel
RNLI	Royal National Lifeboat Institution
ROV	Remotely operated vehicle
RT	Radiographic testing
RTJ	Ring type joint
SALM	Single anchor leg mooring
SAR	Search and rescue
SAW	Submerged arc welding
SBM	Single buoy mooring
SBV	Standby vessel
SCADA	Supervisory control and data acquisition system
SCC	Surface compression chamber
SCE	Safety-critical element
SCR	Safety Case Regulations
SCSSV	Surface controlled subsurface safety valve
SI	International System (of Units)
SI	Statutory Instrument
SMS	Safety management system
SOLAS	Safety of Life at Sea
SPM	Single point mooring
SPS	Surface process shutdown
SSIV	Subsea isolation valve
SSSV	Surface controlled subsurface safety valve
SWIS	Site welding instruction sheet
TEG	Triethylene glycol
TEMPSC	Totally enclosed motor propelled survival craft

TIG	Tungsten inert gas, welding
TLP	Tension leg platform
TLQ	Temporary living quarters
TP	Test pressure
TPS	Total platform shutdown
TR	Temporary refuge
TRA	Total risk analysis
TSR	Temporary safe refuge
UEL	Upper explosive limit
UHF	Ultra high frequency
UKCS	United Kingdom Continental Shelf
UKOOA	United Kingdom Offshore Operators Association Ltd.
UL	Underwriters Laboratory (USA)
UNS	Universal numbering system
UPS	Uninterrupted power supply
UT	Ultrasonic testing
UV	Ultra violet
VDU	Visual display unit
VHF	Very high frequency
WPQR	Welding procedure qualification record
WPS	Welding procedure specification
WSE	Written Scheme of Examination

GLOSSARY

Whilst every effort has been made to fully explain the terminology and expressions used within the text of the book, there follows a list of terms designed to further clarify the more ambiguous references, and to provide a brief definition of some of the more frequently encountered expressions.

Accommodation space — Any room used for eating, sleeping, cooking, laundry or recreation, or as an office or sick bay, and any corridor giving access to any of these rooms, and any storeroom in the vicinity of any of these rooms.

Accumulator — A pressure vessel charged with nitrogen gas and used to store hydraulic fluid under pressure for the operation of hydraulic valve actuators.

Actuator — A mechanical device for the remote or automatic operation of a valve or choke.

Annulus — Space between concentric casing strings.

Associated gas — Naturally occurring reservoir gas found in association with oil, either dissolved in the oil or found as a cap or pocket of free gas above the oil.

Azimuthing thruster — Rotatable ducted propeller used in conjunction with a dynamic positioning system to enable ships to maintain position without the use of anchors.

Block — The subdivided areas of the sea for the purpose of licensing to a company for exploration or production rights. In the UK a block is one thirtieth of a quadrant (one degree by one degree) and is approximately 200–250 square kilometres.

Blowout — Uncontrolled release of well fluids from the well bore during drilling operations.

Bell nipple — Receptacle attached to the top of the blowout preventer or marine drilling riser which directs the drilling mud returns to the shale shaker or mud pits.

BOP — Device attached to the casing head during drilling operations that allows the well to be sealed to confine the well fluids in the well bore.

Brown book — The Department of Trade and Industry's annual publication contains a host of facts and figures relevant to oil and gas production in the UK and is available from Her Majesty's Stationary Office.

Bubble point — Point at which dissolved gasses begin to vaporise from a liquid. Dependant on temperature, pressure and gas/liquid composition.

Caisson — Length of pipe extending vertically downwards from an installation into the sea as a means of disposing of waste waters, or for the location of a sea water pump.

Casing — Pipe used to line and seal the well and prevent collapse of the borehole. A number of casing strings (lengths) are used in ever decreasing diameters.

Cathodic protection — Corrosion protection system which relies on sacrificial anodes or impressed current to protect submerged steel components from corrosion by electrolytic action.

Condensate — Volatile liquid consisting of the heavier hydrocarbon factions that condense out of the gas as it leaves the well, a mixture of pentanes and higher hydrocarbons.

Conductor — The first, and largest diameter pipe to be inserted (spudded) into the seabed when drilling a well. It keeps the hole open, provides a return passage for the drilling mud and supports the subsequent casing strings.

Choke — A valve like device with a fixed or variable aperture specifically intended to regulate the flow of fluids.

Christmas Tree — An assembly of valves attached to the wellhead and used to control well production.

Crossover — Item used to connect one component to another differing in size, thread type or pressure rating.

Dew point — Temperature at which liquids condense from a gas.

Duster — Dry well drilled during exploration.

Dynamic positioning — Satellite monitoring system used to control the action of thruster propellers to maintain a vessel on location without deploying anchors.

Emergency shutdown valve (ESDV) — An automatically operated, normally open, valve used for isolating a sub-sea pipeline.

Enhanced oil recovery — Means used to assist in the extraction of oil either by installing equipment into the production tubing or by injecting water or gas into the reservoir.

Fire loop — A pneumatic control line containing temperature sensing elements (fusible plugs, synthetic tubing, etc.) which, when activated, will initiate a platform shutdown.

Fiscal metering — The measurement of oil, gas or condensate flow rate for taxation purposes.

Flare stack — Elevated tower containing a pipe used for the discharge and burning of waste gases.

Flash drum — Pressure vessel used to lower the pressure of oils and other liquids involved with the production process in order to encourage the vaporization of dissolved gases.

Flowline — Piping which directs well fluids from wellheads to manifold or first process vessel.

Fluid — A generic term meaning a gas, vapour, liquid or combinations thereof.

Grout — Mixture of cement and water (no sand) used to secure and seal attachments such as piles into jacket legs.

HIPPS — High Integrity Pipeline Protection System, a pressure sensing system with voting logic that activates a fast acting isolating valve to protect pipelines which are not designed to withstand maximum upset pressure conditions.

Header — That part of a manifold which directs fluid to a specific process system.

Hydrate — Solid, ice like compound consisting of molecules of water and hydrocarbon gasses.

Hydrocylone — Separation device utilising centrifuging principles to remove oils from water, or as a multiclycone to remove liquids and solids from a gas stream.

Heat affected zone (HAZ) — That potion of the base metal which has not been melted, but whose mechanical properties or microstructure has been affected by the heat generated during the welding process.

Intrinsically safe — Electrical equipment which is incapable of igniting a flammable gas mixture or combustible materials.

Installation — May be fixed or mobile and used directly or indirectly for the exploration or production of mineral resources.

Installation, fixed — A fixed offshore structure involved in the production of oil or gas which may be constructed of steel or concrete. Term used frequently in the UK to describe an offshore installation.

Jacket — Steel support framework used to support platform topsides.

J-T valve — Throttle valve used to reduce the pressure and temperature of a gas stream, associated with the NGL removal process.

Knock out — Removal of liquids from a gas stream within a pressure vessel.

Lower explosive limit (LEL) — The lowest concentration by volume of combustible gasses in mixture with air that can be ignited at ambient temperature conditions.

LNG — Liquefied natural gas, gaseous at ambient temperatures and pressures but held in the liquid state by very low temperatures to facilitate storage and transportation in insulated vessels.

LPG — Liquefied petroleum gas, essentially propane and butane held in the liquid state under pressure to facilitate storage and transportation.

Manifold — An assembly of pipes, valves, and fittings by which fluid from one or more sources is selectively directed to various process systems.

Marine drilling riser — Pipe extending from the blowout preventer on the seabed to the drilling rig on the surface, to permit the return of the drilling mud.

Microwave — High frequency multi-channel radio communications system designed to carry information between two points linked by line of sight transmission.

Mobile installation — One which can be moved from place to place without major dismantling or modification.

Module — Self-contained liftable package forming part of the topside facilities of an offshore installation. e.g. accommodation module, compressor module, drilling module etc..

Mudline — Seabed.

Multilateral — Multiple boreholes drilled from an existing single bore well.

Multiphase — Practise of flowing unstabilised well fluids (oil with high gas content) in a single pipeline by boosting the pressure to prevent vaporisation of the dissolved gasses.

Natural gas — Hydrocarbon gas occurring naturally from underground reservoirs both on and offshore.

NGL — Natural gas liquids, a mixture of hydrocarbon liquids which include butane and ethane obtained from natural gas. May be produced from condensate reservoirs but more probably produced as a by-product of oil production.

Nipple — A section of threaded or socket welded pipe used as an appurtenance that is less than 12 inches in length. Often used to describe any short length of open ended pipe.

Nozzle — Flanged inlet or outlet connection on a pressure vessel.

Packer — Device for sealing one casing string from another, or from the production tubing.

Pedestal — Large diameter, vertical tube or tub onto which a crane is attached.

Photogrammetric — The use of still photography to capture dimensional information for transposing into drawings.

Pig — Spherical device inserted in to a gas subsea pipeline to sweep the line of deposits of rust, scale and condensed liquids. May also be used to clean oil pipelines of wax and may be "intelligent", that is containing measuring and inspection equipment.

Pig trap — A pressure chamber permitting the entry or removal of equipment into the subsea pipeline, normally pigs.

Pipeline — Piping used to convey fluids between platforms or between a platform and a shore facility.

Pipe spool — Single length of pipe with flanged ends.

Platform, offshore — A fixed offshore structure involved in the production of oil or gas which may be constructed of steel or concrete. Term used frequently by Americans to describe an offshore installation.

Pressure vessel — Container, normally cylindrical used to contain internal, or occasionally external pressure.

Produced water — Formation water removed from the oil and gas in the process pressure vessels.

Production separator — Main process vessel used primarily for the separation of gas, oil (and condensate) and water.

Production tubing — Pipe used in wells to conduct fluid from the producing formation into the Christmas tree. Unlike the casing the tubing is designed to be replaced during the life of the well, if required.

Purge — Maintain gas flow in an over rich, or lean concentration so as to avoid the build-up of oxygen and an explosive mixture.

Quality assurance — A sequence of planned and systematic actions necessary to provide adequate confidence that a product or service will satisfy given requirements of quality.

Quality control — The operational techniques and activities that are used to ensure that a quality product or service will be produced.

Rig — A term normally associated with drilling equipment, that is to say a drilling rig. Also a slang term used extensively to describe any of the structures and vessels associated with oil and gas exploration and production.

Riser — The vertical portion of a subsea pipeline (including the bottom bend) arriving on or departing from a platform.

Rock dumping — Deposition of rocks onto subsea pipelines to provide protection against anchors and fisherman's nets when burying the pipe is impractical. Rocks and gravel may also be dumped around subsea wellheads and jacket legs to repair scour damage.

Seismic survey — The use of artificially generated sound waves to determine the type of rock formations below the ground or under the seabed by monitoring the reflected sound wave signals.

Scour — Removal of the seabed in the vicinity of a jacket, subsea wellhead or pipeline by tidal action.

Scrubber — Pressure vessel containing equipment designed to remove or scrub liquids from a gas stream.

Shuttle tanker — Moderate sized oil tanker used to transport oil from larger vessels into port.

Skid — Steel framework used to contain equipment, may be transportable.

Stress corrosion cracking — The cracking which results from a combination of stress and corrosion.

Slew ring bearing — Large ball or roller bearing which connects crane to pedestal and permits rotation.

Slug — An accumulation of water (may also be sand or condensate) in a gas pipeline.

Spudding — A term used to describe the insertion of the conductor into the seabed when drilling a well. May also be used to describe the process of setting the legs of a jack-up into the seabed.

Stinger — Tubular steel support frame attached to the stern of a pipelay barge to control the bending of the pipe as it enters the water.

Stripping — The removal or replacement of drill pipe or tubing strings from a well under pressure using a stripping BOP.

Stripping gas — Gas, normally process gas used to assist in the purification of a liquid by reducing the partial pressure of gaseous contaminants to encourage vaporisation.

Swabbing — The lowering of the hydrostatic pressure in the hole due to the upward movement of the drill pipe and/or tools. Also the use of wireline equipment to clean a well by scooping out liquids.

Sulphide stress cracking — Cracking of metallic materials due to exposure to fluids containing hydrogen sulphide.

Telemetry — System for the collection, collation and transmission of information to a remote source using radio, satellite, fibre optics or cable links. Also associated with the remote control of process equipment.

Third party gas — Term used to describe gas sold direct from oil company to parties other than British Gas, the previous monopoly holders.

Throttle — Regulation of fluid flow by a throttling valve or fixed orifice.

Topsides — Upper part of a fixed installation which sits on top of the jacket and consists of the decks, accommodation and process equipment.

Vam — Trade name for casing thread produced by the Vallourec company of France.

Vent — A pipe or fitting on a vessel that can be opened to the atmosphere.

Vent stack — Open ended pipe and support framework used to discharge vapours into the atmosphere at a safe location above the installation, without combustion.

Wellhead — Permanent equipment used to secure and seal the casings and production tubing and to provide a mounting place for the Christmas trees.

Wireline — Equipment used to introduce tools into the well bore under pressure.

Workover — Re-entry into a completed well for modification or repair work.

Workover rig — Normally a smaller, portable version of the main drilling derrick which can be used to carry our work over operations on installations which do not have a permanent derrick.

STANDARDS, GUIDANCE NOTES AND CODES OF PRACTICE

Provided herein is a list of standards and specifications, most of which are referenced in this book. The list is not exhaustive but identifies the standards and guidance notes which are in most frequent use within the offshore industry, and which provide valuable reference material.

BRITISH STANDARDS

BS 2600	Radiographic examination of fusion welded butt joints in steel. (withdrawn).
BS 2910	Radiographic examination of fusion welded circumferential butt joints in steel pipes. (withdrawn).
BS 3351	Specification for piping systems for petroleum refineries and petrochemical plants. (withdrawn).
BS 3923	Ultrasonic examination of welds. (withdrawn).
BS 4137	Guide to selection of equipment for use in division 2 areas.
BS 4360	Weldable structural steels. Superseded by BS 7668, BS EN 10210-1, BS EN 10155 and BS EN 10155.
BS 4416	Penetrant testing of welded or brazed joints in metals.
BS 4515	Welding of steel pipelines on land and offshore.
BS 4683	Electrical apparatus for explosive atmospheres.
BS 4870	Specification for approval testing of welding procedures.
BS 4871	Specification for approval testing of welders working to approved welding procedures.
BS 5135	Metal arc welding of carbon and manganese steels.
BS 5169	Specification for fusion welded air receivers.
BS 5289	Code of practice. Visual inspection of fusion welded joints. (withdrawn).
BS 5345	Classification of hazardous areas and selection of equipment for use in hazardous areas.
BS 5490	Specification for degrees of protection provided by enclosures.
BS 5500	Unfired fusion welded pressure vessels.
BS 5501	Electrical apparatus for potentially explosive atmospheres.
BS 5750	Quality systems.
BS 6072	Methods for magnetic particle flaw detection.
BS 6443	Penetrant flaw detection. (withdrawn).
BS 6755	Part 2 Specification for fire type testing requirements (valves).
BS 8010	Code of practice for pipelines.

Offshore Engineering

BS EN 287	Approval testing of welders for fusion welding.
BS EN 288	Specification and approval of welding procedures for metallic materials
BS EN 571-11997	Non-destructive testing. Penetrant testing.
BS EN 9701997	Non-destructive examination of fusion welds. Visual examination.
BS EN 14351997	Non-destructive examination of welds. Radiographic examination of welded joints
BS EN 17141998	Non-destructive examination of welded joints. Ultrasonic examination of welded joints.

EUROPEAN EQUIPMENT MANUFACTURERS AND USERS ASSOCIATION — EEMUA

EEMUA No. 107	Recommendations for The Protection of Diesel Engines in Zone 2 Hazardous Areas.
EEMUA No. 150	Steel Specification for Offshore Structures.
EEMUA No. 158	Construction Specification for Fixed Offshore Structures in the North Sea.

GUIDANCE NOTES PROVIDED BY THE HEALTH AND SAFETY EXECUTIVE (HSE)

Prevention of fire and explosion, and emergency response on Offshore Installations. Approved Code of Practise (ACoP) and Guidance

A guide to the Offshore Installations (Safety Case) Regulation 1992.

A guide to the Offshore Installations and Pipeline Works (Management and Administration) Regulation 1995.

A guide to the integrity, workplace environment and miscellaneous aspects of the Offshore Installations and Wells (Design and Construction etc.) regulations 1996.

A guide to the well aspects of the Offshore Installations and Wells (Design and Construction etc.) regulations 1966.

A guide to the Pipelines Safety Regulations 1996.

Provision and Use of Work Equipment Regulations 1992. Guidance on Regulations.

The under listed Department of Energy guidance notes have been withdrawn but are still widely used as reference documents by the offshore industry:

Department of Energy Offshore Installations: Guidance on Design Construction and Certification.

Department of Energy Offshore Installations: Guidance on Fire Fighting Equipment.

Department of Energy Offshore Installations: Guidance on Life Saving Appliances.

Department of Energy Offshore Installations: Guidance on Emergency Pipe-Line Valve.

GUIDANCE NOTES — PETROLEUM INDUSTRY

Institute of Petroleum Code of Safe Practice Part 1, 1965.

Guidance on Permit to Work Systems in the Petroleum Industry (Health and Safety Commission/Oil Industry Advisory Committee, 1991).

AMERICAN STANDARDS — VARIOUS

ANSI/ASME B31.3	Chemical Plant and Petroleum Refinery Piping.
ANSI B16.5	Steel Pipe Flanges and Flanged Fittings.
ANSI B16.9	Wrought Steel Butt Welded Fittings.
ANSI B16.11	Forged Steel Fittings, Socket Welded and Threaded.
ASME IX	Boiler and Pressure Vessel Code.
AWS D1.1	Structural Welding Code.
MSS SP-44	Steel Pipe Line Flanges.
MSS SP-75	Specification for High Test Wrought Butt Welding Fittings.
NACE MR-01-75	Sulphide Stress Cracking Resistant Material for Oilfield Equipment.
NACE RP-01-76	Corrosion Control on Steel, Fixed Offshore Platforms Associated with Petroleum Production.

AMERICAN PETROLEUM INSTITUTE (API) SPECIFICATIONS AND RECOMMENDED PRACTISES

API Spec 5L	Specification for Line Pipe.
API Spec 6A	Specification for Wellhead and Christmas Tree Equipment.
API Spec 6AF	Specification for Fire Test for Valves.
API Spec 14A	Specification for Subsurface Safety Valve Equipment
API Spec 14D	Specification for Wellhead Surface Safety Valves and Underwater Safety Valves for Offshore Installations.
API Std 1104	Standard for Welding Pipelines and Related Facilities.
API RP 2A	Recommended Practise for Planning, Design and Construction of Fixed Offshore Platforms.
API RP 2G	Recommended Practice for Production Facilities on Offshore Structures. (now discontinued)
API RP 6AR	Recommended Practise for Repair and Re-manufacture of Wellhead and Christmas Tree Equipment.
API RP 14B	Recommended Practise for Design, Installation, Repair and Operation of Subsurface Safety Systems.
API RP 14C	Recommended Practise for Analysis, Design, Installation and Testing of Basic Surface Safety Systems for Offshore Production Platforms.
API RP 14D	Recommended Practise for Wellhead Surface Safety Valves and Underwater Safety Valves for Offshore Service.
API RP 14E	Recommended Practise for Design and Installation of Offshore Production Platform Piping Systems.
API RP 14F	Recommended Practise for Design and Installation of Electrical Systems for Offshore Production Platforms.

0API RP 14G	Recommended Practise for Fire Prevention and Control on Open Type Offshore Production Platforms.
API RP 14H	Recommended Practise for Installation, Maintenance, and Repair of Surface Safety Valves and Underwater Safety Valves Offshore.
API RP 14J	Recommended Practise for Design and Hazard Analysis for Offshore Production Platforms.
API RP 16E	Recommended Practise for Design of Control Systems for Drilling Well Control Equipment.
API RP 53	Recommended Practise for Blowout Prevention Equipment Systems for Drilling Wells. (now discontinued).
API RP 64	Recommended Practise for Diverter Systems, Equipment and Operation.
API RP 500	Recommended Practice for Classification of Areas for Electrical Installations at Drilling Rigs on Land and on Marine Fixed and Mobile Platforms.
API RP 520	Recommended Practice for the Design and Installation of Pressure Relieving Systems in Refineries — Parts I and II.
API RP 521	Guide for Pressure and Depressuring Systems.
API RP 1111	Recommended Practice for Design, Construction, Operation and Maintenance of Offshore Hydrocarbon Pipelines.

AMERICAN SOCIETY FOR TESTING AND MATERIALS (ASTM) STANDARDS

ASTM A105	Specification for Forgings, Carbon Steel, for Piping Components.
ASTM A106	Specification for Seamless Carbon Steel Pipe.
ASTM A193	Specification for Alloy-Steel and Stainless Steel Bolting Materials for High-Temperature service.
ASTM A194	Specification for Carbon and Alloy Steel Nuts for Bolts for High-Pressure and High-Temperature Service.
ASTM A234	Specification for Piping Fittings of Wrought Carbon Steel and Alloy Steel for Moderate and Elevated Temperatures.
ASTM A312	Specification for Seamless and Welded Austenitic Stainless Steel Pipe.
ASTM A320	Specification for Alloy Steel Bolting Materials for Low-temperature Service.
ASTM A333	Specification for Seamless and Welded Steel Pipe for Low-temperature Service.
ASTM A790	Specification for Seamless and Welded Ferritic/Austenitic Stainless Steel Tube for General Service.
ASTM A860	Specification for High-Strength Butt-Welding Fittings of Wrought High-Strength Low-Alloy Steel.

TABLE OF COLOUR PLATES

Front Cover — The Conoco/Chevron Britannia installation
Picture reproduced with kind permission of Conoco (U.K.) Limited and Chevron U.K. Limited.

Rear Cover — The Dockwise Transshelf and the Rowan Gorilla V jack-up.
Picture reproduced with kind permission of Dockwise N.V.

Tension Leg Platform (TLP), the Conoco Hutton (now owned by Oryx).

Shell/Esso Kittiwake oil installation.

Plates 1 to 8 are between pages 16 and 17

1. Gas platforms. Shell/Esso Leman 49/26 Alpha.
   Inset — Amoco Leman 49/27 Delta.

2. The Shell/Esso Brent Delta concrete gravity base structure (GBS).

3. The Shell Stadrill semi-submersible drilling vessel.

4. The Castoro Sei semi-submersible pipelay vessel.

5. The Neddrill 3 jack-up.

6. The Putford Artemis standby boat, semi-submersible crane vessel DB101.
   Inset — A fast rescue craft.

7. The FPSO Petrojarl 1.

8. The Shell/Esso 49/19a Clipper platform and temporary support vessels.

Plates 9 to 16 are between pages 124 and 125

9. Flange types (Courtesy of RGB Stainless Limited).

10. Piping systems, typical colour coding plan.

11. Diving support vessels, the Deepwater 1.
    Inset — The Smit Manta.

12. Air diving operations.
    Inset — Deck decompression chamber.

13. The Offshore Hyball Remotely Operated Vehicle (ROV).
    (Courtesy of Hydrovision Limited).
    The Trojan Remotely Operated Vehicle (ROV).
    (Courtesy of Slingsby Engineering Limited).

14. Konsberg Offshore A.S. HOST GL diverless wellhead system.

15. Konsberg Offshore A.S. HOST GL diverless wellhead system (continued).

16. Installation of a sub-sea drilling template.
    Inset — Transportation of a sub-sea wellhead protection frame.

Offshore Engineering

Plates 17 to 24 are between pages 228 and 229

17 The Galveston Key jack-up.

18 The Ocean Benarmin jack-up and the Mobil Camelot 53/1a unmanned gas producing installation..

19 The drill floor.
    Inset — Travelling block hook, gooseneck, rotary hose and swivel.

20 Varco BJ top drive drilling motor.

21 Blowout preventer (BOP) stack.
    Inset — Drillers console.

22 The Arch Rowan jack-up under tow of the Wrestler.

23 The Big Orange XVIII well service vessel.
    Inset — Sphere launcher.

24 Cameron Iron Works (Cooper Oil Tools) emergency shutdown valve (ESDV) and actuator.
    Inset — Solid block Vectco Gary Christmas tree.

Plates 25 to 32 are between pages 308 and 309

25 Piling of the Shell/Esso 49/26 Foxtrot jacket by the Heerema Balder.
    Inset — Clipper jacket and foundation piles during transportation.

26 The Conoco/Chevron Britannia jacket — Construction roll-up and loadout.

27 Barge launch of deep water jacket.
    Transportation of shallow water jacket.

28 Oil and gas deposits always seem to be found at locations where it is either very hot or cold.

29 The Lowering of one of the Shell/Esso 48/19a Clipper jackets.
    Inset — Installation of the jacket complete.

30 Transportation of the Shell/Esso 49/19a Clipper topside structure to the Heeremac DB102.

31 The Amoco North West Hutton oil installation and the semi-submersible crane vessel (SSCV) Herrema Balder.

32 The Conoco/Chevron Britannia topside installation by the Heeremac DB102.

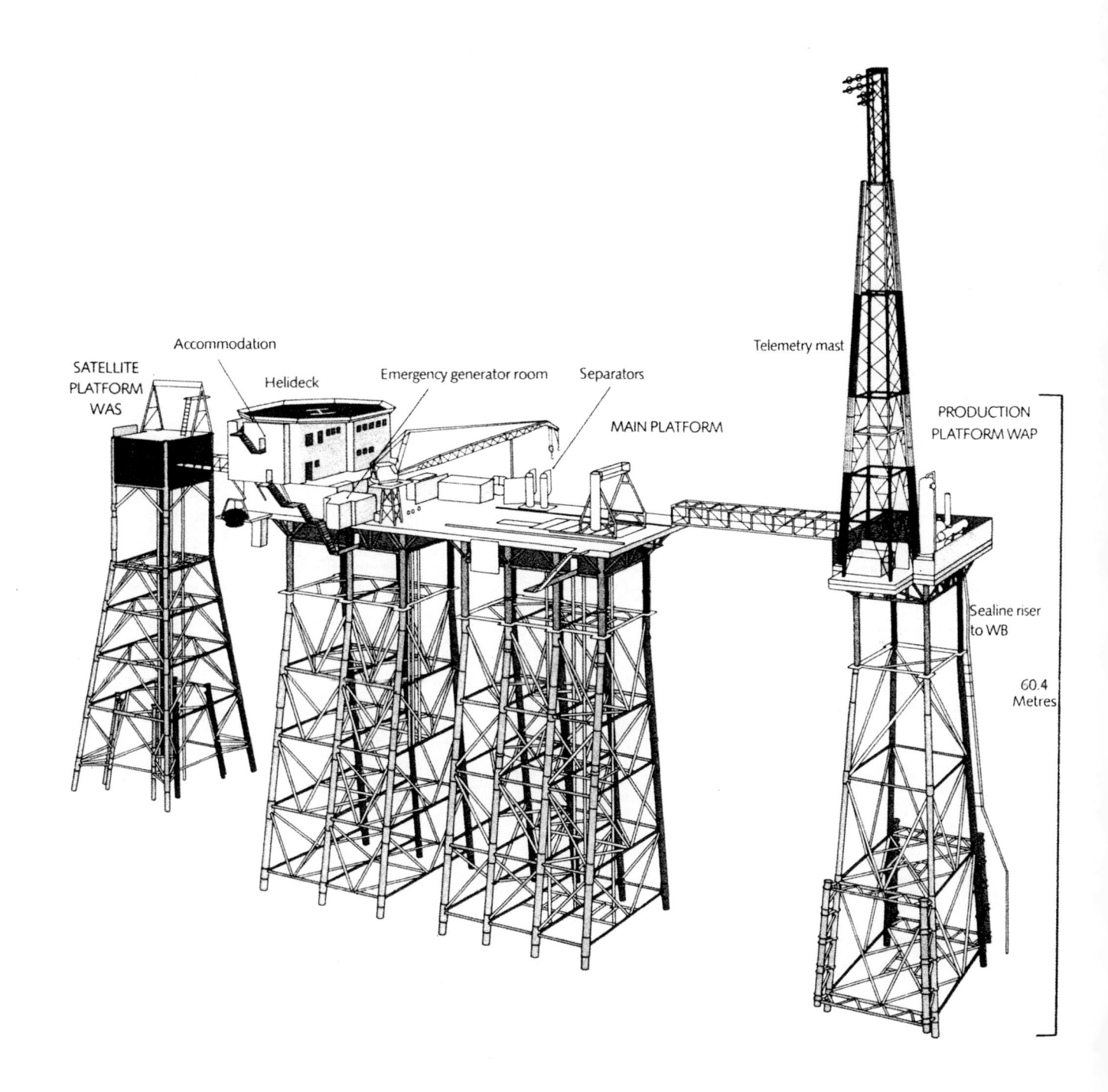

GAS INSTALLATION

Chapter One

OFFSHORE STRUCTURES AND SUPPORT VESSELS

PART 1. OFFSHORE STRUCTURES

PART 2. SUPPORT VESSELS

PART 3. OFFSHORE INSTALLATIONS – DESCRIPTION

PART 4. FIXED STEEL STRUCTURES – INSTALLATION

Offshore Engineering

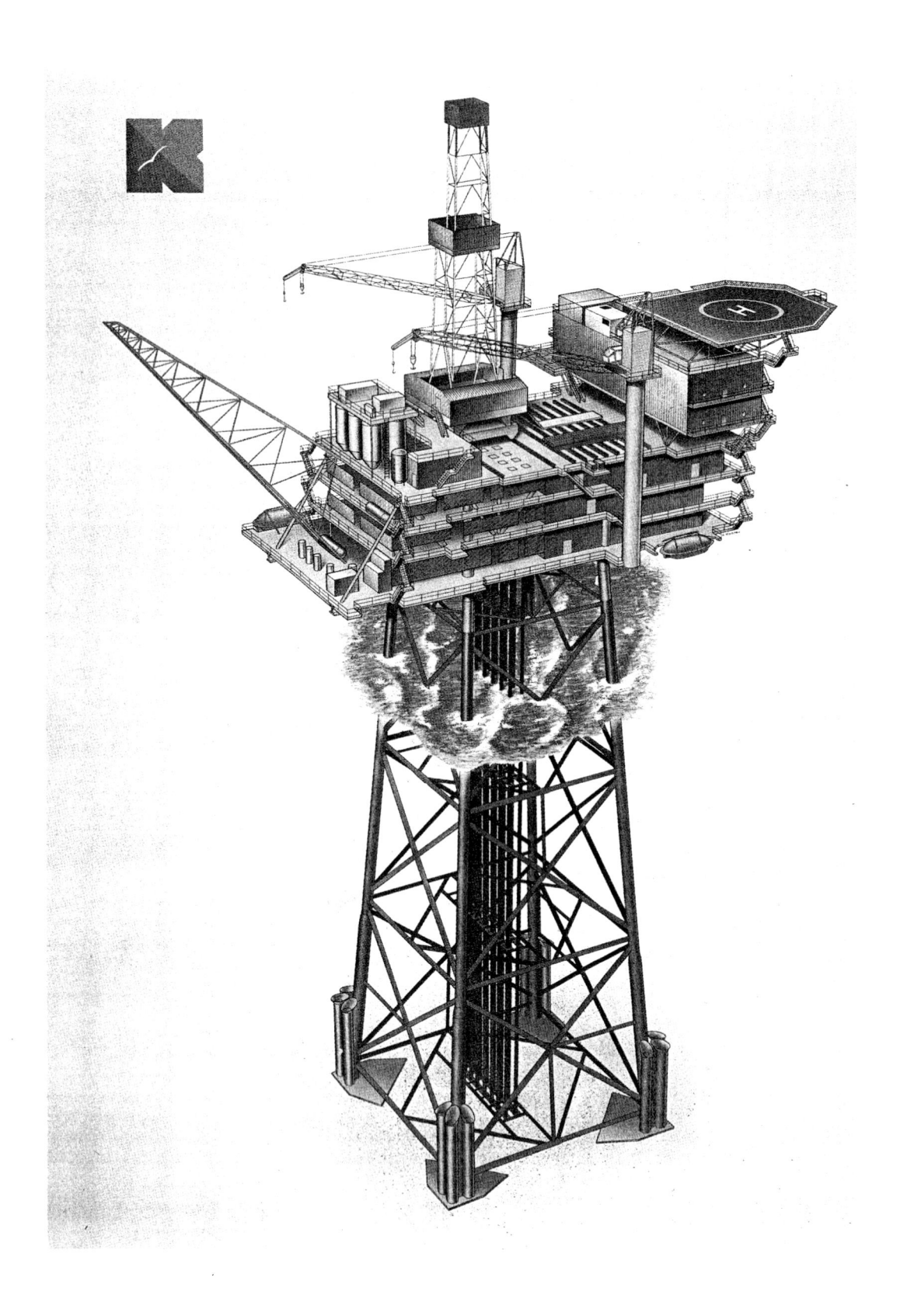

FIXED STEEL INSTALLATION

Part 1. OFFSHORE STRUCTURES

INTRODUCTION

The structure shown opposite is instantly recognisable as an offshore rig. This type of structure, more correctly described as a fixed steel offshore installation or platform forms the backbone of the offshore industry and there are in excess of 7,000 such structures dotted about the oceans of the world.

Not so familiar are the structures and vessels shown elsewhere in this chapter which provide assistance to, or are in competition with the fixed steel structure. They could all be loosely described as oil rigs (or gas rigs) which gives an indication as to the ambiguity of the expression and hints at the complexity of the offshore industry.

The various rig/vessel types shown in the sketches are:

1. Fixed steel structure
2. Concrete gravity base structure
3. Tension leg platform
4. Semi-submersible vessel
5. Floating production systems
6. Self-elevating jack-up
7. Single point mooring

A brief description of each installation and its position within the offshore industry follows on the next few pages.

1. FIXED STEEL STRUCTURES

The traditional offshore installation shown utilises a welded steel, tubular framework or jacket to support the topside facilities and this arrangement is referred to as a fixed steel structure. The topside facilities will vary slightly depending on whether it is an oil or gas producing installation but they will include hydrocarbon process equipment, power generation, a heli-deck, and accommodation and hotel services designed to cater to the needs of the personnel employed in the operation and maintenance of the installation.

The single jacket installation is typical of the rigs found in deep water environments such as the northern sector of the North Sea. It should be noted that the heli-deck and accommodation facilities are situated as far from the potentially dangerous hydrocarbon processing area as is physically possible.

An installation may consist of any number of bridge linked jackets with modern designs tending to favour a separate jacket to house the accommodation and heli-deck. These multi-jacket installations tend to be restricted to shallow water developments where the construction costs are considerably less than their deep water counterparts.

Offshore Engineering

Whilst the fixed steel structure is likely to remain as the industry's first choice of installation design based on operational requirements, its future use may well be restricted to the development of large fields located in intermediate water depths where a substantial return on the capital sums invested can be guaranteed. The ever changing demands of the offshore industry has spawned a number of competitors more ideally suited to specific applications.

The concrete structure has emerged as a viable alternative to the steel jacket but they both tend to suffer from the same inherent disadvantages when used to develop deep water reserves. The scantling requirements of structures that are capable of withstanding the extremes of weather conditions associated with the North Sea are considerable and this is reflected in the construction costs. Generally speaking the *tension leg platform* (TLP) or the *floating production system* (FPS) will be preferred for the development of deep water reserves. The FPS used in conjunction with sub-sea wells also represents the most cost effective solution for the exploitation of the smaller, marginal fields. The alternatives to the fixed steel structure will now be discussed and further information on fixed steel structures can be found in Part 4.

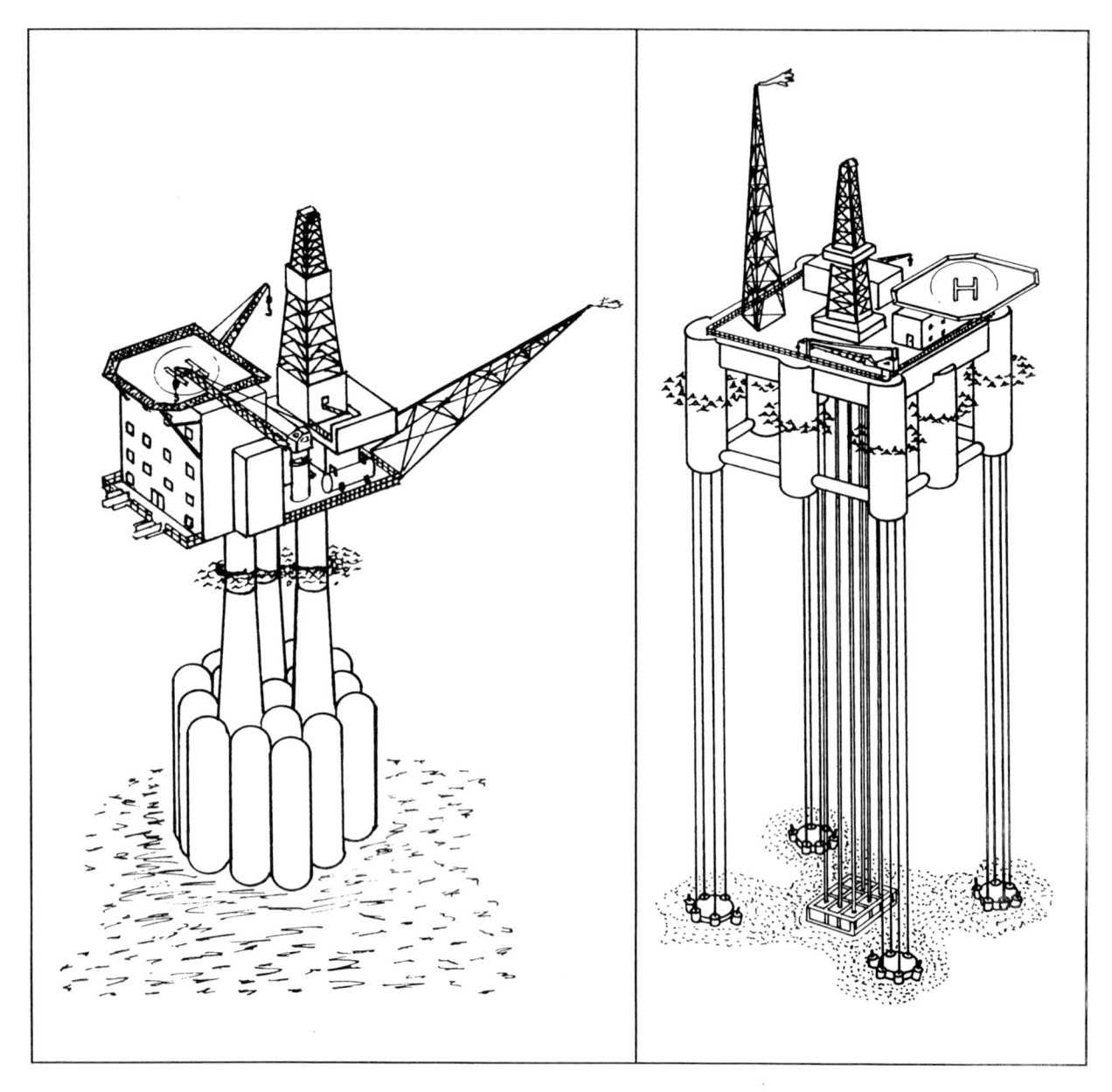

CONCRETE GRAVITY BASE   **TENSION LEG PLATFORM (TLP)**

2. CONCRETE GRAVITY BASE STRUCTURE (GBS)

Whilst the vast majority of fixed offshore platforms employ a tubular steel jacket to support the topside facilities, a number of installations have been constructed using a base manufactured from reinforced concrete. This type of installation was pioneered by the Norwegians who had experienced difficulties in producing the enormous quantities of high quality steel products and the large skilled workforce that are essential ingredients to the success of steel fabrication projects. Also, the presence of the Norwegian Trough, a large undersea valley, made the laying of sub-sea pipelines an expensive and challenging operation so the void spaces which feature so prominently in the design of a GBS provide valuable storage space for crude oil prior to discharge into oil tankers via a *single buoy mooring* (SMB).

The first concrete structure to be installed in the North Sea was constructed by the Norwegians in 1973 and used to develop the *Ekofisk* field. Since 1973, the Norwegians have installed a steady string of concrete structures and it came as no surprise when the Norwegian government elected to develop the *Troll* field with a concrete structure. The Troll field, which stands in 350 metres (1,150 feet) of water is currently the largest offshore oil and gas field in Europe and at 1,270,000 tonnes, the structure is the largest concrete platform in the world.

The construction of a concrete installation base normally commences in a dry-dock. The design of the base includes void spaces or caissons suitably dimensioned to provide the structure with a natural buoyancy which will enable it to be floated clear of the dry-dock for finishing off prior to being towed to its final destination. Once on location the void spaces are flooded to facilitate positioning of the base on the sea bed whilst the topside modules are lifted into place. The void spaces may then be pumped dry and used as storage compartments for crude oil, or filled with permanent iron ore ballast. The colossal weight of the concrete structures obviates the need to install foundation piles, hence the name gravity base structure.

In Europe, technological developments and improved construction techniques have lead to the use of concrete for the building of tension leg platforms, floating production systems and semi-submersible vessels. Further afield, the concrete structure has been selected in preference to other designs because of its ability to withstand impact from icebergs, a feature of vital importance for installations operating in the more northerly waters of Europe, Russia and Canada. In addition to the threat from icebergs, these remote locations frequently prohibit the laying of a sub-sea pipeline and as was found in Norway, the facility to store the oil within the structure prior to transfer to an oil tanker has proven a major factor in determining the viability of the development of the fields.

Before moving on to a description of *tension leg platforms* it is worth noting that not all gravity base structures are made of concrete. In 1982 the Phillips Petroleum Company installed a steel gravity base structure over the *Maureen* field.

The Maureen platform sits on three steel storage legs or tanks and under normal operating conditions the tanks are filled with varying quantities of crude oil and water, the water being used to displace the oil without the aid of pumps. Up to 50,000 barrels can be stored in this way, prior to being transferred into tankers.

As the Maureen field reaches the end of its commercial life Phillips are actively seeking a buyer for the platform with a view to simply refloating her and towing her to a new destination.

3. TENSION LEG PLATFORM (TLP)

The tensions leg platform concept was devised by the Conoco Oil Company as an alternative to the *fixed steel structure* and *floating production system* (FPS) for the development of deep water oil and gas fields. The world's first TLP was built by Conoco in 1984 and used to develop the *Hutton* field in the North Sea. Located in 148 metres (500 feet) of water, the Hutton field could quite easily have been developed using a traditional fixed steel installation but the harsh environment associated with the North Sea was judged by Conoco to provide the ideal test bed for the TLP design prior to venturing into deeper waters.

The TLP fulfils a role midway between the fixed offshore installation and the floating production system. It combines the initial cost saving benefits associated with floating production systems with the operational benefits attributed to the fixed offshore installation.

The TLP consists essentially of a floating production facility which is tethered to the seabed by a number of tensioned legs. The legs are typically constructed from large diameter (250 mm to 750 mm) steel pipes or wire ropes (125 mm diameter). One end of the leg is secured in a foundation template piled into the seabed whilst the other end is connected to a tensioning winch within the floating superstructure. Whilst the legs remain in tension the platform can resist vertical wave induced motion.

The TLP foundation design may appear elaborate when compared to the traditional ship type anchorage systems employed by floating production systems but the advantages are considerable. The restriction on vertical movement permits the use of fixed installation type wellhead equipment and rigid steel conductor-riser assemblies. In contrast, the floating production systems must rely on flexible riser pipes emanating from sub-sea wellheads. These incur a considerable penalty in terms of installation costs and maintenance availability.

The tension leg platform is still a relatively rare bird, but oil companies are beginning to appreciate the advantages offered by this unique design as they are forced to look further afield for their oil and gas. In the Gulf of Mexico, Conoco commissioned their *Jolliet* TLP in a water depth of 575 metres (1,900 feet) and Shell similarly elected to use a TLP to develop the Auger Field (872 metres–2,900 feet), a world record depth in 1993. This record was subsequently eclipsed with the installation of the *Mars* TLP in 1996 in a water depth of 882 metres (2,940 feet), then by the *Ram-Powell* at 965 metres (3,214 feet). Shell's *Ursa* development, brought on stream in 1999 takes the depth record to 1,200 metres (4,000 feet) and the race is sure to continue as technology is continually improved.

Closer to home, Conoco embarked on a project with the Norwegian Government and in 1995 constructed the world's first concrete TLP to develop the *Heidrun* field. Located in 325 metres (1,100 feet) of water the field is one of the largest currently under development in the North Sea and demanded a structure of truly epic proportions. The four concrete foundation units which secure the Heidrun's tethers each weigh 21,000 tonnes, are 48 metres (160 feet) in diameter and are 26 metres (85 feet) high.

In the not too distant future we may well see further refinements of the TLP design which will permit the use of the tension legs as wellhead conductor casings thus saving on the costs of several thousand metres of steel pipe. The ingenuity of the offshore engineers appears to know no bounds.

4. SEMI-SUBMERSIBLE VESSEL (SEMI-SUB)

As can be seen from the illustrations, a semi-sub is essentially an enormous box section barge supported on twin hulls. It has a colossal capacity to consume water ballast and with minimal water plane area and displacement of up to 170,000 tonnes, presents an immovable object to wave action in anything but the most severe weather conditions.

The design of a semi-sub does not permit the installation of an effective main propulsion plant. It is thus heavily dependant on the assistance of support ships for towage to its destination and for the

OFFSHORE STRUCTURES AND SUPPORT VESSELS

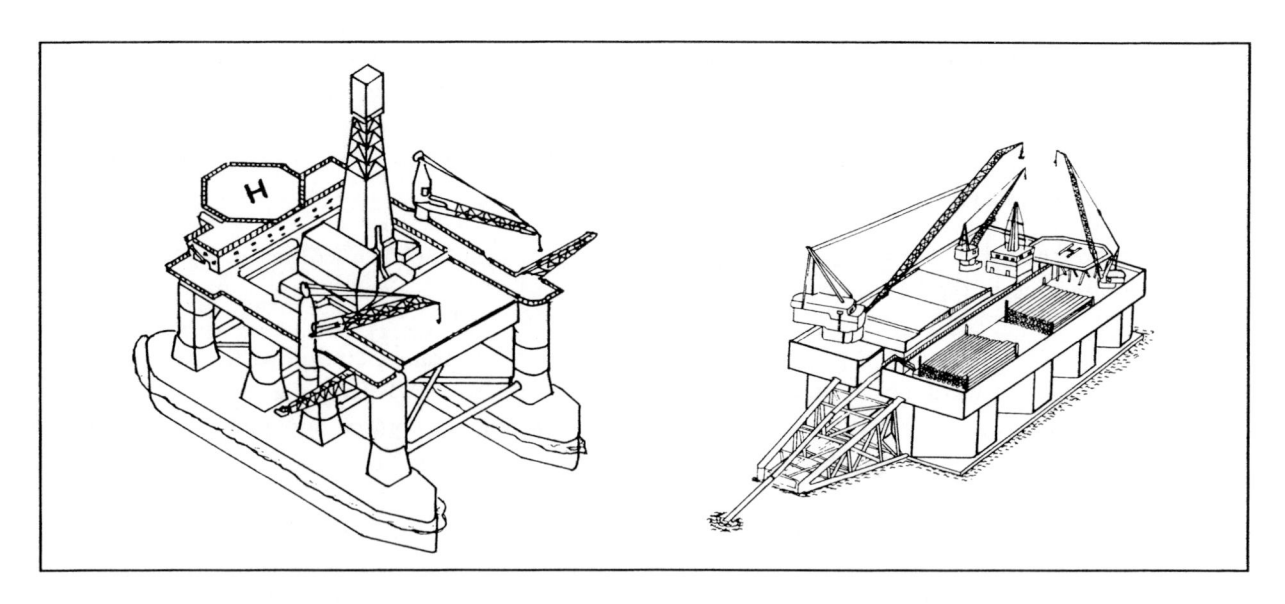

SEMI-SUBMERSIBLE DRILL SHIP

SEMI-SUBMERSIBLE PIPE LAY BARGE

deployment of anchors (typically eight). However, some vessels are fitted with computer controlled azimuthing thruster units which provide an accurate manoeuvring capability, particularly when operated through a satellite navigation system. This facility enables the semi-sub to hold station in water depths in excess of 650 metres (2,200 feet) where the laying of anchors may prove impractical.

The semi-sub represents the ideal choice of vessel for performing operations where accurate station keeping and exceptional stability are prerequisite to success. Depending on the equipment fitted to the main deck the vessel may perform any one of five roles.

i) HEAVY LIFT

The semi-sub will most frequently be encountered in its role as a heavy lift crane barge used for the installation of offshore platforms. Over the years the lifting capacity of the cranes has steadily increased from 2,000 tonnes to 7,000 tonnes and the more modern vessels are fitted with tandem cranes which are capable of lifts of up to 14,000 tonnes. Sophisticated rapid acting ballast systems counteract the huge weight transfers which occur as structures are lifted and manoeuvred into position.

ii) ACCOMMODATION

Some of the older, smaller heavy lift semi-subs are now used for modification or repair projects where the facility to provide extra accommodation and storage space take precedence over the lifting capacity of the crane. They have been joined by purpose built accommodation vessels or flotels with up to 800 beds which can remain in close attendance to fixed structures for months at a time. Transfer of personnel may be by bridge or by helicopter and again the inherent stability of the semi-sub ensures that the workforce sleep soundly, even during a winter in the North Sea.

iii) DRILLING EXPLORATION VESSEL

Depending on water depth, the drilling of exploration wells will be carried out by either a *self-elevating jack-up*, a *semi-submersible vessel* or a *monohull drill ship*. The jack-up is limited to operations in water depths of approximately 120 metres (400 feet) but no such restrictions apply to the semi-sub. The current world water depth drilling record stands at a formidable 2,250 metres (7,520 feet) at a field in the Gulf of Mexico.

The combination of satellite positioning, stability and an abundance of deck space for the storage of drill pipe and well test equipment make the semi-sub the first choice of vessel for deep water exploration projects.

iv) PIPELAY BARGE

The sketches show a typical pipelay barge of the type which has installed nearly 10,000 kilometres of sub-sea pipeline in the North Sea during the past 25 years.

Once started, the laying of a sub-sea pipeline becomes a non-stop operation as the lay barge slowly winches its way forward on its anchors. A modern vessel can lay between 2 and 4 kilometres of pipeline a day, azimuthing thruster units occasionally replacing the anchor spread in very deep water locations.

Sub-sea pipelines vary in diameter from 3″ to 42″ with 30″ and 36″ lines being favoured for the main transmission routes. The pipe is supplied to the barge in 12 metre lengths, ready painted to provide corrosion protection, gas pipelines being coated in concrete to provide the negative buoyancy required to keep them on the sea bed.

On the barge the lengths of pipe are *double jointed*, that is, two lengths of pipe being welded together prior to being lowered into the *firing line*, the inside of the barge where the double joint lengths are welded to the pipeline as it enters the sea. The pipeline is welded using a part manual, part automatic welding process. The welds are then X-rayed and if free from defects are given a protective coating of bitumen before being launched into the sea from the *stinger* framework situated on the stern of the vessel. The stinger prevents buckling of the pipe as it enters the water.

The completed pipeline will be subjected to a hydrostatic pressure test to prove its structural integrity and may then be *trenched* (buried) or *rock dumped* (covered in rocks) to protect it from hazards such as fishermen's nets and ships anchors. Alternately, it may simply be left to rest on the sea bed.

v) FLOATING PRODUCTION FACILITY

The use of the semi-submersible vessel as a floating production facility has been dealt with in the next section.

5. FLOATING PRODUCTION SYSTEMS (FPS)

INTRODUCTION

Floating production systems (FPS) have been with us for many years. In fact, the first oil field in UK waters, the Argyll field, was developed using a floating production system in 1975. During the last five years there has been a considerable growth in the number of floating production systems constructed for service in the North Sea, West Africa, Australia and Brazil.

A floating production system is in effect a floating oil rig. It contains all the equipment associated with a fixed installation and is used in conjunction with sub-sea wellheads to exploit moderate to deep water oil fields.

Basically, floating production systems are selected for field developments for one of three reasons:

OFFSHORE STRUCTURES AND SUPPORT VESSELS

i) they can only be developed economically by a re-useable asset such as

ii) established pipeline infrastructure does not exist,

iii) deep water where it would not be possible to install a conventional

on systems lies in the fact that they can simply lift anchors when production reaches a commercially unprofitable level.

FLOATING PRODUCTION SYSTEMS (FPS)

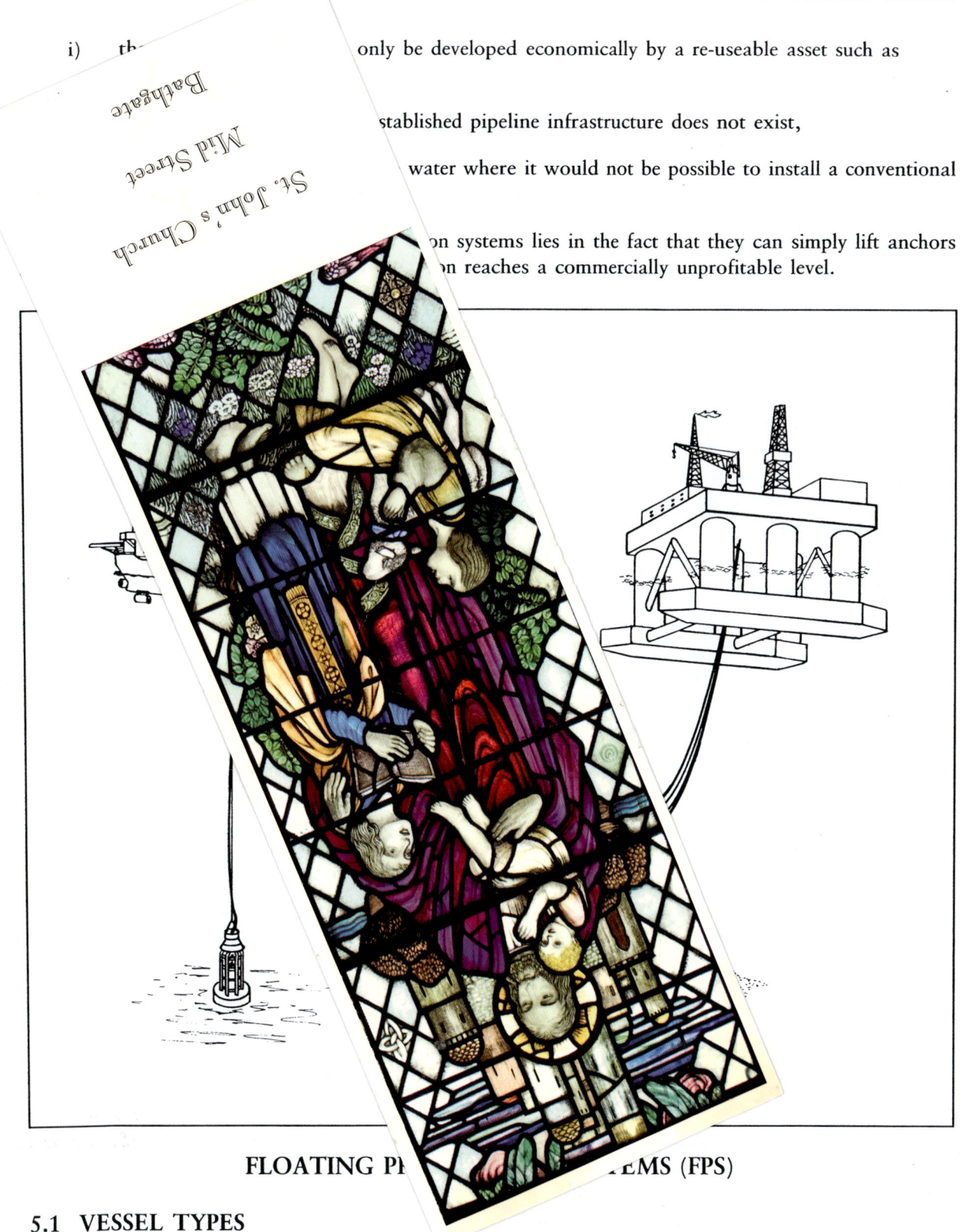

5.1 VESSEL TYPES

Conventional ship shaped *monohull* vessels and *semi-submersible* vessels may both be used as the basis for floating production systems and there are advantages and disadvantages associated with both options. The monohull designs are perhaps more suited to coping with strong winds and tides than semi-submersibles, have ample onboard storage capacity for the stabilised crude oil and are well suited to off-loading duties into shuttle tankers. Whilst semi-submersible production units may require a support vessel for oil storage and to facilitate off-loading (FSU or FSO), they are inherently more stable and offer a greater production capacity than a monohulled vessel, accommodating more than double the number of risers, up to 100.

In the North Sea and seas West of Shetland, monohull floating production systems have in the main been the preferred option whilst the Brazilians, who are the undisputed leaders in deep water floating production technology, have tended to favour semi-submersible vessels. As far back as 1977, Petrobras, the national oil company, broke the 1,000 metre (3,250 feet) water depth barrier with the *Marlim Sul* field development and they currently operate 17 FPSs with a further 17 scheduled to enter service by the year 2002, predominantly in the 800–1,400 metre (2,660–4,660 feet) water depth range. They are also currently developing the giant *Roncodor* field, which at 2,600 metres (8,660 feet) is a world record depth.

5.2 VESSEL CONFIGURATIONS

Whilst the FPS exists as a production installation in its own right, the term *floating production system* is frequently used generically to describe any vessel engaged in the oil production or oil storage process and includes the FSU, FSO and FPSO. A brief description of each type is provided below but it is worth noting that there are many variations on these basic themes, and it is the most frequently used arrangements which have been described.

i)   FSO

The *floating, storage and off-loading unit* (FSO or FSU), is essentially a storage tanker into which processed oil is pumped from a fixed platform (or semi-submersible FPS) which has no storage capacity of its own and is not connected to a sub-sea pipeline. The oil is subsequently off-loaded from the FSU into shuttle tankers for transportation to a refinery.

ii)   FPS

In most cases, the term *floating production system* is used to describe a production vessel which is connected to a sub-sea pipeline, rather than one which has the capability to discharge oil into shuttle tankers. Normally, a semi-submersible vessel, the Rob Roy field development illustrated in Chapter 6 being an example of this type.

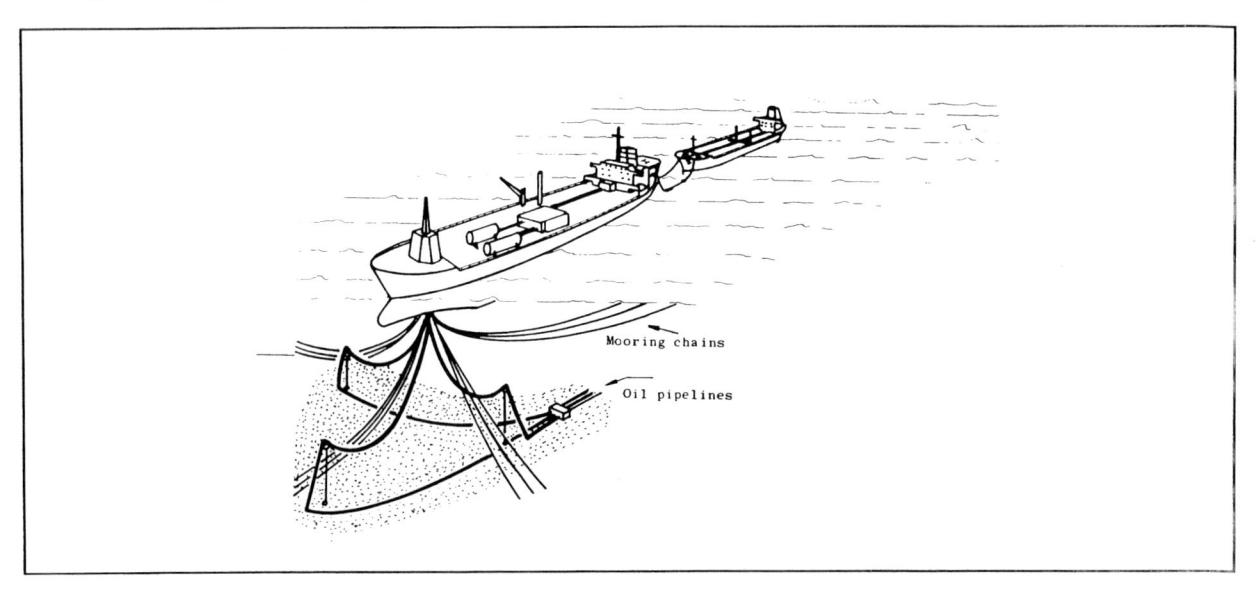

FLOATING PRODUCTION STORAGE AND OFFLOADING SYSTEM

iii) FPSO

Like the FPS, the *floating, production, storage and off-loading* vessel (FPSO) replaces a conventional platform in its entirety. The FPSO differs from the FPS because it is not connected to the beach by a sub-sea pipeline. The oil, which is again extracted via sub-sea wells, is processed and stored onboard the vessel, normally a monohull design, prior to being off-loaded into shuttle tankers.

iv) FPDSO

Currently, no monohull FPSOs are fitted with drilling derricks so drilling and workovers have to be carried out by drill ships. The *floating, production, drilling, storage and off-loading* vessel (FPDSO) is therefore the latest development in the FPS family. Still at the conceptual stage, the FPDSO will only provide advantages for compact field developments, the wells being to far apart for drilling to be carried out from a single location in many cases.

v) PSV

Pioneered by British Petroleum (BP) with the *Seillean*, the Production Storage Vessel design is unique. It is a purpose built vessel designed for the exploitation of small fields which are located in moderate water depths. Essentially it is a floating production system and shuttle tanker in one. A *dynamic positioning* capability enables the vessel to remain on station over the wells without the need to deploy anchors whilst producing oil. Once the vessel's storage tanks are full (typically 20 days production), the vessel simply disengages from the wellhead and proceeds to a shore terminal to discharge the cargo.

5.3 PRINCIPLE FEATURES — MONOHULL FPSO

The floating production system is, as previously stated, a floating oil platform which may be equipped with gas compression and water injection systems, in addition to oil processing equipment. After processing, the stabilised crude oil is stored in the ships cargo tanks prior to being transferred into shuttle tankers via a mooring hawser and hose, reeled from the stern of the vessel. Gas produced as a by-product of oil production is normally used as fuel for the gas turbine powered electricity generators, or where significant volumes are produced and an export pipeline is not available, it may be re-injected into the formation, or flared when production exceeds demand.

There are some significant design challenges which have had to be overcome to make the FPSO a viable proposition for long term field development goals. Whilst these relate primarily to the mooring system, turret and the flexible risers, the basic hull and the topside processing equipment all must be designed to a higher specification than those of a conventional tanker or offshore structure. These items will now be described briefly as they relate to a monohulled vessel, permanently moored on station via an internal turret.

i) HULL

A number of existing tankers have been converted for service as FPSOs and in some cases they have retained their original propulsion system. Where operators have opted for a new purpose built vessel, main propulsion systems are generally omitted, the vessel relying on the electrically powered thruster units to assist in station keeping. The hulls are built in accordance with rules provided by the ship *classification societies* and are designed to remain on station for a period of 10–20 years without dry-docking.

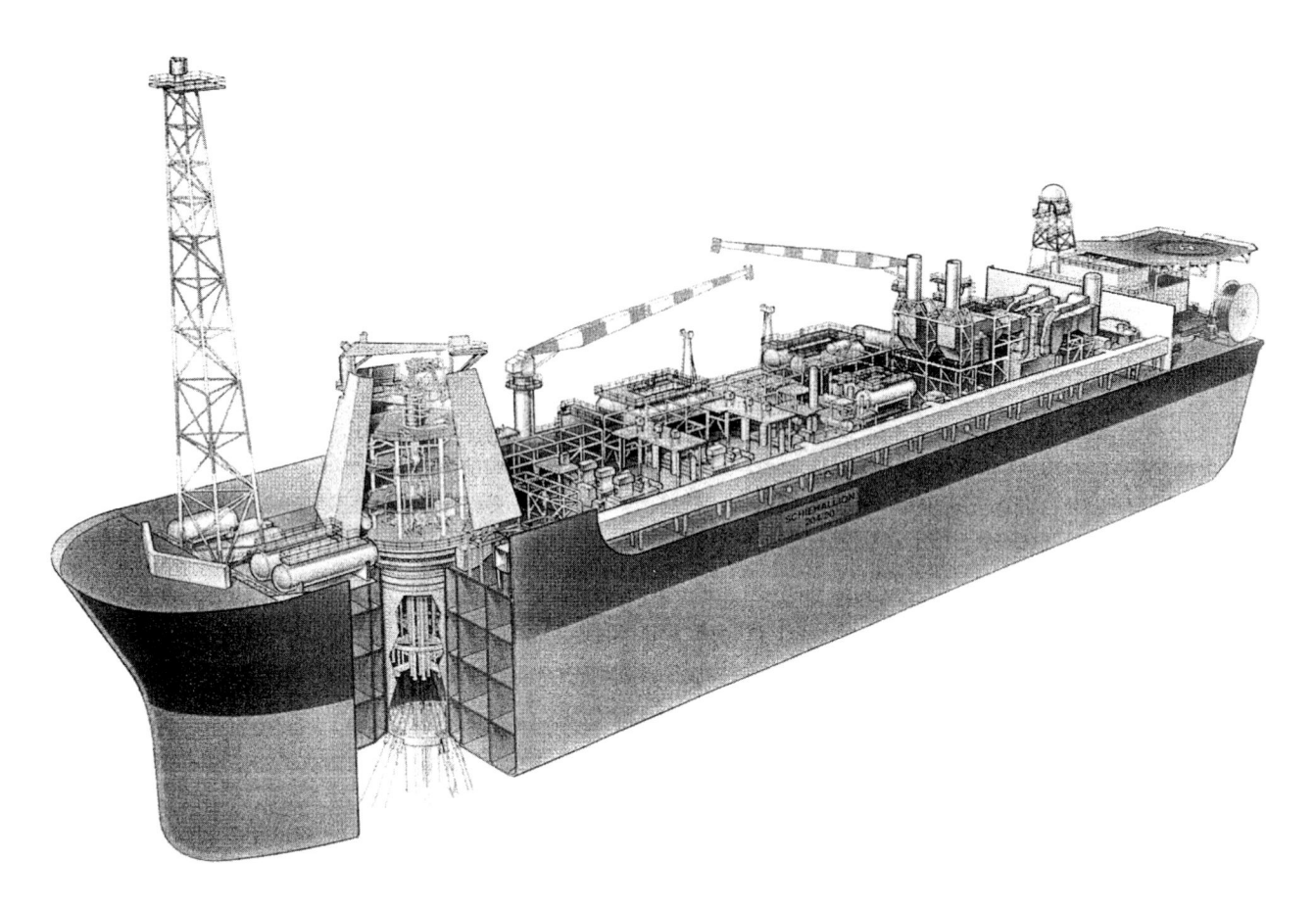

B.P. SCHIEHALLION

ii) TOPSIDES

The topside oil and gas processing equipment is designed and constructed in accordance with fixed offshore platform and refinery standards, specifications and recommendations such as those produced by the American Petroleum Institute (API), American Society of Mechanical Engineers (ASME) and British Standards Institute (BSI). However, the codes and standards must be adapted to take account of vessel motion. The additional loadings due to wave action can be considerable and have a serious effect on the fatigue life of the equipment, particularly the process pressure vessels and pipework.

The process equipment is normally pre-assembled into skids, the skids being mounted on a framework or pillars 3 metres above the main deck of the vessel, the air gap providing some protection against *green seas*, large waves breaking on the deck, and to provide adequate hazardous area boundaries.

iii) TURRET

The turret consists of a large diameter (16–32 metre), vertical cylinder which sits within the hull of the vessel, mounted on heavy duty roller bearings. The mooring wires and the flexible sub-sea risers are attached to the turret which prevent it from rotating. The vessel is thus free to *weathervane* (rotate through an angle of typically 270°) around the fixed turret under the influence of wind, waves and currents, the extent of rotation being controlled by the vessel's azimuth thruster units.

iv) MOORING

A spread of 8 to 14 anchors ensures that the FPSO remains on location whilst producing oil, the anchor spread typically consisting of a combination of wires and chains which are tensioned by winches within the turret. Conventional anchors, suction anchors or piles may be used to make the connection with the sea bed. For the *ultra* deep waters of the *Campus Basin* in Brazil, Petrobras are experimenting with a fibre rope mooring system and twelve, 711 tonne braking strength, *Superline Polyester* ropes will be used to secure an FPSO in 1,400 metres of water at the *Marlim South* field.

v) RISERS

To permit the FPSO to move both vertically and laterally, the connections to the sub-sea wells are made through flexible, steel reinforced risers. The risers are designed to absorb any wave induced motion which might affect the position of the vessel, often being draped over submerged *mid-line buoys* which help in reducing the loads on both the risers and the turret, the oil being transferred from the turret to the process equipment through a swivel stack.

It is the swivels, and to a lesser extent the risers which are currently the limiting factors in the development of ultra deepwater FPSs, gas export lines in particular being difficult to install successfully and limited to a diameter of less than 16 inches (0.406 metres) at operating pressures of 340 bar (5,000 psi). Another problem which increases with water depth is the risk of hydrostatic collapse of the risers, a typical maximum water depth for a steel-reinforced riser of 8-10 inch (0.2-0.25 metres) diameter being approximately 800-1,000 metres (2,660-3,330 feet).

5.4 FUTURE DEVELOPMENTS

Current predictions indicate that floating production systems will continue to feature prominently in the future development plans of the leading oil companies. A total of 134 oil and gas fields are forecast to be developed worldwide in water depths greater than 300 metres (1,000 feet) between 1998 and 2002 and 90% of these are likely to be developed using floating production systems. They offer considerable cost saving benefits when compared with fixed structures and used in conjunction with the diverless sub-sea wellhead, there is virtually no limitation to the water depth in which they can operate.

5.5 SHUTTLE TANKERS

Basically, there are two types of tanker which are equipped for offshore loading operations. Crude oil carriers which transport oil over long distances are provided with fairly basic loading and mooring equipment because the time spent loading is minimal when compared to the duration of the subsequent sea voyage. In comparison, shuttle tankers spend a large proportion of their service life loading, discharging and manoeuvring in, and out of port.

Prior to the boom in FPSOs, shuttle tankers tended to be converted product tankers to which suitable mooring equipment was added. However, the shortage of suitable conversion tonnage has lead to a ship building boom and most major shipyards now have a shuttle tanker within their design portfolio. Approximately 20% of the UK's oil production and 60% of Norway's oil is transported by shuttle tanker rather than by pipeline.

The advanced specification of the new shuttle tankers tends to reflect the industries awareness of the hazards associated with oil spills caused by grounding or collisions. A far greater proportion of the vessels life will be spent loading and discharging when compared to a traditional VLCC (*very large, crude carrier*). Consequently, the latest designs tend to specify double hulls under the cargo tanks and fuel oil bunker spaces, segregated engine rooms and redundancy in steering, manoeuvring and dynamic

Offshore Engineering

positioning equipment. These ships tend to be in the 100 000 dwt to 125 000 dwt tonnage range, approximately half the size of a VLCC.

The manoeuvrability of a shuttle tanker is extremely important because of the long periods spent holding station whilst loading oil. Typically, three 1,750 kW bow thrusters and two 1,750 kW stern thrusters are provided in addition to the main propulsion system. The dynamic positioning and mooring systems are designed to cope with maximum wave heights of 6.0 metres and mean wind speeds of up to 72 knots, although typical loading restrictions are applied at wave heights of 4.0–4.5 meters for connecting operations, with oil transfer ceasing when wave height exceeds 4.5 metres.

The next generation of multi-purpose shuttle tankers (MST) are likely to be designed to facilitate drilling, as well as for operation in the shuttle tanker mode, and for conversion to FSO or FPSOs with the minimum of work should the need arise.

6. SELF-ELEVATING JACK-UP

The self-elevating jack-up is one of the stalwarts of the offshore industry, its origins dating back over 60 years when it could be found searching for oil in the muddy swamps of Louisiana.

The self-elevating jack-up consists of a triangular shaped (sometimes rectangular), box section barge fitted with three (sometimes four) moveable legs which enable the vessel to stand on the sea bed in water depths of up to approximately 120 metres (400 feet).

The shape of the vessel imposes design restrictions similar to those encountered with semi-submersible vessels which prevent the fitting of a main propulsion unit. Transportation over short distances is effected under the tow of *tugs* or *anchoring handling vessels* whilst it is both quicker and safer to accommodate a jack-up on the back of a *submersible heavy lift ship* to undertake long sea passages.

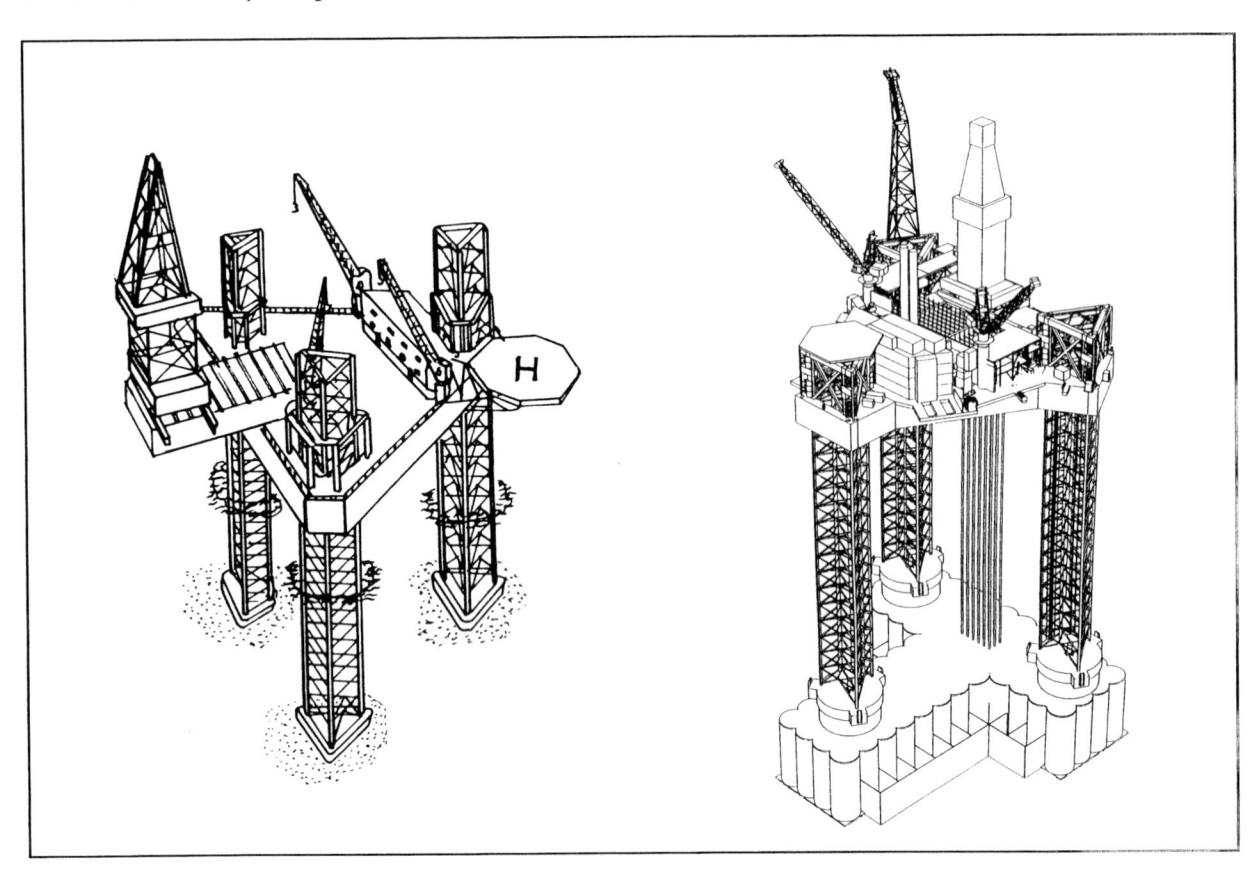

SELF-ELEVATING JACK-UP

Once on location, thruster units enable the jack-up to maintain position whilst the legs are lowered to the sea bed and the hull is jacked into position, clear of wave action. The base of each leg is fitted with a *spud can* which consists of a plate or dish designed to spread the load and prevent over penetration of the leg into the sea bed. High pressure jets of water or compressed air may be used to remove loose debris in the vicinity of the spud cans whilst the legs are manoeuvred into position, this process being referred to as *spudding in*. The legs are raised and lowered by means of a *rack and pinion* arrangement, the racks being attached to the corners of each leg, running from top to bottom whilst the pinions are driven by electric motors via reduction gearboxes, typically 12 motors for each leg. The operation of the electric motors is synchronised to ensure that the hull of the jack-up is raised and lowered on an even keel.

Jack-ups are primarily used for drilling operations but a number have been constructed to act specifically as *accommodation support* vessels to provide assistance to fixed installations during construction, modification or repair programmes. The jack-ups used for drilling operations fulfil two completely different roles. The majority of jack-ups are used for exploration (wildcat drilling) purposes and they lead a solitary existence searching for oil or gas in some of the remotest corners of the globe.

The other role in which jack-ups will be encountered is in the drilling of wells for permanent installations, particularly gas producing platforms such as those located in the southern sector of the North Sea. The gas production process does not require the same level of ongoing drilling activity and well modification work that is associated with the production of oil. Consequently derricks are not normally fitted to gas producing installations and all well modifications and drilling operations must be carried out by a jack-up sited alongside. The drilling derrick is fitted to rails so that it can be cantilevered into a position which provides direct access to the wellhead area.

The future of the self elevating jack-up looks assured with old rigs being replaced with designs capable of operating in ever deepening, harsh water environments. In fact the jack-ups look set to increase their share of the offshore market with a new generation of rigs being designed to operate as process facilities for the development of marginal oil fields using both conventional, and sub-sea wells. The *BP Harding* development provides a good example of these new trends and is a particularly innovative concept.

The BP Harding field development which was brought on stream in 1996 employs a jack-up as a production installation. The jack-up stands on a concrete gravity base in 110 metres (400 feet) of water, the base doubling as a storage facility for 580,000 barrels of oil. The gravity base is an open storage system, the produced crude oil displacing sea water from the base of the compartments where it is stored prior to export to a shuttle tanker via a submerged turret loading buoy (STL). This is an extremely large structure, the jack-up weighing 27,000 tonnes and the concrete base 84,000 tonnes.

7. SINGLE POINT MOORING (SPM)

The single point mooring concept originated in the 1960's as a solution to the problem of transferring crude oil from an onshore reception facility or refinery into very large crude oil tankers (VLCC — very large crude carrier) which were physically too big to enter port. The tankers are moored to a large buoy located at some considerable distance from the coast, often several miles, the buoy itself being secured by a spread of anchors. The oil is transferred from a sub-sea pipeline, through a swivel connector in the buoy into floating loading hoses attached to the tanker, the tanker thus being free to weathervain around the buoy independent of both wind and tide.

Single point moorings of the tethered buoy and flexible riser type similar to those just described, the *catenary anchor leg mooring* (CALM) buoys, are still used extensively and a number of more sophisticated designs such as the *buoyant* or *submerged turret loading* (STL) system have been developed over the years to meet the ever demanding needs of the oil industry.

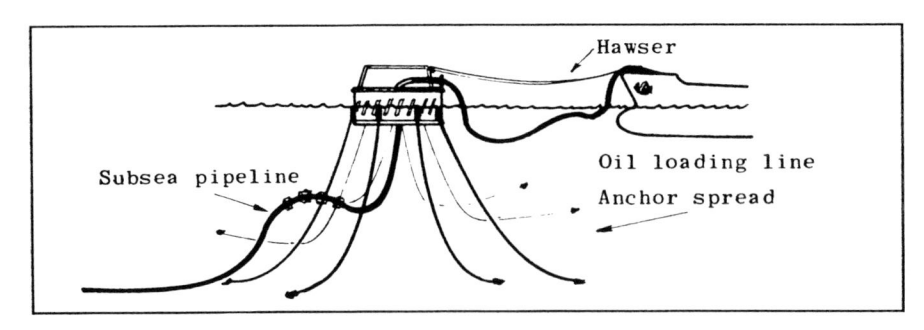

CALM BUOY

There are a number of variations on the single point mooring buoy theme which employ a rigid structure or riser column to convey crude oil from the sea bed to the surface. Examples of this particular breed are the *single anchor leg mooring* (SALM) and the *articulated loading platform* (ALP). Regardless of design they all attempt to achieve the same end result which is to provide a safe mooring whilst the tankers load oil in wind speeds of 80km/hour (50 mph) and wave heights of 5.5 metres (18 feet).

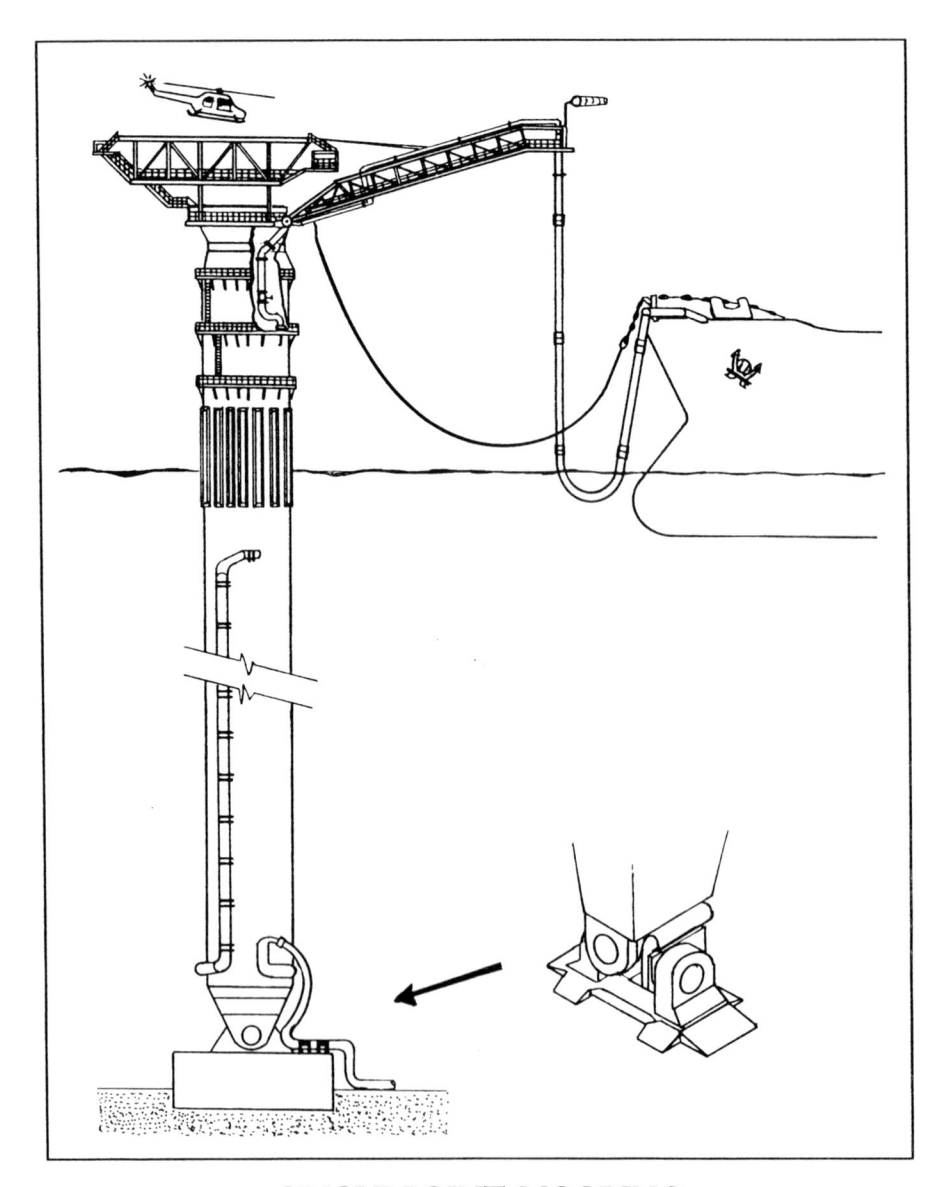

SINGLE POINT MOORING

1 – Gas platforms. Shell/Esso Leman 49/26 Alpha multi jacket installation typical of the gas gathering and compression facilities found in the Southern North Sea.
Inset – Amoco Leman 49/27 Delta. Note the "piled barge" production platform.

2 – *The Shell/Esso Brent Delta concrete gravity base structure (GBS).*

3 – The Shell Stadrill semi-submersible drilling vessel involved in wildcat drilling operations.

*4 – The Castoro Sei semi-submersible pipelay vessel.
The pipeline is guided into the sea from the "stinger" framework attached to the vessel's stern.*

5 – *The jack-up Neddrill 3 testing an exploration well. The gas is being flared from a temporary overboard vent boom.*

*6 – The Putford Artemis standby boat and the DB101 semi-submersible crane vessel (SSCV) at work installing a topside module.
Inset – a fast rescue craft.*

7 – *The FPSO Petrojari 1 discharging in to a shuttle tanker. The Petrojarl 1, now operated by Talisman, formerly operated by ARCO is on location over the Blenheim field.*

8 – The Shell/Esso 49/19a Clipper installation. In attendance the Safe Lancia accomodation support vessel and the Santa Fe Monarch jack-up.

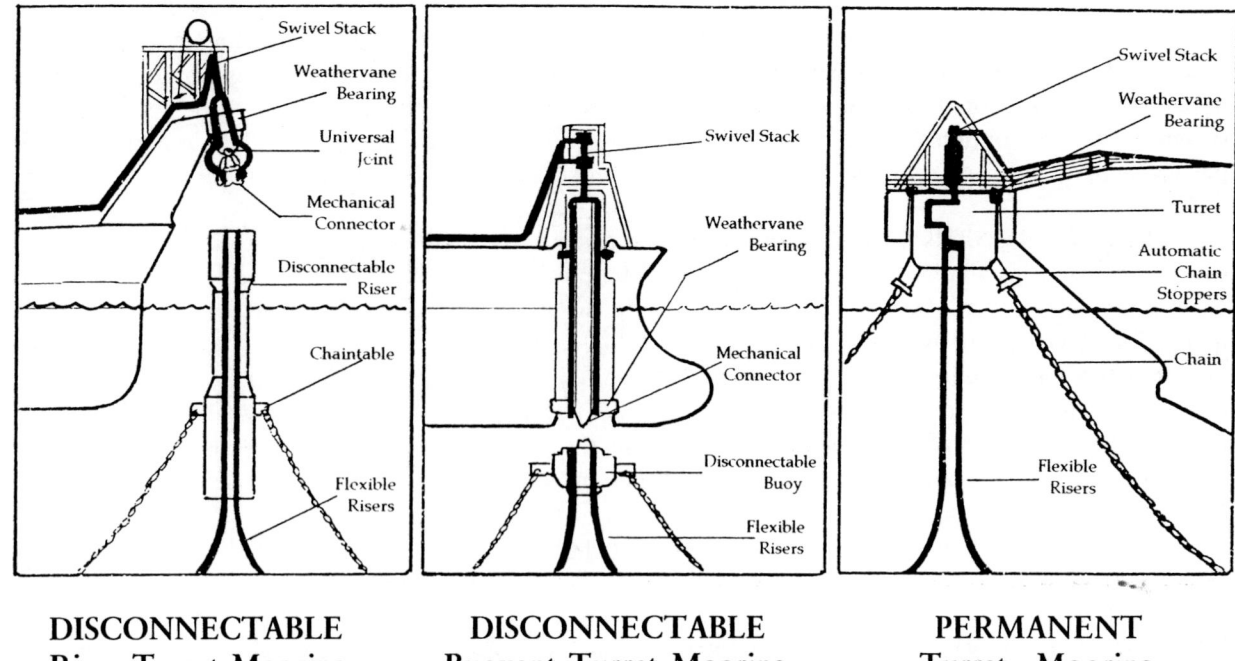

DISCONNECTABLE
Riser Turret Mooring

DISCONNECTABLE
Buoyant Turret Mooring

PERMANENT
Turret Mooring

A brief account of a fixed column SPM will now be given which is representative of the type used as a loading terminal by *Statoil* for their *Statfjord B* development.

This particular SPM (or SALM) consists of three main components:

i) **GRAVITY BASE**

The gravity base shown measures 20 × 20 × 8 metres (65 × 68 × 27 feet), weighs 950 tonnes and contains 4,500 tonnes of iron ore ballast to assist the forces of gravity in keeping the column firmly rooted to the sea bed.

ii) **COLUMN**

The steel column boasts dimensions equally as impressive as those of the gravity base measuring 170 metres (560 feet) in height, 9 metres (30 feet) in diameter and weighing 4,000 tonnes. To assist emplacement the column is filled with 2,400 tonnes of iron ore in addition to the permanent water ballast. A Cardan type articulated joint provides the means of attachment to the base a feature that also permits a degree of lateral movement. The crude oil is transferred from the manifold at the gravity base up to the rotating head by two 36 inch diameter (750 mm) pipelines routed up the outside of the column.

iii) ROTATING HEAD

The rotating head contains the quick release mooring and loading attachments, temporary accommodation and a helideck and is designed to act as a moving pivot for a ship to rotate around under the influence of wind and tide. The oil is transferred from the column into the loading hoses through a swivel coupling at discharge rates of up to 57,000 barrels an hour (8,000 tonnes) and this particular SPM has the facility to load ships ranging in size from 80,000 to 150,000 tonnes.

Riser column SPMs such as those just described have two main applications when used in support of fixed installations. Connected to a sub-sea pipeline, the riser towers provide the ideal means of loading tankers for whilst a pipeline is certainly the most cost effective means of conveying large quantities of crude oil over long distances, the SPM enables tankers to load oil destined for global export without the inconvenience and expense of entering port. They are also used in support of fixed installations located in isolated environments where the laying of a sub-sea pipeline would be impractical or prohibitively expensive. Processed oil from the platform is pumped into tankers via the riser tower which is installed at a safe distance from the installation, connected by a short sub-sea pipeline.

The other area where the SPM has grown in popularity in recent years is in the sub-sea market. When a field is developed exclusively by sub-sea wells, the oil is normally processed by a floating production system (FPS) prior to export. The wells may be connected to the floating production system by individual flexible risers attached to a turret within the hull of the vessel, or alternatively they may be manifolded on the sea bed and connected to the FPSO by a detachable single point mooring tower or buoy, often a more suitable arrangement in locations where hurricanes and typhoons occur.

Courtesy of STATOIL

OFFSHORE STRUCTURES AND SUPPORT VESSELS

Part 2. SUPPORT VESSELS

The photograph gives an indication as to the variety of vessels required to ensure the continued good health of the offshore industry. Whilst appearing to have little in common with one another apart from perhaps a heli-deck, they can loosely be divided into construction or support vessels. The larger vessels tend to be employed on the installation, maintenance and repair of sub-sea pipelines and fixed platforms and were introduced, if not fully discussed in chapter one. The smaller vessels consist primarily of diving support (DSV), survey, supply and standby boats.

The offshore supply and standby boats deserve a special mention as they will be encountered repeatedly in almost daily attendance to the production installations and construction fleets, the DSV and survey vessels are discussed in subsequent chapters.

1. THE OFFSHORE SUPPLY BOAT

The distinctive profile of the offshore supply boat cannot be confused with any other vessel. The high bow and forward accommodation are designed to withstand the severest of weather conditions and permit 360° of unrestricted vision from the wheelhouse whilst the long flat afterdeck provides an ideal platform for the stowage of containers, drill pipe and the occasional 20 ton anchor. Below deck, a refrigerated cargo hold facilitates the transportation of perishable food stuffs and potable water, diesel fuel, cement and barytes are carried in purpose built tanks.

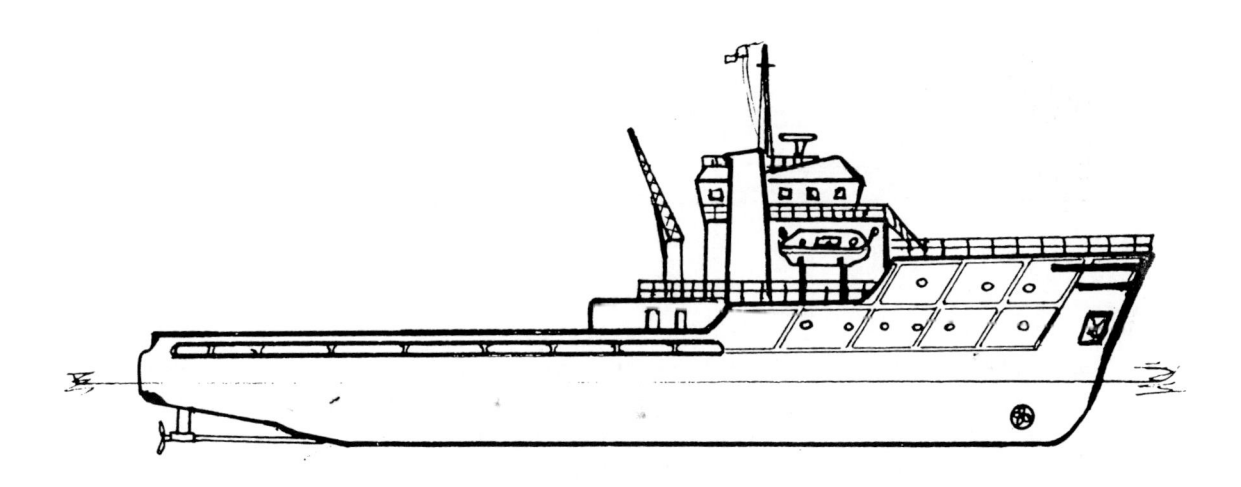

SUPPLY BOAT

More than anything else the supply boat personifies power and manoeuvrability. A twin engine, twin propeller arrangement provides propulsive power and a degree of mechanical redundancy whilst twin rudders and bow thrusters provide the manoeuvrability. The entire operation is controlled from a wheelhouse which contains duplicate controls facing forward and aft. The helmsman has literally finger tip control over the vessel to enable him to maintain close attendance to an offshore installation whilst the stores are plucked from the after deck by the platform crane.

To ensure regular employment in a fiercely competitive market the supply boat has had to develop into a *jack of all trades* and in addition to the basic function of delivering stores the more modern vessels will be encountered towing rigs, handling the anchors of semi-submersible vessels and providing fixed installations with fire fighting cover during major repair programmes.

Offshore Engineering

It would be true to say that the offshore industry would rapidly grind to a halt were it not for the services provided by the supply boat fleet.

2. STAND-BY VESSELS

Under Statutory Instrument No. 1542, *Offshore Installations (Emergency Procedures)* 1976, now replaced with the PFEER regulations, *Offshore Installations (Prevention of Fire and Explosion, and Emergency Response) Regulations, SI 1995/743S 743*, all manned offshore installation located in the UK sector of the North Sea had to have in attendance a stand-by vessel. This vessel was to be capable of accommodating the entire compliment of the installation and of providing first aid facilities.

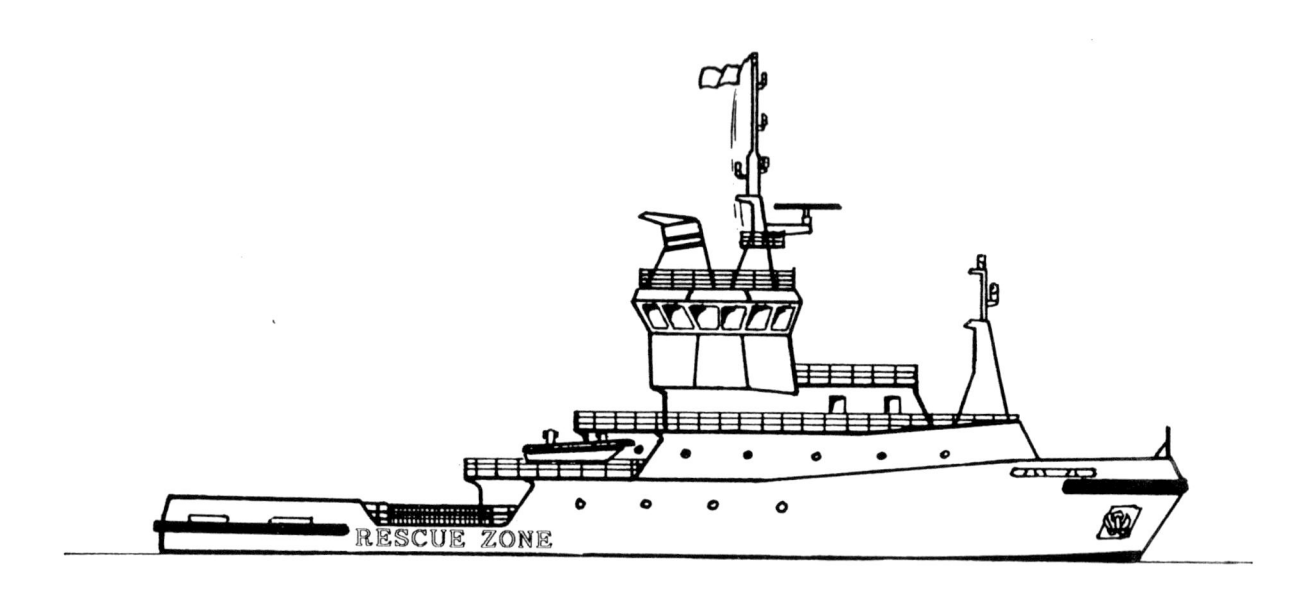

STAND-BY BOAT

The Regulations permitted one stand-by vessel to cover any number of installations located within a five mile radius. In practise, the operators of offshore installations tended to employ a number of vessels considerably in excess of this minimum requirement, particularly when engaged in operations such as underdeck scaffolding, painting and inspection programmes where there was a possibility that individuals may fall into the sea. In the last 15 years over 125 people have fallen into the sea of which 90 were recovered safely (these figures do not include those rescued following the Piper Alpha disaster).

Prior to the *Piper Alpha* disaster in 1988, the majority of the 160 strong stand-by fleet operating within the UK sector of the North Sea consist of converted trawlers. However, the disaster highlighted deficiencies in their operational capabilities and they were severely criticised during the *Cullen Inquiry*.

In response to the Cullen Report, the *Department of Transport* and the *Health and Safety Executive* (HSE) formulated a code of practise for stand-by vessels aimed primarily at improving their manoeuvrability and speed of response. New vessels must be at least 11 metres in length, capable of a speed of 10 knots, and provide 360° unrestricted vision of the surrounding seas from the wheelhouse. They must be of either twin screw (propeller), or single screw assisted by a 360° thruster powered propulsion unit, and be fitted with a bow thruster. They must also carry two fast rescue craft available for immediate deployment.

OFFSHORE STRUCTURES AND SUPPORT VESSELS

A certain latitude existed as to the acceptance of older vessels which did not comply fully with the new recommendations but the days of the converted trawler were numbered and there are few left now in service. Under the PFEER regulations the requirements for stand-by vessels will be determined on an installation by installation basis, governed primarily by the results of the *assessment* required by regulation five, and the requirement under regulation 17 to provide, *effective arrangements for recovery and rescue*, which only, *a suitable vessel standing by will provide*. The purpose of the assessment is to identify which events could give rise to a major accident involving fire or explosion, or the need to evacuate the installation.

Further information

i) Assessment of The Suitability Of Stand-by Vessels Attending Offshore Installations published by HMSO.
ii) Guidelines for The Management Of Emergency Response for Offshore Installations, published by UKOOA.

OFFSHORE SUPPORT VESSELS (see overleaf)

1	Multi-functional support vessels (MSVs)	10	Heavy transportation vessels
2	Accommodation units	11	Anchor handling, tug, supply vessels
3	Fallpipe dumping vessels	12	Diving support vessels
4	Pipeline bury barges	13	Well service vessels (see also 18)
5	Seismic survey vessels	14	Multi purpose vessels (see also 1)
6	Flexible pipelay vessels	15	Reel pipelay vessels
7	Derrick lay barges	16	Pipelay barges
8	Pipelay ships	17	Standby vessels
9	Derrick and crane barges	18	Well stimulation vessels

(Reproduced by permission of Oilfield Publications Limited
Homend House, 15 The Homend, Ledbury, Herefordshire HR8 1BN)

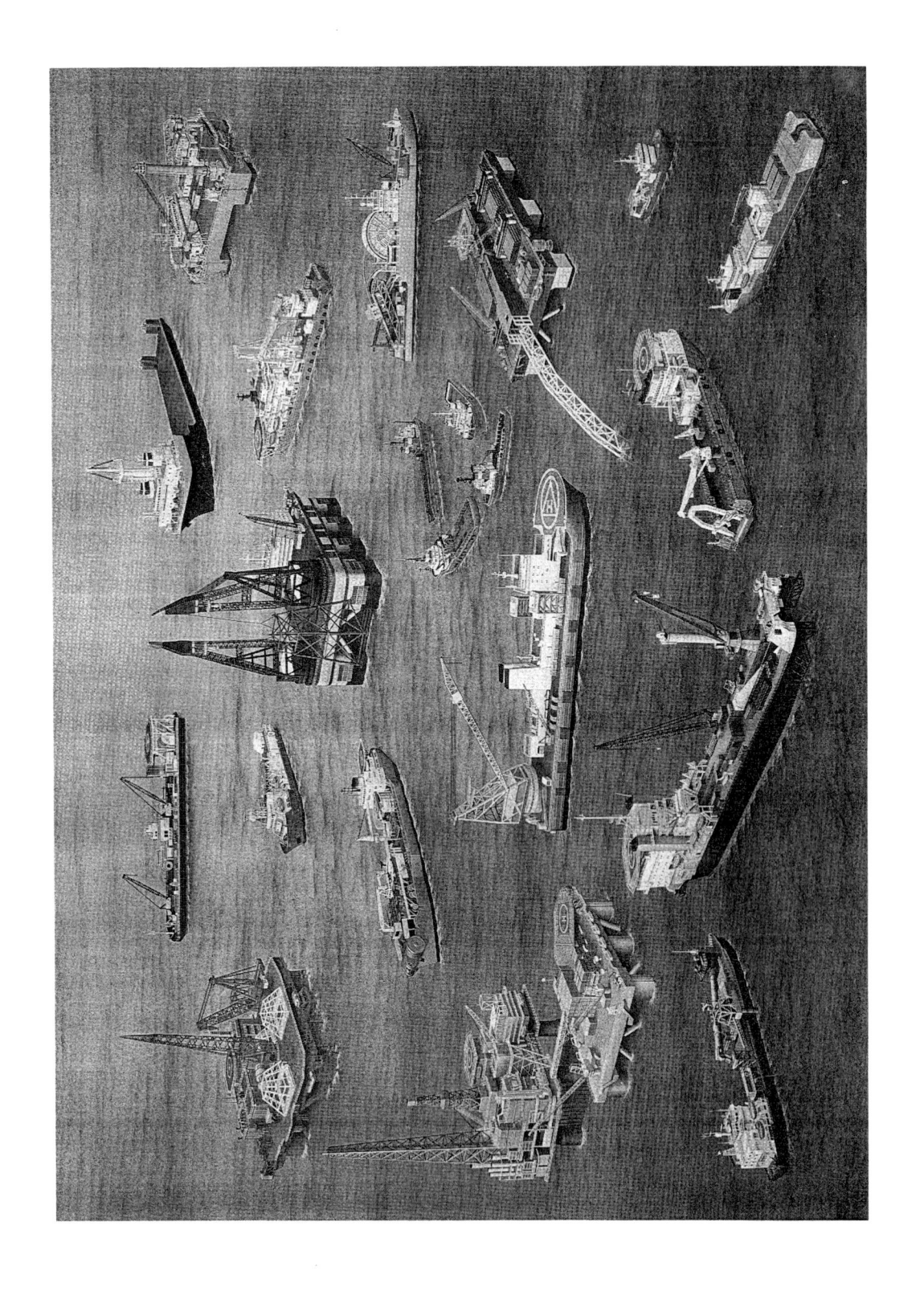

Offshore Engineering

OFFSHORE SUPPORT VESSELS
(Reproduced by permission of Oilfield Publications Limited
Homend House, 15 The Homend, Ledbury, Herefordshire HR8 1BN)

Part 3. OFFSHORE INSTALLATIONS DESCRIPTION

The sketch shows a typical offshore installation. The main component parts may be described as:

1   Accommodation

The accommodation block provides a full range of hotel services designed to cater to the needs of the personnel employed in the operation and maintenance of the installation.

On new installations the accommodation may be designated as the Temporary Refuge (TR) and will be designed to provide maximum protection to personnel during an emergency situation.

2   Wellhead Area

The wellhead area contains the Christmas trees which regulate the flow of hydrocarbon products from individual wells to the process equipment.

3   Process Area

The process area contains the pressure vessels and associated equipment required to remove impurities and bi-products from the oil or gas prior to their discharge into the sub-sea pipeline.

4   Power Generation

The majority of offshore installations are located a considerable distance from the coast and as such must be self-sufficient in all aspects, including the generation of electricity. The alternators may be driven by reciprocating diesel or gas fuelled engines, or by gas turbines.

5   Heli-deck

Helicopters provide the means by which personnel are transported to and from offshore installations and they are used as the primary means of evacuation in the event of an emergency. In UK waters, helicopters and heli-decks come under the jurisdiction of the Civil Aviation Authority (CAA).

6   Lifeboats

In the event of an emergency which necessitates abandonment of the installation, the lifeboats provide a means of escape to the sea in the absence of helicopter assistance.

7   Radio Mast

A steel tower which accommodates communication components such as satellite and telemetry dishes.

8   Vent Stack – (gas producing installations only)

Vertical, open ended discharge pipe through which process gas may be expelled to atmosphere in order to depressure and make safe the gas process equipment.

9   Flare Stack – (oil producing installations only)

The flare stack provides a safe, remote location for the disposal by burning of unwanted gaseous hydrocarbon bi-products produced during the oil refining process.

10   Drilling Derrick – (normally oil producing installations only)

The drilling derrick is used throughout the life of the installation to drill new wells, to drill wells for enhanced oil recovery, and to modify and repair existing wells.

11 Pedestal Cranes

The cranes are used to assist in maintenance operations and to facilitate the unloading and loading of stores from supply boats.

12 Cellar Deck

Lowermost deck in process area.

13 Spider Deck

Walkway located above the high water line which facilitates inspection and maintenance of the jacket structure. It also provides an emergency escape route to the sea.

14 Jacket

Tubular steel support structure for the topside modules.

15 Conductor Guide Frame

Guide frames are located at regular intervals both above and below sea level to restrain the wellhead conductors against lateral movement.

16 Conductor

Section of pipe extending from the sea bed to the wellhead area. The conductor supports the wellhead and Christmas tree and contains the casing strings and production tubings which conduct oil or gas from the reservoir to the installation.

17 Riser

Section of the sub-sea pipeline extending from the sea bed to the emergency shutdown valve (ESDV) on the installation.

18 Riser Clamp

Clamp or clamps used to secure the riser to the jacket.

19 Caisson

Tubular steel pipes or casings extending to a position below the lowest sea water level. They may accommodate deep well pumps for fire fighting and service water facilities, or provide a disposal route for drainage waters.

20 Mud Mats

Steel plates attached to the base of each leg to prevent over penetration of jacket into a soft sea bed.

21 Pile Clusters

Fitted on deep water jackets to house the foundation piles.

22 "J" Tubes

Open ended "J" shaped pipe attached to the jacket structure and extending from cellar deck to sea bed. Provides protection for flexible flowlines and umbilicals emanating from sub-sea wells.

OFFSHORE STRUCTURES AND SUPPORT VESSELS

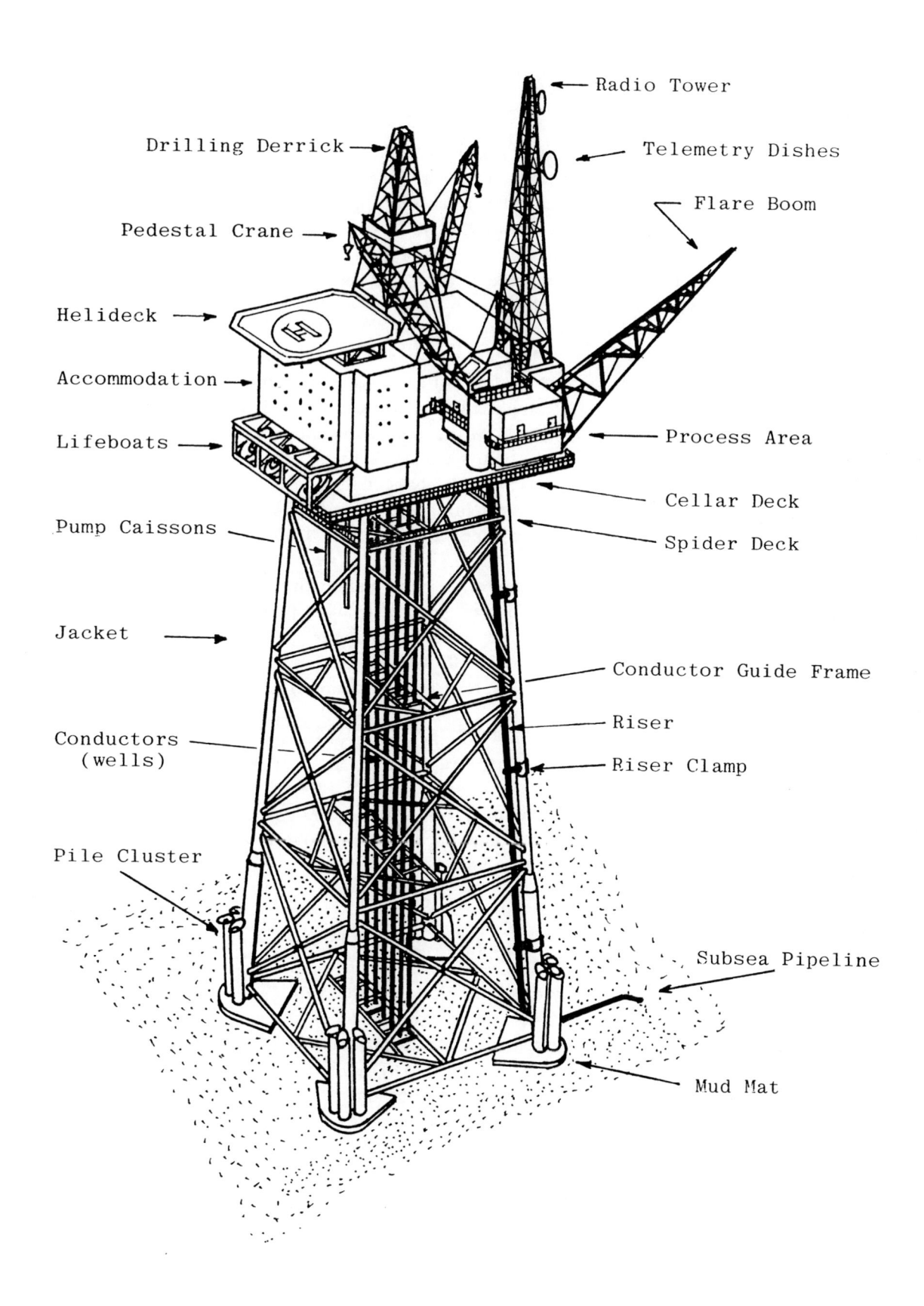

FIXED STEEL STRUCTURE

Part 4. FIXED STEEL STRUCTURES – INSTALLATION

INTRODUCTION

The North Sea represents one of the most hostile marine environments in the world and a considerable amount of specialised technology has been developed over the years to cope with the tantrums of mother nature. The bulk of the offshore engineering prior to 1965 was carried out in the relatively calm waters of the Gulf of Mexico and it was Gulf technology which was used to construct the early North Sea installations. In fact some of the first process modules were actually built in the USA and shipped over to the UK. Since those early days the technology developed to cope with the North Sea's extremes of weather has elevated British offshore engineering into a position as a world leader.

For the purpose of this chapter we are dealing with fixed steel structures which represent the vast majority of offshore installations. They vary considerably in size and on the UKCS (UK Continental Shelf) there is a very definite North-South divide created by the tremendous differences in water depths, weather conditions, and platform complexity. The southern sector installations stand in relatively shallow water depths of 12 to 40 metres (40 to 135 feet) and the basic nature of the gas processing equipment permits the construction of small, lightweight structures. The support structures or jackets weigh in the region of 250–1,500 tonnes and the topsides or superstructures from 1,500–10,000 tonnes.

In the northern sector jacket weights tend to be in the region of 5,000–20,000 tonnes although the giant rigs installed in the boom years are much bigger. The heaviest jacket belongs to the *BP Magnus* installation which weighs 35,000 tonnes. These colossal structures are required to support topside weights of up to 40,000 tonnes in water depths of 100 to 160 metres (325 to 550 feet) and reflect the considerable quantity of process and drilling equipment required to produce and process oil.

The jacket and topside dimensions quoted for the North Sea are representative of the size of installations found at all geographic locations where water depths are similar.

Offshore structures are constructed on the mainland and their design must reflect how they will be installed offshore. They are assembled in building brick fashion as can be seen from the sketches. The various stages of installation will now be explained.

1. SUB-SEA TEMPLATE

The first installation operation involves the citing of the sub-sea template on the sea bed. The template is piled into the sea bed in a location considered to be the most favourable for reaching the hydrocarbon deposits. The main objective of the template is to provide a guide frame through which wells can be drilled prior to the arrival of the jacket. Drilling wells takes a considerable amount of time and it is advantageous to enlist the services of a drilling vessel to pre-drill some, or all of the wells whilst the jacket is under construction on the mainland.

In addition to well guides, guides are provided in the template to assist in the accurate positioning of the jacket over the template. Sub-sea templates are unlikely to be deemed necessary on small installations situated in shallow waters.

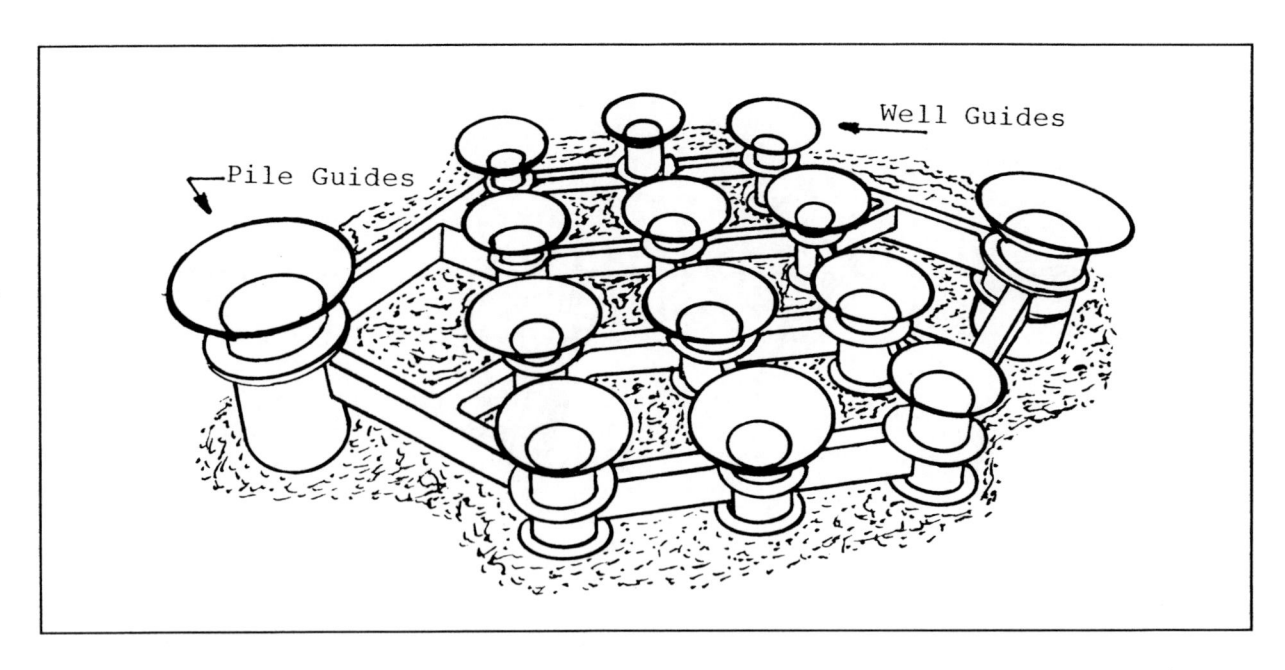

SUB-SEA TEMPLATE

2. JACKET

Tubular steel jackets are completely fabricated onshore prior to transportation to site by dumb barge. The smaller jackets may be lifted in place by a floating crane whilst the largest jackets may employ flotation devices to assist in their installation once launched from the barge The flotation devices are sequentially flooded to enable the jacket to sink slowly into its final resting place. Once located on the sea bed the jackets are normally secured by foundation *piles*.

Jacket size dictates the method of pile installation as can be seen from the sketches. On the smaller jackets tubular piles are driven through the legs to a pre-determined depth of 30–50 metres (100 to 165 feet), or inserted into pre-drilled holes and grouted (cemented) when the formation resists conventional piling techniques. The piles are fitted with spacers to ensure that they locate centrally within the jacket leg and on completion of piling, steel shims are inserted between the top of the pile and the jacket leg and welded. The pile is then cut level, just above the top of the jacket and grout is pumped into the base of each jacket leg until it flows from a vent hole cut in the top of the leg. Filling the annulus space formed by the jacket leg and pile in this manner ensures an extremely rigid structure is produced. This arrangement has been used to secure jackets in water depths of up to 100 metres (335 feet).

The immense size of the deep water jackets makes piling through the legs impractical so a pile cluster is fitted to the base of each leg. The piles can then be driven through the cluster guides to the required depth of 60–90 metres (200 to 300 feet) before being grouted into position. Over 6,000 tonnes of piles were used to secure the *BP Magnus jacket*.

An alternative to piling has recently been developed by Statoil and used to secure the *Europipe* riser jacket. The jacket legs sit within *skirted mud mats*, essentially large inverted buckets, 12 metres (40 feet) in diameter and 6 metres (20 feet) deep which penetrate the sea bed under the weight of the jacket, the jacket legs being filled with water ballast during installation to assist the penetration process. Suction provides further penetration prior to the addition of 2,200 tonnes of permanent grout ballast which was poured into the top of the buckets. This novel concept offers considerable savings on installation time and costs and is sure to be more widely used in the future.

Offshore Engineering

Deep water launch

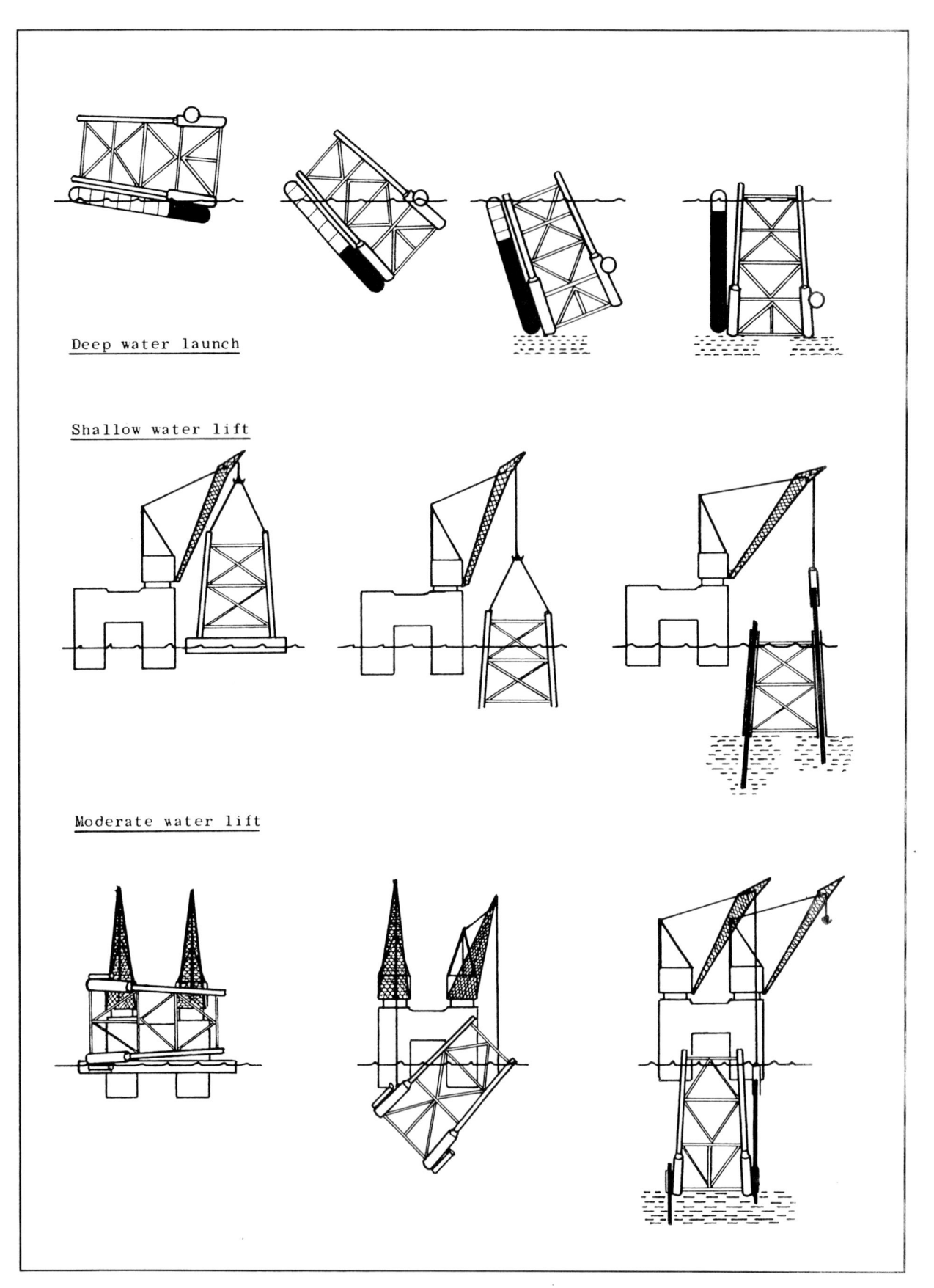

Shallow water lift

Moderate water lift

JACKET INSTALLATION

OFFSHORE STRUCTURES AND SUPPORT VESSELS

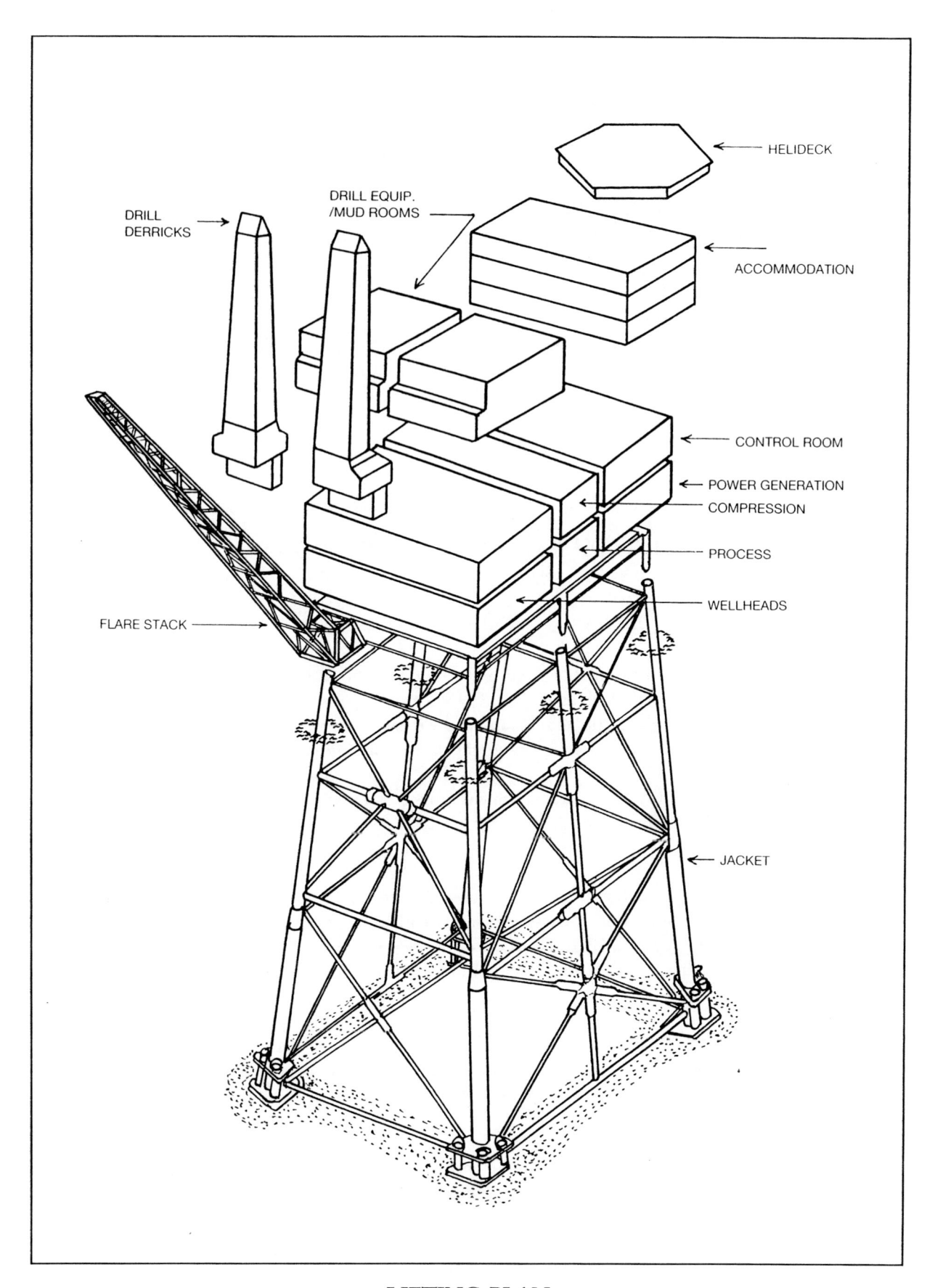

LIFTING PLAN

Offshore Engineering

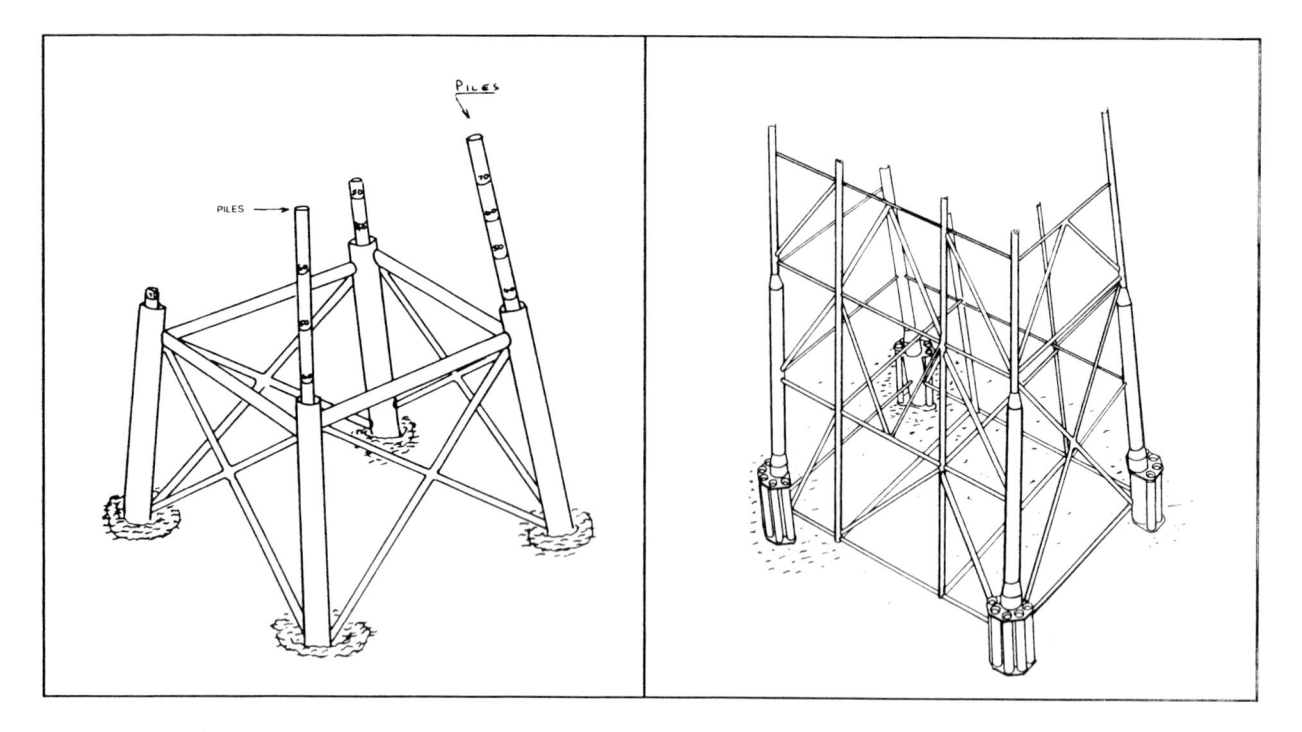

PILING – THROUGH JACKET   PILE CLUSTERS – DEEP WATER

3. TOPSIDE INSTALLATION

The topside weight of an installation may be as low as 1500 tonnes for an unmanned installation in the southern North Sea or it may be as high as 35,000 tonnes for one of the northern sector monsters. Whilst a floating crane can lift 1,500 tonnes on to a jacket, a single lift of 35,000 tonnes is clearly beyond the capacity of any crane. Consequently, the larger platforms must be constructed within the constraints of the lifting facilities available, which to date is approximately 14,000 tonnes. This entails constructing the topsides in liftable packages which can be installed and secured by welding one at a time, the world record lift being the integrated deck of the Conoco-Chevron *Britannia* installation which weighed 11,000 tonnes.

The heavy lift cranes used for installing rigs are mounted on semi-submersible barges which have been purpose built for offshore work. These enormous catamaran type vessels with deck capacities of upto 11,500 tonnes often accommodate the smaller topsides modules as deck cargo for transportation to the site. Where this is not practical the modules are loaded onto flat top barges and towed to their destination.

To assist installation the modules and jackets are fitted with male and female type location devices referred to as stab ins or bucket guides examples of which can be seen in the installation photographs. On shallow water jackets with through-leg piles, the topside structure is actually welded to the top of the pile, rather than to the jacket.

Assembly of an offshore installation in this manner creates tremendous competition within the module fabrication industry. Until recently a module destined for service in the UK sector of the North Sea would have been constructed almost entirely in the UK. However, the advent of the common market has seen the emergence of the *Euro rig* which can contain a collection of modules fabricated in the UK, Spain, France, Italy and Portugal and the Britannia installation completed in 1998 is a good example of this type of practise.

30

The Britannia is the largest gas platform to be installed in the UK waters of the North Sea this decade and will supply 8% of the UK's gas. The jacket weighs 21,000 tonnes and was constructed in Dragados in Spain whilst the topside structures which weigh 28,000 tonnes were built in the UK. The jacket is secured by the longest piles in the North Sea, 123 metres (400 feet) in length and 2.74 metres (9 feet) in diameter.

4. ABANDONMENT

De-commissioning and abandonment of offshore installations is rapidly becoming a growth industry, particular in locations such as the US Gulf of Mexico where there are currently in excess of 4000 installations. In Europe there are considerably fewer installations but they tend to be much larger, there being approximate 220 structures in the UK sector of the North Sea and about 70 structures in the Norwegian sector. Further afield there are nearly 100 offshore installations in Thailand, 200 in Malaysia, over 400 installations in Indonesia and 700 in the Middle East and they will all require removal one day.

There are three main options available for the removal of offshore installations that are considered to be commercially unviable propositions. In all cases the topside equipment and decks are removed which basically leaves the jacket structure to be disposed off.

The first option involves removing the jacket in it's entirety which is the favoured option for jackets located in shallow waters, the jacket simply being cut at the sea bed and lifted onto a barge. Alternatively, the upper jacket structure may be removed with the lower section left in place, cut at a depth below the surface where it will not prove a hazard to shipping. The third option is to use explosives to topple the jacket onto the sea bed where it will be left to form a natural reef.

Oil companies are keen to adopt what is perceived by the general public to be the correct environmental option when decommissioning offshore installations and considerable thought is given to the removal and recycling processes. However, some of the very large structures are simply to big to permit their safe removal and the artificial reef is likely to prove the only viable solution to the problem.

Chapter Two

THE NORTH SEA HISTORY AND LEGISLATION

SECTION ONE — THE NORTH SEA
PART 1.   THE STORY SO FAR

PART 2.   OIL AND GAS DISTRIBUTION

SECTION TWO — LEGISLATION
PART 1.   HISTORY

PART 2.   NEW LEGISLATION

PART 3.   THE SAFETY CASE

PART 4.   STATUTORY INSTRUMENTS

PART 5.   ASSOCIATED INFORMATION

Offshore Engineering

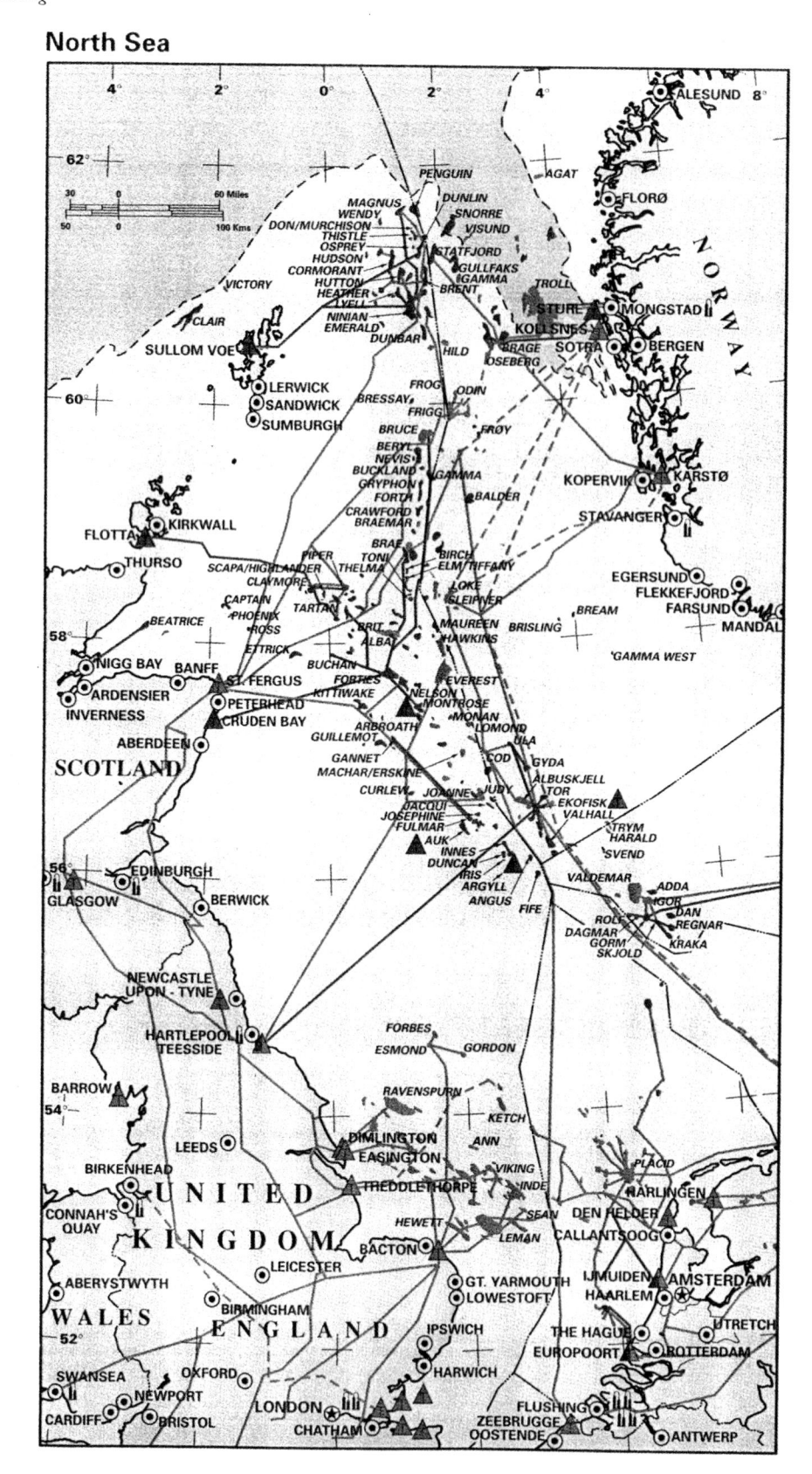

North Sea

THE NORTH SEA U.K. OIL AND GAS FIELDS
(Reproduced by permission of Oilfield Publications Limited
Homend House, 15 The Homend, Ledbury, Herefordshire HR8 1BN

Part 1. THE NORTH SEA THE STORY SO FAR

INTRODUCTION

The object of this chapter is to provide the reader with a brief history of the North Sea oil and gas industry and to place the UK's involvement in European and world wide hydrocarbon production into perspective.

GAS

The UK Offshore gas industry dates back to 1963 when exploration of the southern North Sea commenced following the discovery of a large onshore gas field at Groningen in Holland. The first offshore gas field, the West Sole situated off the Humber Estuary was discovered in 1965 and this was quickly followed by the Leman gas field approximately 50 miles off the coast of East Anglia. For many years the Leman field was the largest offshore gas field in the world and it is still one of the largest producers of gas for the UK, supporting in excess of 35 installations.

Today there are over 75 gas producing fields located in the southern North Sea supporting over 100 installations which produce 250 million cubic metres of gas a day, making the UK by far the largest producer of offshore gas in Europe. However, as recently as 1990 the industry was regarded as being in decline with production figures predicted to be halved by the year 2000. There has since been a dramatic about turn initiated largely by the liberalisation of the UK gas market following the privatisation of British Gas, and by the recent European Union Directive which opened up national markets. These changes have stimulated the development of both new and existing fields resulting in a return to record levels of production.

OIL

The offshore oil industry has experienced far greater periods of growth and restraint than the gas industry because historically, oil has always been more of an international commodity than gas and has consequently been subjected to significant price fluctuations following periods of political instability. It was not until 1969 that the colossal Ekofisk oil field was discovered in the Norwegian sector of the North Sea and it was a further two years before it went into production. The Argyll field heralded the start of oil production from the UK sector and the subsequent development of over 111 oil and condensate fields elevated Britain into the position of leading European oil producer, a position it held until recently.

During the latter half of the 1980s, oil production, like the gas, was in decline having thought to have peaked in 1985, but, a resurgence in field developments has once again seen production running at record levels. Today, approximately 2.7 million barrels of oil a day are produced from the UK sector compared to 3.1 million barrels from the Norwegian Continental Shelf. This increase in production can largely be attributed to the drive towards increased efficiency and the reduction in operation and development costs brought about by the CRINE initiative, the *cost reduction in the new era*. The cost of producing a barrel of oil (including development costs) has fallen from $24 in 1980 to $12 in 1998.

Offshore Engineering

NORTH SEA MAJOR OIL AND GAS FIELDS — 1999

OIL FIELDS	UK				NORWAY			
Field Operator	Year	Original Reserves mnbbls	Daily Production barrels		Field Operator	Year	Original Reserves mnbbls	Daily Produce barrels
1. BRENT Shell	1976	1,800	125,000		1. OSEBERG Norsk Hydro	1988	1,460	390,000
2. NELSON Enterprise	1994	458	115,000		2. GULFAKS Statoil	1986	1,321	363,000
3. SCOTT Amerada Hess	1994	450	115,000		3. STATFJORD Statoil	1979	3,240	250,000
4. FORTIES BP	1975	2,470	95,000		4. EKOFISK Phillips	1971	1,700	220,000
5. ALBA Chevron	1994	375	89,000		5. TROLL B Norsk Hydro	1994	1,200	210,000
6. HARDING BP	1996	185	85,000		6. DRAUGEN Shell	1992	410	205,000
7. FOINAVEN BP	1997	230	85,000		7. HEIDRUN Statoil	1995	750	200,000
8. MILLER BP	1992	300	83,000		8. SNORRE Saga	1992	770	165,000
9. MAGNUS BP	1983	645	73,000		9. NORNE Statoil	1998	450	120,000
10. BERYL Mobil	1976	835	73,000		10. VALHALL Amoco	1982	247	95,000

GAS FIELDS	UK				NORWAY			
Field Operator	Year	Original Reserves Bcm	Daily Production Mcm		Field Operator	Year	Original Reserves Bcm	Daily Produce Mcm
1. MORECAMBE Bay–Centrica	1985	173	56.5		1. TROLL Statoil	1994	1,290	80
2. BRITANNIA Chevron/Conoco	1998	88	19.8		2. SLEIPNER Statoil	1993	186	19
3. BRENT Shell Expro	1976	119	17.4		3. SNORRE Saga	1992	7	9.7
4. BRUCE BP	1993	74	16.0		4. STATFJORD Statoil	1979	73	9.7
5. BRAE AREA BP	1983	63	12.8		5. MURCHISON Conoco	1980	2	9.6
6. ARMADA British Gas	1997	34	11.5		6. EKOFISK Phillips	1971	143	8.5
7. LEMAN Shell Expro	1969	162	10.5		7. GULFAKS Statoil	1986	14	5.5
8. SOLE PIT Shell Expro	1990	50	9.9		8. HEIMDAL Elf	1986	34	3.8
9. LEMAN BP-Amoco	1969	160	7.2		9. VALHALL BP-Amoco	1982	21	2.7
10. INDE Shell/Amoco	1971	134	5.0					

NORTH SEA OIL AND GAS PRODUCTION SUMMARY

OIL PRODUCTION			GAS PRODUCTION		
Barrels/day	Number of Producing Fields		Million scf/day	Number of Producing Fields	
	UK	Norway		UK	Norway
0 – 10,000	27	7	0 – 50	50	8
10,000 – 20,000	18	3	50 – 100	17	2
20,000 – 30,000	10	1	100 – 200	37	2
30,000 – 40,000	9	0	200 – 300	4	0
40,000 – 50,000	9	0	300 – 400	1	4
50,000 – 60,000	3	2	400 – 500	3	0
60,000 – 70,000	3	1	500 – 600	2	0
70,000 – 80,000	3	2	600 – 700	0	1
80,000 – 90,000	3	1	700 – 800	0	1
100,000 – 200,000	4	1	800 – 900	0	1
200,000 – 300,000	0	3	2,010	1	0
300,000 – 400,000	0	5	2,857	0	1
400,000 – 500,000	0	2			

Note: 20,000 barrels/day = 1 million tonnes a year

As can be seen from the table, there are considerably more oil and gas fields located in the UK sector of the North Sea than in the Norwegian sector, however, the Norwegian fields tend to be much larger in size.

The UK is the largest producer of offshore gas in Europe producing twice as much gas as Norway. Norway is the largest producer of oil in Europe producing approximately 20% more oil than the UK.

The purpose of the tables in this chapter is to provide a picture of activity levels in the North Sea, and to give an indication of field size and daily production figures for comparison with other geographic locations. In terms of oilfield size, an oil field with recoverable reserves of 400–500 million barrels is still regarded as being quite a respectable development proposition with the vast majority of new finds now ranging from 20–70 million barrels. As far as production rates are concerned there are relatively few fields which produce in excess of 100,000 barrels a day, 15,000–50,000 barrels being more the norm.

As far as the gas is concerned, a field containing reserves of 70–80 billion cubic metres is regarded as being a sizeable find, 15–20 billion cubic metres currently representing the average size discovery, whilst fields of 1.5–2 billion cubic metres are still considered to be worth developing. Whilst there are some prodigious gas producers, the vast majority of gas fields produce less than 100 mscf/day (2.8 mcm) once the initial pressure peak has subsided.

Offshore Engineering

WORLD'S TOP TEN

OIL RESERVES (1,000 tonnes)			GAS RESERVES (billion cu ft.)		
Total World	139,742,923	100.00%	Total World	4,945,362	100.00%
1. Saudi Arabia	35,479,452	25.39%	1. Russian Federation	1,700,000	33.42%
2. Iraq	15,342,466	11.03%	2. Iran	810,000	15.92%
3. Kuwait	12,876,712	9.21%	3. Qatar	300,000	5.90%
4. Iran	12,739,726	9.12%	4. Saudi Arabia	190,000	3.74%
5. Abu Dhabi	12,630,137	9.04%	5. Abu Dhabi	189,000	3.72%
6. Venezuela	8,887,397	7.03%	6. USA	166,474	3.27%
7. Russian Federation	7,808,219	4.76%	7. Venezuela	143,078	2.81%
8. Mexico	6,684,384	3.92%	8. Algeria	130,600	2.57%
9. Libya	4,041,096	2.89%	9. Nigeria	114,852	2.26%
10. China	3,287,671	2.35%	10. Iraq	109,800	2.16%
Others	21,326,227	15.26%	Others	1,232,665	24.23%
OIL PRODUCTION (1,000 tonnes)			**GAS PRODUCTION (billion cu ft.)**		
Total World	3,283,255	100.00%	Total World	78,472	100.00%
1. Saudi Arabia	395,000	12.03%	1. USA	19,249	24.53%
2. USA	320,000	9.75%	2. Russian Federation	18,744	23.89%
3. Russian Federation	295,000	8.98%	3. Canada	5,535	7.05%
4. Iran	180,500	5.50%	4. United Kingdom	3,071	3.91%
5. China	160,500	4.89%	5. Indonesia	2,436	3.10%
6. Norway	158,750	4.84%	6. Algeria	2,383	3.04%
7. Venezuela	158,750	4.84%	7. Netherlands	2,369	3.02%
8. Mexico	151,250	4.61%	8. Norway	1,649	2.10%
9. United Kingdom	131,000	3.99%	9. Uzbekistan	1,620	2.06%
10. Nigeria	113,500	3.46%	10. Saudi Arabia	1,550	1.98%
Others	1,119,005	34.08%	Others	19,866	25.32%

Note: 1 trillion = one thousand billion (10^{12})   1 billion = one thousand million (10^9)
1 billion cubic metres of gas = 0.83 million tonnes of oil equivalent.

Note: The figures quoted above are for the year ending 1997. Whilst the figures will vary from year to year it is unlikely that table positions will change significantly and they will thus provide a clear indication of the general order of the main contenders.

THE NORTH SEA – HISTORY AND LEGISLATION

FIELD DEVELOPMENTS — THE PRESENT

The oil and gas fields currently under development tend to be much smaller than those that represent the golden years of the offshore industry. The Norwegian Troll field may appear to contradict this statement as it is nearly four times the size of the original Leman Field. However, the field was actually discovered in 1974 but development was delayed due to the difficulty in tapping the gas bearing formations. Advances in drilling technology have since permitted full exploitation of the fields reserves.

A more realistic picture of current development trends within the UK sector may be obtained by considering the reserves contained within the BP-ETAP, *Eastern Trough Area Project*, development. The ETAP development is one of the largest of the decade and will eventually provide nearly 10% of the UK's oil. It contains estimated oil reserves of 435 million barrels which equates to an output of 216,000 barrels a day. The development, which comprises seven small reservoirs, also contains significant quantities of gas and *natural gas liquids* (NGL). Similarly, the Britannia field, operated by Chevron and brought on stream in 1998, is the largest gas field currently under development and with estimated reserves of 81 billion cubic metres of gas and large reserves of condensate, it will meet almost 10% of the UK's gas demand.

It is worth noting at this point that whilst the BP-ETAP development is operated by BP there are numerous other oil companies listed as licensees or joint owners, these being Shell, Esso, AGIP, Mitsubishi Oil, Murphy and Total. This form of sharing the development and operational costs, and subsequent profits is referred to as 'partnering' and is used extensively all over the world on virtually every development.

The fields currently under development in the UK may appear to be small when compared to the Leman and Forties fields that heralded the birth of the oil and gas industries but they compare favourably with field sizes elsewhere in the world. For instance, the Mars field is the largest find in the Gulf of Mexico for 20 years and at 500 million barrels is comparable with the previously mentioned ETAP development. Similarly, the Norne field, with reserves of 450 million barrels, is the largest discovery on the Norwegian continental shelf in more than a decade. They represent the norm within the offshore industry and are far more numerous than the giant discoveries that tend to make the headlines.

An air of optimism still prevails in the North Sea. The figures for proven reserves show that only one third of the UK's reserves of oil and gas have been produced to date. Potential exists for the development of a further 125 oil and 30 gas fields which will require up to 50 fixed platforms, 30 floating production systems and 60 sub-sea installations, in addition to the 450 fixed structures already operating in north European waters. This should ensure the survival of the offshore industry for at least another 30 years.

THE FUTURE

Both Britain and Norway rely heavily on the oil and gas industries to maintain acceptable balance of trade figures. However, to put the industry into perspective we must appreciate that the North Sea contains less than 5% of proven world hydrocarbon reserves. Saudi Arabia, Iran, Iraq and Kuwait own over 50% of existing reserves and the Former Soviet Union (FSU or the Commonwealth of Independent States — CIS) also possess huge reserves of oil and gas.

Whilst the UK does not possess reserves which enable it to compete with the FSU and the OPEC nations, what it can provide is the technology and expertise developed from 30 years experience in the North Sea, technology and expertise that has recently tamed the Atlantic Ocean frontier provinces, West of Shetland.

As the North Sea is now considered to be a mature province the majority of oil companies are looking for development opportunities elsewhere in the world and the Caspian Sea area of the FSU is a prime target. For years their oil and gas fields have been in decline due to lack of investment and it is to the well established international oil companies that the new governments have been looking for assistance in reviving their oil and gas industries.

The four main countries in the Caspian region are Azerbaijan, Kazakhstan, Turkmenistan and Uzbeckistan and oil majors such as Agip, BP Amoco, British Gas, Chevron, Statoil and Texaco have taken a large stake in the various consortiums which have been formed to share the costs, profits and risks associated with these new ventures. The risks are great because the ownership of the oil is often in doubt as the various countries are in dispute over their territories, particularly offshore where there are problems over the establishment of maritime boundaries.

Transportation is another problem area with potentially troublesome countries bordering the new states and all these considerations, allied to a falling oil price have somewhat tempered the enthusiasm of western companies of late. However, the rewards can be considerable with the Tengiz field in Kazakhstan having estimated reserves of between 6 billion and 9 billion barrels of oil and the Karachaganak field has estimated reserves of 2.4 billion barrels of oil and 16 trillion cubic feet of gas.

OIL AND GAS PRICES

NORTH SEA CRUDE OIL PRICES (average per year in $US per barrel)							
1979	18.0	1984	28.6	1989	18.6	1994	14.9
1980	34.0	1985	27.5	1990	23.7	1995	16.6
1981	35.8	1986	14.2	1991	19.6	1996	20.8
1982	32.8	1987	18.5	1992	19.4	1997	18.5
1983	29.5	1988	15.2	1993	16.9	1998	12.9

NORTH SEA GAS PRICES

Traditionally, North Sea gas prices have not fluctuated as dramatically as oil prices because of the long term contracts signed between the major suppliers and British Gas. For several years the operators received approximately 20 pence per therm for the gas, a figure which equates to £70,000 per million cubic metres, or £2,000 per million cubic feet of gas.

Since the privatisation of British Gas the price has fluctuated wildly with spot market prices at periods of peak demand rising to as much as 70 pence a therm. However, generally the price of gas sold on long term contracts has fallen to around 14 to 16 pence a therm which has forced Centrica (formerly British Gas) to re-negotiate its contracts with the major suppliers. In this market, small fields command a premium price, there being a wider and more competitive market for small volumes than for large fields. Whilst a 50 billion cubic feet (1,400 million cubic metres) gas field is considered small, it still represents sales revenues of £100 million when sold at 20 pence a therm, a realistic spot market price. Further changes are sure to occur as the oil majors become gas suppliers to the public.

Whilst every effort has been taken to ensure the accuracy of the figures reproduced in this chapter and indeed the book as a whole, it should be noted that figures quoted by the oil and gas industries are notoriously inconsistent and vary by as much as 400%. As an example, in spite of the sophisticated measuring equipment available the UK proven oil reserves are ambiguously quoted as being somewhere between 610 and 2015 million tonnes with the undiscovered reserves estimated at between 285 and 2,680 million tonnes. Gas reserves are similarly quoted as being somewhere between 630 and 1985 billion cubic metres for proven reserves and 500 to 1,700 billion cubic metres for estimated undiscovered reserves. However, these results will provide a basis for general comparisons to be made with the global picture.

FACTS AND FIGURES PERTAINING TO THE UK OIL AND GAS INDUSTRY

1 Tonne = 7.5 barrels (crude oil).
1 cubic metre = 35.31 cubic feet.
1 tonne of oil equivalent is defined as having a calorific value of 397 therms
Current world oil demand 66.4 million barrels a day.
Average UK oil production 2.69 million barrels a day.
Total UK oil production 128,200,000 tonnes a year.
Total UK oil exports 74,300,000 tonnes.
Total UK oil imports 42,600,000 tonnes.
UK gas production 250 million cubic metres a day.
Total UK gas production 91.3 billion cubic metres a year.
Annual UK oil and gas revenue £19 billion pounds (based on £87 tonne).
Average oil production costs from North Sea fields $4 to $6 barrel.
Combined overall average cost (development and production) per barrel of oil $16.
New field, overall average cost (development and production) per barrel of oil $12.
Overall average cost (development and production) per therm of gas 13 pence.

The various tables in this chapter have been prepared so that future developments within the oil and gas industry can be compared with the established milestones of the preceding 30 years.

Oil field size is defined in terms of millions of barrels (mnbbls) of recoverable oil. The daily production figures are quoted in thousands of barrels of oil a day (bbls/d) and give an indication as to the size of the platform required to exploit a particular reservoir.

Gas field reserves are generally defined in billions of cubic metres (bcm), although they are still regularly quoted in units of standard cubic feet (scf). Daily output figures are measured either in millions of cubic metres a day (mcm/d), or millions of standard cubic feet a day (mscf/d).

Offshore Engineering

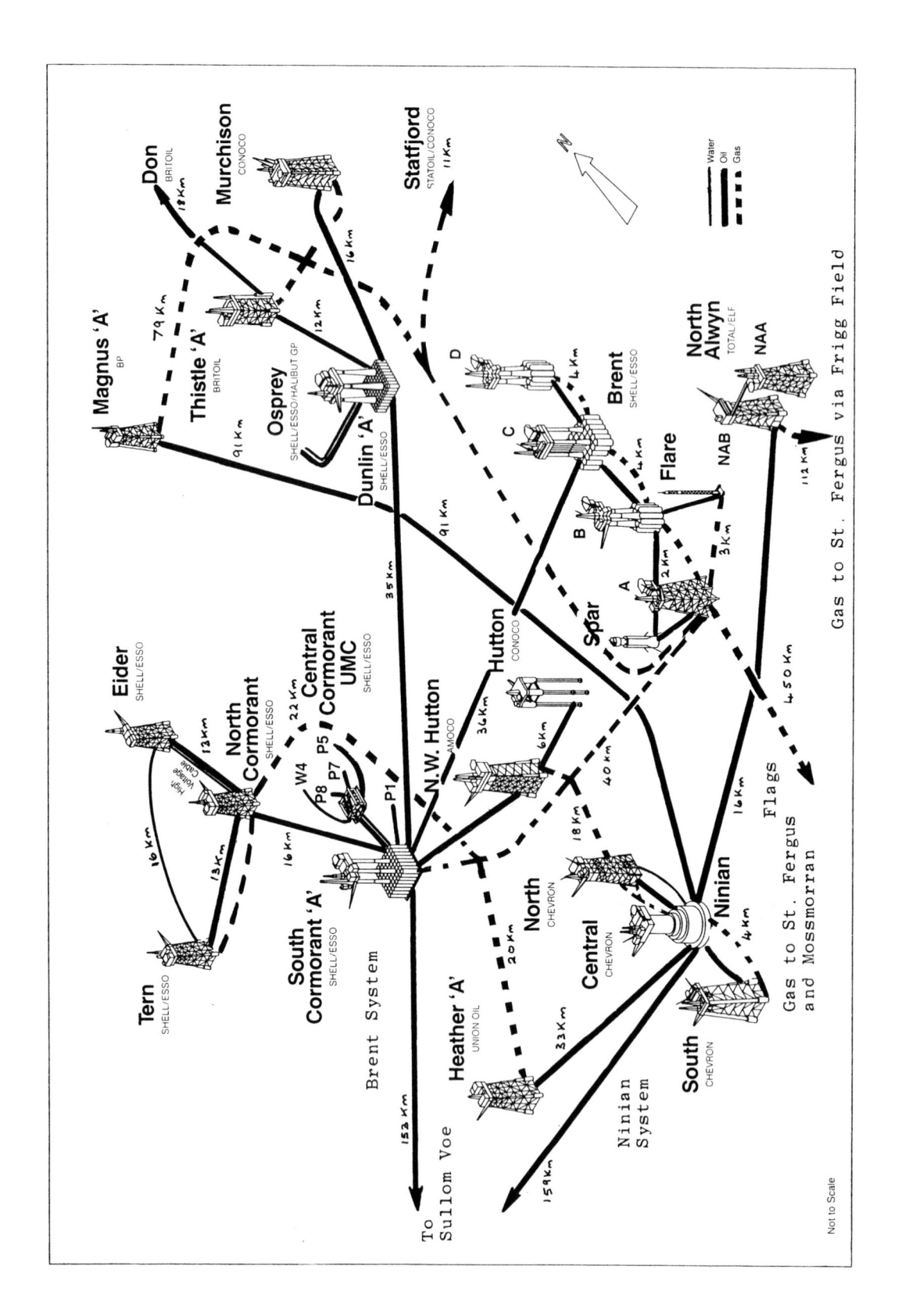

NORTHERN NORTH SEA PIPELINE SYSTEMS

Part 2. OIL AND GAS DISTRIBUTION

If the map for onshore and offshore oil and gas distribution is studied, a complex picture involving several thousand kilometres of pipeline emerges. It is estimated that the UK's onshore gas distribution network contains over 5,320 km (3,325 miles) of main trunk line, whilst the combined oil and gas sub-sea pipeline systems in the North Sea amount to over 8,700 km (5,450 miles) with almost 1,000 km (625 miles) of new pipeline being added each year.

In the UK the entire onshore gas distribution system is owned and operated by B.G. Transco (formerly British Gas) but the situation prevailing with offshore oil and gas pipelines is far more complex. The colossal cost of installing a sub-sea pipeline has created a situation whereby companies lease space from one another, or share the construction costs. This approach minimises capital investment and ensures that the pipeline operates at maximum capacity and hence profitability. These arrangements are particularly beneficial to the northern sector installations which frequently require two sub-sea pipelines, one for the transportation of oil and one for gas.

OIL DISTRIBUTION

The UK is currently the ninth largest producer of oil in the world and it is all produced from the northern sector of the North Sea, an area also responsible for supplying 15% of the UK's gas. The gas is produced as a by-product of oil production and whilst it is a valuable commodity it's existence considerably complicates the oil distribution process as can be seen from the map of the Northern North Sea Pipelines Development.

The production of oil does not suffer from the same extremes of seasonal demand that affects the gas industry and because the production of associated gas is tied to the production of oil, the quantities of gas produced during the summer months frequently exceed demand. Traditionally the excess gas was simply burnt off at the flare stack but government legislation has since deemed this practice as being environmentally unacceptable. The problem was eventually solved by the conversion of part of the ageing British Gas *Rough field* into a storage facility. During the summer months the excess gas is pumped into the reservoir for recovery at a later date.

An examination of the main distribution map will show that the majority of offshore pipelines terminate at *St. Fergus* in Scotland, *Flotta* in the Orkneys and *Sullom Voe* in the Shetlands with Teeside receiving oil and condensate direct from the Norwegian Ekofisk development. Whilst the UK is self sufficient in oil in terms of quantity, a thriving import-export industry exists in order to obtain the different grades of crude required for blending purposes. Sullom Voe and Flotta operate as major refining and redistribution centres, the deep waters surrounding the Islands providing natural berths for the supertankers used to transport oil to worldwide destinations.

GAS DISTRIBUTION

Whilst the UK is self sufficient in gas and likely to remain so for the foreseeable future, it has since 1976, imported approximately 15% of the gas consumed each year from Norway via the *Frigg* field pipeline. The gas imports were permitted under the terms of the Frigg Treaty signed between the Norwegian and UK governments due to the fact that one third of the Frigg field lies in UK waters. Over a similar period, gas was being exported to Holland.

Gas consumption in the UK is split fairly evenly between domestic and industrial usage and the demand from summer to winter varies tremendously. To accommodate large fluctuations in demand B. G. Transco use underground reservoirs located onshore in North Humberside and offshore in the Rough field near Easington. Gas is pumped underground during the summer months and recovered during periods of peak consumption in the winter.

Nearly 85% of the gas produced in the UK originates from the southern sector of the North Sea and the production and distribution processes could not be more straightforward, particularly when compared with the complexity of the oil distribution network. The smaller offshore installations supply gas to infield terminal platforms or gathering stations which compress the gas for onward transmission to onshore reception facilities at *Bacton* in Norfolk, *Theddlethorpe* in Lincolnshire and *Easington* in Humberside. The compression platforms ensure that the gas is delivered to the beach at a pressure in excess of 1125 psi (72 bar), the operating pressure of the national grid.

Prior to sale the gas requires very little in the way of preparation. The calorific value is checked and the characteristic gas smell is added to the naturally occurring odourless gas to assist the public in the detection of leaks.

Looking further afield, the European gas industry is experiencing a period of growth unseen since the early days of exploration. Norway is in the enviable position of owning some of the largest oil and gas fields in the North Sea and looks set to become the major European producer of both products. Prior to the discovery of oil and gas Norway was self-sufficient in hydro-electric power and with a population of only 4 million people there exists an unrivalled opportunity for exporting energy.

Until recently the European Economic Community (EC) restricted both the carriage of gas from one country to another, and its use for the generation of electricity. The relaxation of these restrictions has lead to a tremendous growth in the construction of gas fired powered stations and the completion of the *Interconnector* in 1998, a large 40-inch diameter pipeline linking the UK with Belgium will further increase the trading in gas. New pipelines have also permitted Norway to export to countries as far apart as Spain and Italy, who are in turn linked by pipeline to Algeria. Further links are likely to be established with the states which once formed Russia and with Iran, making gas a truly international trading commodity.

SECTION TWO – LEGISLATION

PART 1. HISTORY

1. Initial Legislation
2. Certification
3. The Cullen Inquiry

PART 2. NEW LEGISLATION

1. Summary – Applicable Legislation
2. The Safety Case
3. Goal Setting
4. Verification

PART 3. THE SAFETY CASE

1. The Safety Case – Introduction
2. Safety Management System
3. Risk Assessments
4. Verification Schemes

PART 4. STATUTORY INSTRUMENTS

1. Safety Case
2. Amendments to The Safety Case
3. Prevention of Fire, and Explosion and Emergency Response
4. Design and Construction
5. Management and Administration
6. Pipeline Safety Regulations
7. Provision and Use of Work Equipment

PART 5. ASSOCIATED INFORMATION

1. The Health and Safety Executive
2. UKOOA

SECTION TWO

Part 1. HISTORY

INTRODUCTION

The offshore industry in the UK is a relative newcomer to the industrial scene especially when compared with the likes of the steel, mining and shipbuilding industries which date back to the days of the industrial revolution and beyond. For all that, the large sums of money invested and the potentially huge returns have ensured that the offshore industry has maintained a high public profile, particularly when so many traditional sources of engineering are in decline.

In terms of legislation, the offshore industry started with a clean sheet of paper and a golden opportunity to develop regulations specific to the requirements of what is a very specialised industry. In practise the industry has suffered at the hands of more than one master and in the past produced disjointed, conflicting regulations, mainly in response to disasters involving considerable loss of life, culminating in the destruction of the *Piper Alpha* installation in 1988. The Piper Alpha proved to be the catalyst for a radical change in the way the industry was both certified and regulated.

To fully appreciate the significance of the post Piper legislation and the move towards *Verification*, a basic understanding of the previous regulatory regime under *Certification* is essential and will thus be described prior to embarking on a full explanation of the new Verification regulations. It is hoped that this will emphasise the shift in responsibilities that lie at the heart of the regulatory changes. An appreciation of Certification will also prove of value to readers involved with offshore installations at geographic locations, other than the UK Continental Shelf (UKCS), where, governments have adopted and still operate, schemes of certification which are similar to those originally instigated by the United Kingdom.

1.0 INITIAL LEGISLATION

The first significant offshore industry legislation was enacted as a result of the 1958 *Geneva Convention On The Continental Shelf* which recognised the right of countries to exploit the natural resources within their coastal waters.

The UK Government ratified the 1958 Convention agreements and incorporated them into domestic law under the *Continental Shelf Act* (CSA) of 1964. That same year, the first gas was discovered offshore in the UK following a massive onshore discovery in Groningen in Holland. The first oil in the North Sea was discovered five years later in 1969.

The Geneva Convention and Continental Shelf Act may be regarded as legislation pertaining to property. Legislation pertaining to the safety of offshore installations and their personnel followed with the introduction of the *Mineral Workings (Offshore Installations) Act* in 1971 and the *Health And Safety At Work Act* (HSWA) in 1974.

1.1 MINERAL WORKINGS (OFFSHORE INSTALLATIONS) ACT 1971 (MWA)

The Mineral Workings Act was enacted as a consequence of the loss of 13 lives when the drilling jack-up, the *Sea Gem*, capsized and sank in 1964. The Act empowered the Secretary of State to ensure that offshore installations were certified and remained *fit for purpose* in compliance with the Regulations. In order to enforce the requirements, the Secretary of State for Energy passed the *Offshore Installations (Construction and Survey) Regulations, Statutory Instrument Number 289* in 1974 and this marked the start of certification regulations for offshore structures in the UK.

The Construction and Survey Regulations decreed that all offshore installations located in the territorial waters of the UK, be in possession of a *Certificate of Fitness* before producing hydrocarbons and by this means, the requirements for certification referred to in the MWA would be met.

The Secretary of State decreed that *Certifying Authorities* would be established who would issue the *Certificates of Fitness* for offshore installations. The operational effectiveness of the Certifying Authorities was to be monitored by the *Petroleum Engineering Division* (PED) of the *Department of Energy*.

1.2 HEALTH AND SAFETY AT WORK ACT (HSWA) 1974

The Heath and Safety at Work Act of 1974 marked a transformation in the approach to safety at work in as much as it made the individual more responsible for his or her safety and the safety of their working environment. It attempted to remove the considerable burden created by excessive and ineffective legislation.

The HSWA empowered the *Secretary of State for Employment* to provide for the occupational health, safety and welfare of individuals at their place of work. The Secretary of State created the *Health and Safety Committee* (HSC) whose function was to formulate policies and regulations pertaining to the Heath and Safety at Work Act. Policing of the regulations would be carried out by inspectors of the newly formed *Health and Safety Executive* (HSE), a body responsible to the Health and Safety Commission.

Initially, the Heath and Safety at Work Act was not applicable to offshore workers. However, in 1977 the Act was extended to include the territorial waters of the UK, and thus offshore installations. The *Secretary of State for Energy* was made responsible for offshore safety, a responsibility that was discharged through the *Petroleum Engineering Division* (PED), *Safety Directorate*.

Whilst the Petroleum Engineering Division would monitor the offshore installations to ensure that they complied with the Heath and Safety at Work Act, they would themselves be responsible to the Health and Safety Commission. This arrangement was further complicated by the fact that the Health and Safety Commission could not introduce regulations for offshore work without consulting the Petroleum Engineering Division, and vice versa.

This division of responsibility between the Secretary of State for Employment and the Secretary for State for Energy lead to what many people believed was a dilution of the Heath and Safety at Work Act, and that it created a conflict of interests within the Department of Energy. It was felt that the responsibility for occupational health, safety and welfare in any industry should not be held by the Department with policy making responsibility for that industry.

Whilst the Mineral Workings Act provided ample scope for the development of legislation which reflected the specialist needs of the offshore industry, it appeared to conflict with the Health and Safety at Work Act and created an air of confusion that would persist for nearly 15 years. The situation was not resolved until 1991 and the outcome of the Piper Alpha enquiry when all the subsequent new regulations were introduced under the Health and Safety at Work Act, rather than the Mineral Workings Act.

2.0 CERTIFICATION

2.1 OFFSHORE INSTALLATIONS (CONSTRUCTION AND SURVEY) REGULATIONS, STATUTORY INSTRUMENT No. 1974/289)

Having outlined the organisational arrangements and responsibilities imposed by the Mineral Workings and the Heath and Safety at Work Acts, we must return to the requirements of Certification and the issue of the Certificate of Fitness introduced under SI 289.

2.2 CERTIFYING AUTHORITIES

Whilst a Government department could have fulfilled the functions required of a Certifying Authority, the Secretary of State elected to make use of the existing ship *Classification Societies* in the role of Certifying Authority because they operate similar certification schemes within the shipping industry and it was felt that the offshore industry would benefit from their established organisational framework and experienced personnel. They were adjudged to be both competent and independent and thus ideal for the task in hand.

Six classification societies were authorised by the Secretary of State for Energy to issue Certificates of Fitness:-

- Lloyd's Register of Shipping — LRS.
- Bureau Veritas — BV.
- Det Norske Veritas — DNV.
- Offshore Certification Bureau — OCB.
- American Bureau of Shipping — ABS.
- Germanischer Lloyd — GL.

2.3 CERTIFICATE OF FITNESS (C of F)

The point at which a Certificate of Fitness was issued normally coincided with the location of the drilling template or the installation of the jacket. In order to issue a Certificate of Fitness the Certifying Authority would carry out a design appraisal of the main structure and equipment and monitor the construction, installation and commissioning of the installation.

The issue of the Certificate of Fitness was dependent on the acceptance of the installation by the Certifying Authority, the final decision resting on the shoulders of the attending surveyor for the regulations stated that the installation and it's equipment would be subjected to whatever tests were required *in the opinion of the surveyor*. This placed the Certifying Authorities in a position of great authority and responsibility.

A *Letter of Limitation* and a *Letter of Qualification* accompanied each Certificate of Fitness. These letters enabled the Certifying Authority to highlight operational and environmental constraints, and any deficiencies in operating procedures or equipment. Typically a limitation amounted to a restriction on the use of a particular item of equipment or operating procedure such as a defective crane whilst a qualification related to items of a less critical nature such as a requirement to update the *Statutory Operations Manual* or the *Planned Preventative Maintenance* system.

Once issued, a Certificate of Fitness remained valid for a period of five years subject to annual mechanical, structural and electrical inspections being carried out by the Certifying Authority.

2.3.1 ANNUAL SURVEYS

The Annual Survey consisted of a thorough visual inspection of the installation, a review of equipment certification and maintenance records, and selective function testing of the safety systems. Changes to any item of equipment on the installation required the prior approval of the Certifying Authority.

2.3.2 MAJOR SURVEYS

The fifth annual inspection was referred to as the Major Survey and had to be completed prior to renewal of the Certificate of Fitness. In addition to items examined during the annual survey, a major survey included a more detailed investigation to ensure that no significant changes to approved designs had been made. The major survey included an underwater inspection of the structure of the installation. This was, and still is, normally spread over the five year period as part of a planned maintenance programme carried out by a specialist diving contractor. This subject has been dealt with more fully in chapter six.

Up until July 1998, the Construction and Survey Regulations, or SI 289 as they were more commonly referred to, was the single, most important piece of offshore legislation having been in existence for more than 20 years. The revocation of SI 289 heralded in a further attempt by the regulators to find a suitable means of controlling an extremely hazardous working environment.

With the introduction of the *Design and Construction Regulations* in 1996 and the amendments therein to the *Safety Case Regulations*, the Certificate of Fitness, Annual and Major Surveys, and the role of the Certifying Authority disappeared as *Operators*, or *Duty Holders*, as they are referred to, being tasked with certifying their own installations. However, whilst the use of an approved Certifying Authority is no longer a mandatory requirement of the legislation, all the Certifying Authorities previously listed continue to provide services to the offshore industry in the new role of *Independent, Competent Bodies or Independent, Competent Persons*, a position which is clearly identified within the new regulations.

3.0 THE CULLEN INQUIRY

The explosion and resulting fires which engulfed the *Piper Alpha* installation on the 6 July 1988 claimed the lives of 165 of the 226 persons onboard the platform and 2 of the crew of a rescue boat. The death toll was the highest in any accident in the history of the offshore industry and lead to the biggest shakeup of the offshore safety regime in the UK for 20 years.

The Public Inquiry into the disaster was chaired by the *Hon. Lord Cullen* and lasted for over a year. When the report was published it contained 106 recommendations which proposed far reaching changes in the way that offshore installations would be both certified and regulated.

The central recommendation of Lord Cullen's report was that every operator or owner of an offshore installation should prepare a Safety Case, a decision influenced by the HSE's experience of regulating major hazards onshore under the *Control of Industrial Major Accident Hazards Regulations 1984 (CIMAH) (SI 1984/1902)*. There were to be changes to the regulatory body and the progressive replacement of the existing regulations.

As a result of the Inquiry, the *Offshore Safety Division* (OSD) was created in 1990 as a department within the *Health and Safety Executive* (HSE). By the Spring of 1991, all duties pertaining to the occupational health, safety and welfare of offshore installations were transferred from the Petroleum Engineering Division (PED) of the Department of Energy, to the Offshore Safety Division of the Health and Safety Executive. The following year the first of the legislative reforms took place with the introduction of the *Safety Case Regulations*.

Part 2. NEW LEGISLATION

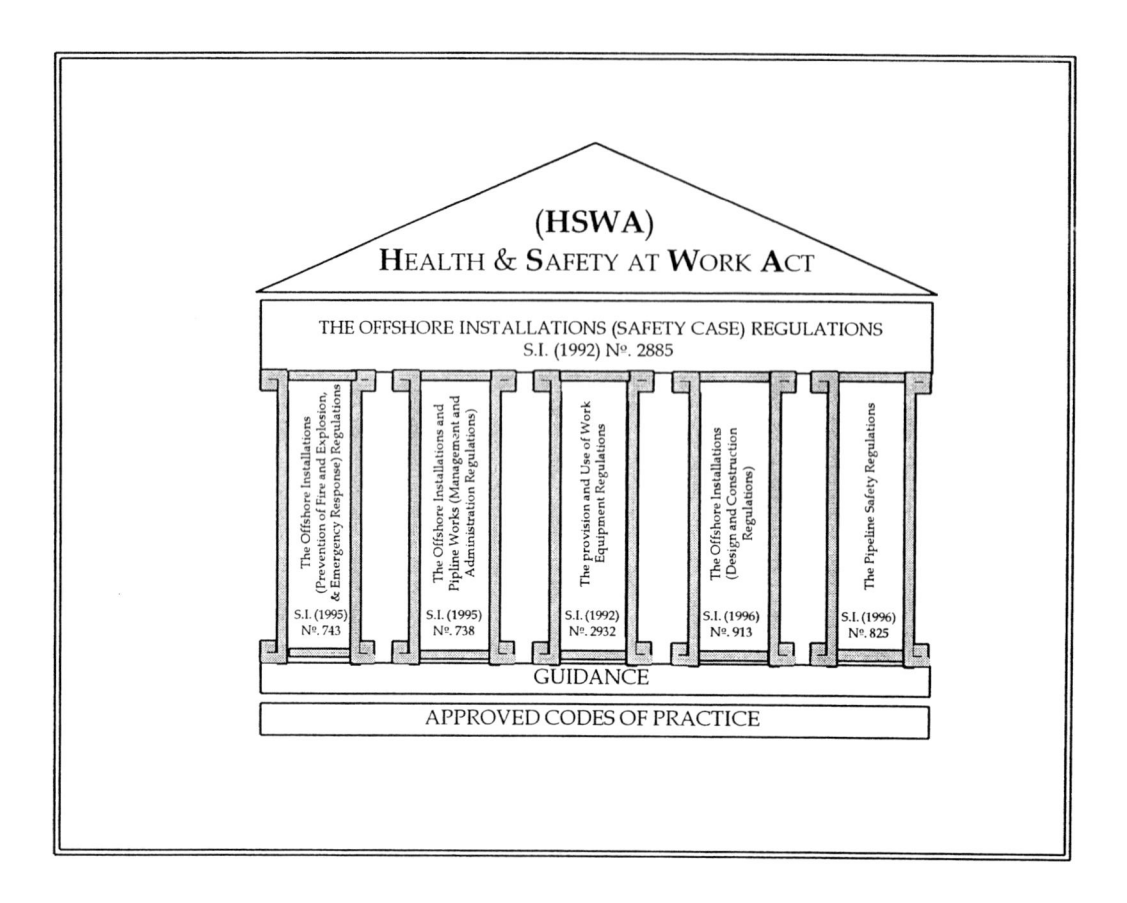

![Diagram depicting a classical temple structure with HSWA as the pediment, The Offshore Installations (Safety Case) Regulations S.I. (1992) No. 2885 as the entablature, and five pillars labelled: The Offshore Installations (Prevention of Fire and Explosion, & Emergency Response) Regulations S.I. (1995) No. 743; The Offshore Installations and Pipeline Works (Management and Administration Regulations) S.I. (1995) No. 738; The provision and Use of Work Equipment Regulations S.I. (1992) No. 2932; The Offshore Installations (Design and Construction Regulations) S.I. (1996) No. 913; The Pipeline Safety Regulations S.I. (1996) No. 825. The base is labelled GUIDANCE and APPROVED CODES OF PRACTICE.]

1.0 SUMMARY OF APPLICABLE LEGISLATION

Health and Safety at Work Etc. Act 1974 (HSWA Act)

The Health and Safety at Work Act places general duties on employers to ensure, so far as is reasonably practicable, the health and safety of their employees, and others may be affected by their undertaking. These general duties are supported by the specific requirements in regulation 3 of the *Management of Health and Safety at Work Regulations* (SI 1992/2051) (MHSWR) for employers to undertake risk assessments for the purpose of identifying the measures which need to be put in place to prevent accidents and protect people against accidents.

Safety Case Regulations (SCR) — SI 1992/2885

The Safety Case Regulations are the key focal point of the new regime of offshore health and safety legislation and they are underpinned by the DCR, MAR, PFEER and PUWER. Key elements of the Safety Case Regulations (SCR) are the identification of hazards with the potential to cause a major accident and the reduction of risks associated with the hazards to levels that are as low as reasonably practicable.

Prevention of Fire and Explosion, and Emergency Response Regulations (PFEER) — SI 1995/743

The PFEER focuses on identifying and preventing fire and explosion hazards and as these hazards have the potential to affect the structure, they are also relevant to complying with Design and Construction Regulations (DCR). Work done to comply with PFEER regulations 18 and 19, requirements for a *written scheme for the systematic examination* of equipment and procedures, may also contribute to the verification of safety critical items requiring verification under the amendments to the Safety Case Regulations introduced by the DCR.

Design and Construction Regulations (DCR) — SI 1996/913

The objectives of the Design and Construction Regulations (DCR) are to replace the certification regime established by the *Offshore Installations (Construction and Survey) Regulations* 1974 (SI 289). The Regulations contain requirements for insuring that offshore oil and gas installations, and oil and gas wells are designed, constructed and kept in a sound structural state for purposes of health and safety. The Regulations deal with the integrity of installations and the provisions in it impose duties on the duty holder relating to ensuring that integrity is achieved and maintained.

Management and Administration Regulations (MAR) — SI 1995/738

One of the major facets of the Safety Case Regulations is the requirement to demonstrate that management systems are adequate to ensure compliance with relevant statutory health and safety provisions. The verification arrangements introduced into the SCR by the DCR will be an important component of such management systems.

Provision and Use of Work Equipment Regulations (PUWER) – SI 1992/2932

PUWER seeks to ensure the safe provision and use of work equipment and covers most offshore plant. Since the hazards associated with plant will, to some degree, affect an installation's structure, and hence integrity, compliance with PUWER is expected to contribute significantly to meeting the requirements of the Design and Construction Regulations, particularly at the design stage.

Pipeline Safety Regulations (PSR) – SI 1996/825

The PSR lays down duties on pipeline operators relating the design, construction, operation, maintenance and decommissioning of pipelines and any associated apparatus or works.

Aviation

The duty holder should ensure that the design of the installation also meets the requirements of relevant aviation legislation. Guidance on this subject may be obtained from the Civil Aviation Authority (CAA) publication CAP 437, Offshore Helicopter Landing Areas.

2.0 THE SAFETY CASE

At first glance the general format of the new regulations appears to be very similar to the old. There are still regulations covering the design and construction of the installation, pipelines, fire fighting and life saving equipment, and the provision of emergency procedures, and there is a good reason for this. Whilst some operators thought that the whole of the offshore industry could be controlled under one set of regulations, the Safety Case Regulations, incorporating a requirement for a Formal Safety Assessment (FSA), this idea was rejected by Lord Cullen. He was of the opinion that *any large system is usually best handled by braking it down into more manageable parts, in some form of hierarchy. These other regulations would compliment the Safety Case by setting intermediate goals and would give the regime a solidity which it might otherwise lack.*

3.0 GOAL SETTING

The intention of the new regulations is to provide a less prescriptive form of legislation and to encourage a more integrated approach to dealing with potential hazards. Each of the new regulations provides a stated objective in the form of a *General Duty* and the duty holder must specify exactly how compliance with these duties will be achieved, effectively setting out what are considered to be acceptable levels of risk, and establishing performance goals. Ensuring that stated objectives are met rather than prescribing the detailed measures to be taken is referred to as *goal setting*.

Lord Cullen gives this example of a goal setting situation in respect of fire pumps, the goal being to ensure that the *fire pump will operate in all accident situations other than if it is destroyed by the accident itself. The duty holder must look at the various threats which could immobilise the fire pump and build in such safety measures as will provide protection to the fire pump*. These measures would typically consider fire and blast protection, a gas release and dropped object protection.

4.0 VERIFICATION

Having installed the necessary equipment and instigated procedures which define how the goals will be achieved, the regulations require the duty holder to prepare a *written scheme for the systematic examination* of the plant and equipment installed under the Prevention of Fire and Explosion, and Emergency Response Regulations (PFEER), and to set in place a *verification scheme* for items deemed to be *safety-critical elements* by the Safety Case Regulations, the object being to ensure that the efficiency of the equipment and the effectiveness of the procedures are being maintained in service. The preparation of the written schemes and the subsequent examinations are to be carried out by, or in conjunction with, an *independent and competent body* selected by the duty holder.

SUMMARY

To summarise, the new regulations set the duty holder three basic tasks;

i) to set goals by which compliance with the regulations will be achieved,

ii) to prepare a written scheme of examination/verification scheme to ensure that these goals are achieved and maintained in practise,

iii) to select *independent and competent person(s)* to carry out the ongoing inspections and examinations associated with the written schemes of examination.

The new regulations completely change the philosophy of offshore legislation, the operator or duty holder now being in a position where they effectively self-certify their installations, rather than being obliged to employ a certifying authority to fulfil this function, as was previously the case.

Part 3. THE SAFETY CASE

1.0 THE SAFETY CASE REGULATIONS – INTRODUCTION

The central recommendation of Lord Cullen's report on the Piper Alpha disaster was that every operator or owner of an offshore installation should prepare a safety case for each installation, both fixed and mobile, old and new. The new Regulations cited as the Offshore Installations (safety case) Regulations came into force on the 30th November 1993 with older installations having until the 30th November 1995 to comply.

The safety case is a written document prepared by the duty holder which sets out comprehensive information on arrangements for managing health and safety, and a means of demonstrating that the hazards on an installation have been identified and assessed and that they are under control. The new Regulations are accompanied by a set of Guidance Notes to assist in the interpretation of the requirements and in the preparation of the safety case, which is by its very nature, a very involved and complex task.

The format of the safety case is similar to the *Installation Operations Manual* required under the SI 289 Regulations. Both are essentially self-contained documents designed to provide factual information such as the geographic, meteorological and oceanographic conditions which affect the installation, and include information and drawings showing the design and layout of the structure and plant. However, in addition to providing factual information, the principal aims of the safety case are to ensure that:

a) the management system is adequate to ensure compliance with statutory health and safety requirements,

b) adequate arrangements have been made for audit and preparation of audits reports,

c) all hazards with the potential to cause major accidents have been identified, evaluated and means taken to reduce the risks to personnel.

It is quite clear from a) and b) that the *Health and Safety Executive* (HSE) place considerable emphasis on the importance of employing sound management practises to minimise the risks to personnel from working activities and the working environment. Consequently, a large part of the *Guidance On Regulations* are dedicated to an explanation of the extremely onerous requirements of *Safety Management Systems* (SMS).

The other departure from previously established offshore practices is the means by which the hazards referred to in paragraph c), will be dealt with, essentially by the introduction of the requirement to carry out *Quantitative Risk Assessments* (QRA). Extensive advice is provided on QRA in the Guidance of Regulations.

Furthermore, the subsequent amendments to the Safety Case Regulations made by regulation 26 of the *Design and Construction Regulations* in 1996 introduced the requirement to verify the effectiveness of items identified as being *safety-critical*.

Risk assessments, safety management systems and verification will now be discussed briefly prior to presenting an abridged account of the new regulations.

2.0 THE SAFETY MANAGEMENT SYSTEM (SMS)

OBJECTIVES

The term *Safety Management System* refers to the elements in the management system which are particularly concerned with health and safety aspects, and with loss control. The duty holder's policies and aims for achieving high standards of health and safety should be described and the plans and programmes which are aimed at delivering these objectives, and safety objectives should be outlined. Safety goals must be set and performance standards established, in order to provide a level against which the systems can be measured.

Furthermore, the Safety Management System should specify how the assessment of risks has been carried out and contain a summary of the methods and criteria used. A clear statement must be made on the companies policy towards the use of Quantitative Risk Assessments (QRA).

OBLIGATIONS

The Safety Management System should describe the duty holder's proposals to ensure that its aims for achieving high standards of health and safety performance are communicated to all concerned. There should be clear unbroken lines of command and accountability from top management down to the lowest operational level. Clear charts must be provided to demonstrate formal acceptance of responsibility for the safety policy at Board level for UK based companies and arrangements for holding individuals accountable for their health and safety performance should be included. Responsibility to co-ordinate and monitor the implementation should also be allocated at this level.

High standards of health and safety performance require a positive and informed commitment from the work force and the SMS should describe the methods for securing this commitment. Safety measures must be discussed with safety representatives to assist in the development of health and safety measures and representatives should have direct access to the responsible senior company officer at Board level. Statutory requirements for formal representation of the workforce in discussions with management on health and safety matters are contained in the Offshore Installations (Safety Representatives and Safety Committees) Regulations 1989.

Line managers and supervisors should be aware of their responsibilities and formally accept them and the SMS should clearly state the responsibilities placed on the *offshore installation manager* (OIM). The responsibility placed on managers may not extend beyond matters which are in practise within their control, the employer remaining responsible for ensuring that managers have clear guidelines.

PROCEDURES

The SMS should indicate the form in which detailed procedures are to be documented and made available to managers and employees and of particular importance are:

(a) written schemes providing for the systematic examination, maintenance and, where appropriate, testing, by a competent person, of all of the parts of the installation and its equipment; for the intervals at which this should be done; and arrangements to ensure that defects are suitably acted upon;

(b) permit to work (PTW) systems, specifying activities covered, the duties of all responsible persons, records required to be kept, and other procedures.

MONITORING

Monitoring, as distinct from auditing, should be carried out at prescribed intervals by personnel responsible for the operations under scrutiny, assisted as necessary by health and safety advisors. Two forms of monitoring should be used because a low incident rate is not a guarantee that risks are being effectively controlled;

(a) **Active Monitoring** — provides feedback on performance, before an incident occurs. Examples of active monitoring are achievements of objectives assigned to individuals or groups, periodic inspection of plant and equipment, and the monitoring of operation of the permit to work system.

(b) **Reactive Monitoring** — provides information on accidents, ill health, dangerous occurrences and near misses and the system for internal reporting and recording of events should be described in the SMS.

AUDIT

Audit is described as the structured process of collecting independent information on the efficiency, effectiveness and reliability of the Safety Management System and for the drawing up of plans for corrective action.

Operators, or duty holders must satisfy themselves by means of regular audits that the Safety Management System is being adhered to. The independent audit constitutes the final stage in the management control cycle and should be carried out by competent persons who are out side of the line management chain for the areas or activities being audited. Staff from different parts of the organisation or external consultants may be used and there should be a facility for conveying constructive recommendations to senior management with a view to making improvements in the management systems.

CONCLUSIONS

The main principles of effective safe management are no different to the sound management practises advocated by proponents of quality and business excellence and they should be monitored in the same way, and with the same vigour.

3.0 RISK ASSESSMENTS

Risk, is a subjective concept and its quantification is universally accepted as being extremely difficult to define and open to subjective judgement. It is defined in British Standard 4778 as, *the process whereby decisions are made to accept a known or assessed risk and/or the implementation of actions to reduce the consequences or probability of occurrence.*

Risk assessments may be qualitative or quantitative studies. The term Formal Safety Assessment (FSA) was used extensively by Lord Cullen during the Piper Alpha Enquiry and whilst an FSA could be an entirely qualitative study, it would normally be a combination of both qualitative and quantitative risk assessments. A qualitative assessment often identifies the more obvious problems which can be rectified immediately, prior to commencing with the more detailed quantitative assessment.

Qualitative assessments — These tend to be based on actual operational experience, engineering standards, and sound engineering judgement.

Quantitative risk assessments — These involve calculations to assist with the identification of the risks and to determine the frequency, magnitude and consequence of hazardous events.

QUANTITATIVE RISK ASSESSMENT (QRA)

QRA has two main functions:

- to assist in the comparison of design and operating philosophy options (with a view to effective and efficient safety management),

- to demonstrate to the Regulatory Authority that specified safety standards and risk targets are being achieved (or maintained).

It was Lord Cullen's wish that quantitative risk assessments would be used extensively as a means of establishing compliance with the new regulations. Whilst the use of QRA is considered to be highly desirable, it's use is only mandatory under the PFEER regulations in relation to the assessment of the survivability of the temporary safe refuge and escape routes. However, it is expected that most duty holders will use QRA extensively and where they choose not to, an acceptable alternative form of engineering assessment must be provided.

QRA ANALYSES

Quantitative Risk Assessments essentially involve an analyses of the installation, equipment and operational activities to identify any hazards which have the potential to cause a major accident such as the escape of hydrocarbons, impact with an aircraft or ship, structural failure, fire or explosion. It normally involves the following processes:

- hazard identification, essentially to determine what could go wrong;

- initiating events analysis, to determine where, and how often things could go wrong;

- event tree analysis, to determine what contributes to an accident;

- consequence analysis, to determine how severe the accident is likely to be, and what the consequences are likely to be.

- risk calculation and sensitivity analysis;

- risk assessment, how do risk levels compare against company criteria;

- identify practices measures to reduce risk.

The setting of an acceptance criteria for risk provides means for evaluating the relative significance of the results obtained from the risk analysis. The acceptance criteria are arrived at by discussion, a company effectively setting what it considers to be acceptable levels of risk, as part of the *goal setting* process. The risk levels should compare favourably with established industry standards and criteria to show that the measures taken to prevent such occurrences happening are adequate. The Health and Safety Executive use the concept of *As Low as Reasonably Practicable* (ALARP) in judging the level of acceptable risk.

ALARP

The guiding principles of ALARP are:

- there is an upper limit above which risk cannot be tolerated on any grounds;

- there is a lower limit below which the risk is acceptable and no further risk reducing measures are necessary;

- in between these two limits there lies a region in which the benefits justify the risks. This is the ALARP region.

The main safety goal is that risks should be as low as reasonably practicable (ALARP). This means that a situation can be achieved when further improvements in safety are not required if their costs are grossly disproportionate to their benefits.

HAZOP AND HAZAN

In discussions involving an investigation of hazards the terms *hazop* and *hazan* will be frequently encountered and a brief explanation of these terms will clarify the subject of hazard analyses. Basically these operations afford a means of producing on paper in a systematic and thorough fashion, and in advance of plant start up, potential hazards to the plant, process and personnel, and ultimately making recommendations to eliminate these hazards.

HAZOP — A hazard operational study involves going through an installation system by system, drawing by drawing, operation by operation to identify what potential hazards may arise. Once identified these potential hazards are subjected to a hazard analyses.

HAZAN — A hazard analyses is a technique for assessing the probability of a hazard occurring and determining the subsequent consequences for comparison with a set target, or known criteria.

THE ALARP PRINCIPLE

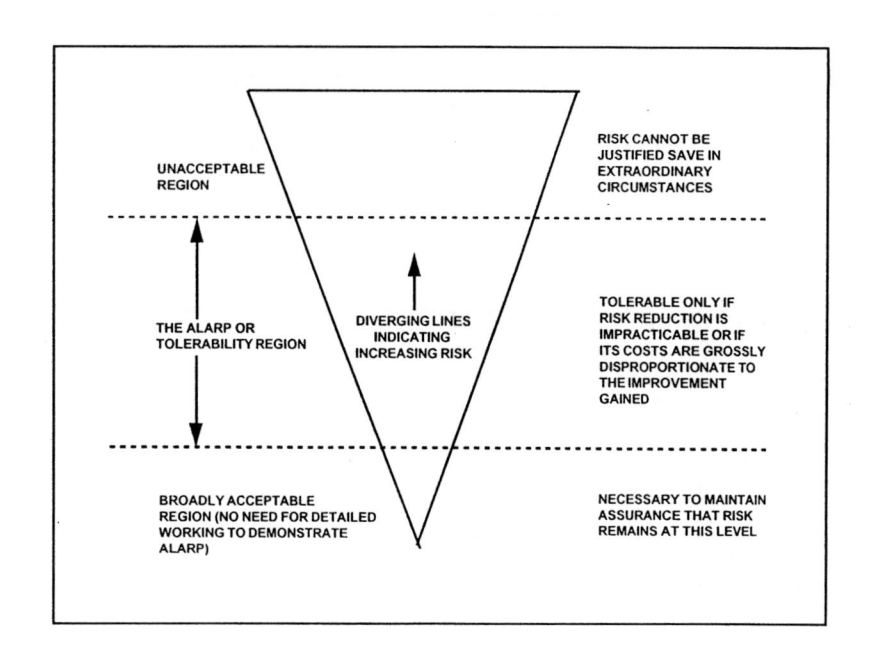

UNACCEPTABLE REGION	RISK CANNOT BE JUSTIFIED SAVE IN EXTRAORDINARY CIRCUMSTANCES
THE ALARP OR TOLERABILITY REGION (DIVERGING LINES INDICATING INCREASING RISK)	TOLERABLE ONLY IF RISK REDUCTION IS IMPRACTICABLE OR IF ITS COSTS ARE GROSSLY DISPROPORTIONATE TO THE IMPROVEMENT GAINED
BROADLY ACCEPTABLE REGION (NO NEED FOR DETAILED WORKING TO DEMONSTRATE ALARP)	NECESSARY TO MAINTAIN ASSURANCE THAT RISK REMAINS AT THIS LEVEL

4.0 VERIFICATION SCHEMES

INTRODUCTION

Regulation 26 of the Design and Construction Regulations (DCR) deals with the amendments to the Safety Case Regulations, in particular, the requirement for a verification scheme for safety-critical elements.

The requirements relating to verification are applicable to the following regulations:

- safety Case Regulations (SCR) — SI 2885

- prevention of Fire and Explosion, and Emergency Response Regulations (PFEER) — SI 743

- design And Construction Regulations 1996 — SI 913

OBJECTIVE

The overall objective of the verification scheme is to set in place independent and competent scrutiny of those parts of an installation which are critical to safety. The verification scheme should be written and provide independent checks to confirm the continuing suitability of *safety-critical elements* throughout the life cycle of the installation.

The verification scheme should address the items contained in Schedule 9 of the Design and Construction Regulations and it should be drawn up by, or in consultation with an *independent and competent person(s)* (ICP), before completion of design work. It may be set out in a single document, or in a series of documents, and it may be recorded electronically, but information about the scheme should be readily accessible and a summary of the scheme is to be included in the safety case.

SCHEDULE 9 OF THE DESIGN AND CONSTRUCTION REGULATIONS

Matters To Be Provided For In A Verification Scheme

1. The duty holder is required to provide particulars of how the independent and competent person(s) will be selected and to specify the appropriate levels of independence.

2. An effective communication system must exist to convey information to those carrying out verification work. They must have a thorough understanding of the implications of all facets of the different stages of the installation's life cycle and how they fit together.

3. The nature and frequency of examination and testing.

4. Arrangements for review and revision of the scheme.

5. The arrangements for the making and preservation of records showing

    - the examination and testing carried out;
    - the findings;
    - remedial action recommended; and
    - remedial action performed.

6. Arrangements for communicating the output of the verification process to an appropriate level in the organisation to ensure that the action required in the light of verification findings is taken.

1.0 THE VERIFICATION PROCESS

The first step, and an essential preliminary to developing a verification scheme, is the accurate identification and listing of the *safety-critical elements* which require verification. *Performance standards* must then be determined, followed by the preparation of a *suitable written scheme*, often referred to as a *written scheme of examination*, to ensure that all the verifiable items are examined to confirm that they remain suitable for their intended purpose.

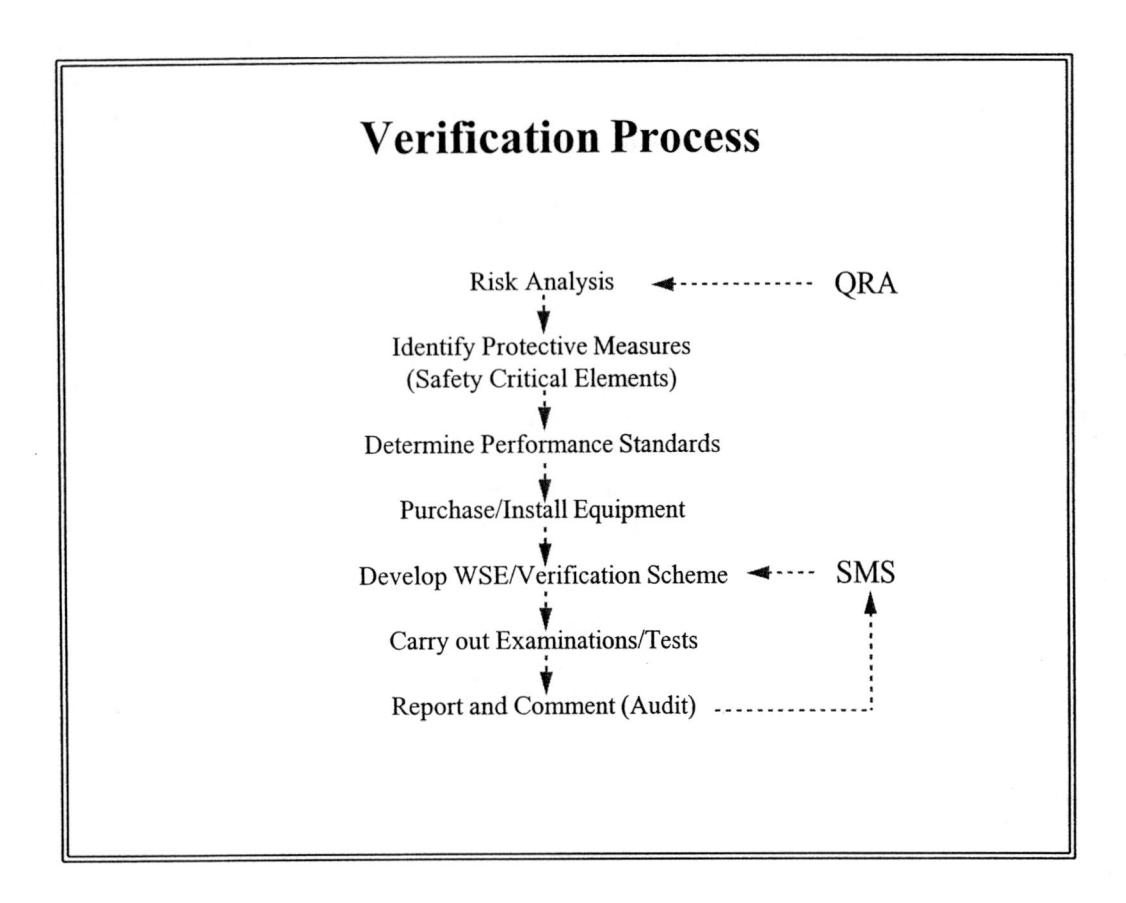

Verification Process

```
Risk Analysis  ◄------- QRA
     ▼
Identify Protective Measures
(Safety Critical Elements)
     ▼
Determine Performance Standards
     ▼
Purchase/Install Equipment
     ▼
Develop WSE/Verification Scheme  ◄---- SMS
     ▼                                  ▲
Carry out Examinations/Tests            |
     ▼                                  |
Report and Comment (Audit) -------------
```

1.1 SAFETY-CRITICAL ELEMENTS (SCE)

Safety-critical elements are defined as *those structures, plant, equipment, systems (including computer software) or component parts whose failure could cause or contribute substantially to a major accident*, and includes equipment which is intended to prevent or limit the effect of a major accident. The definition of a major accident event can be found in regulation 2 of the Safety Case Regulations and in:

- a fire, explosion or the release of a dangerous substance involving death or serious personal injury;

- any event involving major damage to the structure of the installation or plant;

- the collision of a helicopter with the installation;

- the failure of life support systems for diving operations in connection with the installation, or the detachment or trapping of a diving bell;

- any other activity involving death or serious personal injury to five or more persons.

The list of safety-critical elements is likely to include most of the equipment on the platform. If we consider the definition, the jacket structure, pipeline risers and the hydrocarbon process equipment in their entirety are all likely, *to cause or contribute to a major accident*, should they fail, so will be classed as safety-critical elements. The ESD system, fire and gas systems and fire fighting equipment are all designed to, *prevent or limit the effect of a major accident*, so they will also be classed as safety-critical elements. These are just examples to illustrate the potential scope of the scheme, there are many more safety-critical elements besides these.

The safety-critical elements may be established as part of the study carried out to identify hazards in the preparation of the safety case and in the *Assessment Of Major Accident Hazards*, required by PFEER. The safety-critical elements may then be reviewed for completeness against a generic list of SCEs developed by specialist safety engineers for similar installations. This should ensure that there are no obvious anomalies.

Once the list of safety-critical elements is complete, it must be reviewed and commented on by the duty holders nominated independent and competent persons. A detailed assessment by the independent and competent person of the methodology used to identify the safety-critical elements is not required. They are expected to use professional judgement, expertise and experience to reach a view on whether the list is appropriate for the installation in question. Any problems identified by the ICP which cannot be resolved with the duty holder must be listed as reservations and maintained on file.

Having identified and confirmed the completeness of the list of safety-critical elements, the next stage in the verification process is to establish levels of performance, or *performance standards*, with which the equipment must comply in order to fulfil its function.

Note: In addition to the items described as safety-critical elements (SCE) by regulation 26 of the Design and Construction Regulations, most duty holders have elected to adopt an integrated approach to verification and have included the plant and equipment identified under regulations 18 and 19 of the PFEER Regulations in a common verification scheme.

1.2 PERFORMANCE STANDARDS

Performance Standards are most clearly defined in PFEER:

> A performance standard is a statement, which can be expressed in qualitative or quantitative terms of the performance required of a system, item of equipment, person or procedure, which is used as the basis for managing the hazard — e.g. planning, measuring, control or audit — through the life cycle of the installation.

Setting performance standards is a crucial aspect of the verification process They may be described in what is termed as FARSI format, that is functionality, availability, reliability survivability and on how they interact with other systems. They should be measurable and auditable.

Using a fire pump as an example, the pump performance or *functionality* can clearly be specified, typically to supply a calculated volume of water at a particular pressure in order to meet the demands of the largest section of fire main and deluge. As regards the other aspects, it should have a 100% *availability* and operate *reliably* for a period of time as designated in the PFEER assessment. It should be capable of *surviving* in all accident situations, other than if it is destroyed by the accident itself and in order that it is not immobilised by the failure of other systems, it should be fully self contained, that is, it should not rely on *interaction* with other systems which could immobilise it, should they fail.

Once the performance standards have been established, suitable equipment can be purchased and an examination scheme set in place to ensure that the performance of the equipment can be achieved and maintained in practice.

2.0 EXAMINATION PROCESS

The *arrangements in writing for systematic examination*, or *written scheme of examination* as the examination process may be referred to, is the means by which it will be established that all verifiable items are being effectively maintained and remain, suitable for their intended purpose. The scheme may be complimentary to, but not a substitute for routine maintenance programmes.

The written scheme should detail exactly what items are to be examined and the frequency of the examinations. The examinations which will be carried out by *independent and competent persons* will involve careful and critical scrutiny of the safety-critical elements using appropriate techniques, which may include testing. The findings and any remedial action required should be clearly stated, and recorded.

It should be noted that in addition to the examinations which are carried out during production, examinations are also to be carried out during design, fabrication, construction and repair. Information channels must be in place to follow design verification, fabrication and commissioning activities through into production and in to ongoing verification.

3.0 INDEPENDENT AND COMPETENT PERSONS

Duty Holders are responsible for selecting persons to carry out verification work and they will need to justify their selection. This may be done in a number of ways, one of which would be to use a formally accredited organisation such as the National Certification Scheme for, In-service Inspection Bodies.

The *competence* of a person should consider such matters as technical competence and experience and these persons should be available in adequate numbers. Professional judgement will be an important aspect of their work.

The *independent* persons must be sufficiently impartial and objective in their judgement to ensure that safety is not compromised. They should not verify their own work, and their management lines should be separate from those whose work they are checking. Whilst it is acceptable in principle for a duty holder's in-house team to check work done elsewhere in the same organisation, it would influence objectivity if their management chain included the manager responsible for meeting production targets which might be unachievable if plant were shut down on safety grounds.

4.0 RECORDS AND REVIEW OF THE SCHEME

The operator or owner of an installation shall ensure that its verification scheme is kept at an address in Great Britain and that the records are kept for a period of six months after the scheme ceases to exist. The purpose of this requirement is to provide an auditable trail showing what work has been done, its findings, any recommendations made and any work carried out as a result.

The operator, in the case of a fixed installation and the owner, in the case of a mobile installation, shall ensure that, as often as may be appropriate the verification scheme is reviewed and, where necessary, revised or replaced by or in consultation with an independent and competent person. A note is made of any reservations expressed by such persons in the course of drawing it up.

5.0 RESPONSIBILITY

Whilst the preparation of the list of safety-critical elements, the compilation of performance standards and indeed the writing of the verification scheme can be carried out by a safety engineering consultancy of the duty holders choice, the responsibility for all the elements which constitute the scheme, rest with the duty holder.

Part 4. STATUTORY INSTRUMENTS

Introduction To Part Four — The more significant of the new regulations which are currently applicable to offshore installations in the UK sector of the North Sea are covered in this section. They have been dealt with on a regulation by regulation basis with the more salient points from the regulations and the various guidance notes being quoted in order to give the reader an overall view of the intent of the regulations.

1. SAFETY CASE REGULATIONS 1995 — SI 2885

Regulation 1   Citation and Commencement
The Regulations may be cited as the Offshore Installations(Safety Case) Regulations 1992. They came into force on the 31st May 1993.

Regulation 2   Interpretation
Some of the more relevant definitions are explained.

Major Accident means:

    (a) a fire, explosion or the release of a dangerous substance involving death or serious personal injury;

    (b) any event involving major damage to the structure of the installation or plant;

    (c) the collision of a helicopter with the installation;

    (d) the failure of life support systems for diving operations in connection with the installation, or the detachment or trapping of a diving bell;

    (e) any other activity involving death or serious personal injury to five or more persons.

Mobile Installation — means an installation (other than a floating production platform) which can move without major dismantling or modification, whether or not it has motive power.

Floating Production Facilities (FPSs) are considered to be fixed installations by virtue of their specific exclusion from the definitions of a mobile installation.

Quantitative Risk Assessment — means identification of hazards and the evaluation of the extent of risk arising therefrom incorporating calculations based upon the frequency and magnitude of hazardous events.

Operating an installation — means the exploitation or exploration of mineral resources, the storage for recovery of gas under the sea bed, the conveyance of things by pipe and the provision of accommodation for people who work on or from an installation.

Construction Activity — Construction activities are identified as being the installation or dismantling of the platform, or any activity which requires the use of a heavy lift vessel.

Regulation 3   Application
The safety case regulations apply to all installations located in the tidal waters and parts of the sea in or adjacent to Great Britain up to the seaward limits of territorial waters and parts of the Continental Shelf designated by the Act of 1964.

Regulation 4 Safety Cases For Fixed Installations
At least three months before the completion of a design for a fixed installation the operator must submit a design safety case to the HSE for review. This is to enable the HSE to comment on matters pertaining to health and safety so that their comments can be considered in the final design.

Six months prior to commencement of the operation of a fixed installation the operator must send the operational safety case to the HSE for assessment. Operation is defined as the drilling of the first well from the installation, or when hydrocarbons are admitted to the installation through a pipeline, whichever is the earlier and neither may proceed until the HSE have accepted the safety case. Where a sub-sea template is used to drill wells prior to the arrival of the jacket the safety case must still be submitted prior to commencement of drilling.

Detailed advice on the content of the safety case is provided in Schedules 1 and 2 of the Guidance on Regulations. A summary is provided here following regulation 17.

Regulation 5 Safety Case For Mobile Installations
A design safety case is not required for a mobile installation because due to the vessels mobility it may be built or operated at any geographic location where different regulations and international conventions apply. However, a safety case is required for the operation of the installation and this must be submitted to the HSE for approval at least 3 months before the vessel enters the relevant waters of the UK.

Detailed advice on the content of the safety case is provided in Schedule 3 of the Guidance on Regulations.

Regulation 6 Safety Cases For Combined Operations
A combined operations safety case is required where two or more installations (each themselves required to have a safety case when operating remotely) interact. The combined safety case must be submitted at least 6 weeks prior to commencement of combined operations (4 weeks in the case of well workovers) and be approved by the HSE prior to work commencing. Examples of combined operations are:

(a) a mobile installation providing accommodation to another installation;

(b) a mobile drilling unit cantilevered over a fixed wellhead platform;

(c) a mobile drilling unit working on a sub-sea well linked back to a fixed production platform.

A combined operational safety case is NOT required for the placing of jackets and the hook up phases of a new installation unless the installation is producing hydrocarbons. A combined operational safety case is also not required for vessels employed in activities connected with the installations although their activities should be covered in the safety case for the installation for which they are working.

Examples of such activities connected with the installation are the loading and discharging of materials from supply boats, diving activities undertaken within 500 metres of the installation, the loading of shuttle tankers and heavy lifts undertaken from a heavy lift vessel alongside the installation (prior notification of heavy lift activities would need to be made to the HSE).

The primary reason for a combined safety case is to demonstrate that all the risks have been considered, that the co-ordination of management activities has been addressed and that provisions are in hand for dealing with evacuation, escape and rescue in the event of an emergency. Detailed advice on the content of the combined operations safety case is provided in Schedule 4 of the Guidance on Regulations.

THE NORTH SEA – HISTORY AND LEGISLATION

Regulation 7   Safety Case for Abandonment of Fixed Installations
At least 6 months prior to the abandonment of an installation the operator must submit a safety case for the abandonment to the HSE and it must be approved prior to work commencing.

Regulation 8   Management of Health and Safety and Control of Major Accident Hazards
The principle matters to be demonstrated in safety cases are that:

(a) the management system is adequate to ensure compliance with statutory health and safety requirements;

(b) adequate arrangements have been made for audit, and the preparation of audit reports; and

(c) all hazards with the potential to cause a major accident have been identified, their risks evaluated, and measures taken to reduce risks to persons to as low as reasonably practicable.

The safety case must include details of the operators management system to ensure that the relevant statutory provisions are complied with and that all risks and hazards have been satisfactorily evaluated and measures taken to reduce the risks to persons affected by the hazards to the lowest level that is reasonably practicable.

The safety case must include details of the audit arrangements for the management system. The audit must be an independent and systematic assessment of the adequacy of the management system to achieve the requirements of the Regulations.

Regulation 9   Revision Of Safety Cases
The operator of an installation should revise the contents of the safety case as often as may be considered appropriate. Apart form the obvious operational need to keep the safety case up-to-date, careful logging of modifications is essential in order to demonstrate to the HSE that the installation is being operated in accordance with the safety case.

Even apparently minor changes should be assessed but the burden of re-submission of the safety case for reappraisal is unlikely to be justified in relation to trivial details of day-to-day management, maintenance and repair.

The original safety case should set the parameters on which re-submission of the safety case is required and these will in all probability be in respect of substantial modifications, meriting reappraisal of the risk control arrangements.

Examples of changes considered to be significant and likely to warrant a re-submission of the safety case are the introduction of new topside modules, modification of emergency shutdown systems and repairs following significant structural damage as may occur from a dropped object or collision with a ship. The development of additional reservoirs and the change of a major management contractor is also likely to warrant changes to the safety case.

Whilst it may be decided that changes to the safety case are required, no changes can be made until they have been submitted to the HSE for assessment and this normally requires an approval period of three months. Shorter periods may be acceptable to the HSE under their discretionary powers. Periods of 21 days are acceptable for well modifications and six weeks may be acceptable for relatively straightforward changes.

An installation should not be operated beyond a period of 3 years from the acceptance of the safety case unless the safety case has been revised and accepted by the HSE. The revised safety case should be submitted to the HSE for approval 3 months prior to the third anniversary.

Regulation 10 Duty To Conform With Safety Case
A duty holder must ensure that the installation is operated in conformity with the safety case and it is an offence not to do so.

It is a prime purpose of the safety case to demonstrate foreseeable excursions from normal operation conditions which may have the potential to cause a major accident, can be effectively brought under control without injury to people. To achieve this the operator must set operating limits and performance standards for activities which are critical to safety and set acceptable values. A HAZOP may be useful for some of these activities. A regular audit should be carried out to ensure effect adherence to the specified limits and standards.

Regulation 11 Notification Of Well Operations
At least 21 days before drilling a well the operator must send notification of these intentions to the HSE. Detailed advice on the notification of well operations is provided in Schedules 6 of the Guidance on Regulations.

Regulation 12 Notification Of Construction Activities
The HSE are to be notified at least 28 days prior to construction activities commencing on an installation. Construction activities are identified as being the installation or dismantling of the platform, or any activity which requires the use of a heavy lift vessel.

An individual is to be nominated through which arrangements can be made for inspection by an inspector of the HSE. The notification should detail the extent of the work and the hazards which it may involve, the dates on which the work is due to commence and finish and a programme for the work.

Regulation 13 Transitional Provisions
A transitional period for the acceptance of safety cases for existing installations was provided up to the 30th November 1995.

Regulation 14 Co-operation
Every person to which the Regulations apply shall co-operate with the operator and owner of an installation. These will include contractors, operators of other installations during combined operations, operators of installations joined by a common pipeline, persons in control of stand-by, supply and heavy lift vessels.

Regulation 15 Keeping Of Documents
A copy of the safety case must be kept on board the installation and at a nominated company address in the UK. Copies of any audit reports of the safety case, including the findings, recommendations, actions and action time scales are to be kept in an identical manner.

Regulation 16 Amendments to the Offshore Installations (Safety Representatives and Safety Committees)
The Safety Case Regulations require that safety representatives are entitled to see the Safety case, to be supplied with copies of relevant documents if requested and a summary of the main features. The safety representatives are to be consulted during the preparation of the Safety Case and are entitled to view the written audit statement following a review of the Safety Case.

Regulations 17 Exemptions

The HSE may exempt by a certificate in writing any person or installation from the provisions of the safety case regulations, provided that they are satisfied that the health and safety of persons who are likely to be affected by the exemption will not be prejudiced in consequence of it.

The HSE are also at liberty to accept shorter periods of time for the acceptance period of a safety case.

A summary of the requirements contained in Schedule 2 of the particulars to be included in a safety case, are included in chapter three.

2. MODIFICATION OF THE OFFSHORE INSTALLATIONS (SAFETY CASE) REGULATIONS 1992

Introduction — Regulation 26 of the Design and Construction Regulations (DCR) deals with the amendments to the Safety Case Regulations, Schedule 2, in particular the requirement for a verification scheme for safety-critical items.

Schedule 2 — Amendments relating to installation verification.

Paragraph (1) of regulation 2 shall be amended in accordance with this Schedule.

After the definition of "safety case", there shall be inserted the following definition:
"Safety-critical elements" means such parts of an installation and such of its plant (including computer programmes), or any part thereof–

    (a) the failure of which could cause or contribute substantially to; or

    (b) a purpose of which is to prevent, or limit the effect of, a major accident.

After paragraph (7) of regulation 2 there shall be inserted the following paragraphs:

(7A)   Any reference in these Regulations to a verification scheme is a reference to a suitable written scheme for ensuring, by means described in paragraph (7B), that the safety-critical elements

    (a) are or, where they remain to be provided, will be suitable; and

    (b) where they have been provided, remain in good repair and condition.

The verification scheme should provide independent checks to confirm the continuing suitability of safety-critical elements throughout the life-cycle of the installation. The scheme may be complimentary to, but not a substitute for routine maintenance programmes.

(7B)   The means referred to in paragraph (7A) are:

    (a) examination, including testing where appropriate, of the safety-critical elements by independent and competent persons;

    (b) examination of any design, specification, certificate, CE marking or other document, marking or standard relating to those elements by such persons;

    (c) examination of such work in progress by such persons;

    (d) the taking of appropriate action following reports by such persons;

Offshore Engineering

(7C) For the purposes of paragraph (7B) and regulations 15A and 15C a person shall be regarded as independent only where:

(a) His function will not involve the consideration by him of an aspect, of a thing liable to be examined, for which he bears or has borne such responsibility and might compromise his objectivity; and

(b) He will be sufficiently independent of a management system, or of a part thereof, which bears or had borne any responsibility for an aspect, which he might consider, of a thing liable to be examined, to ensure that he will be objective in discharging his function.

After paragraph (2) of regulation 15 there shall be inserted the following paragraph:

(2A) The operator or owner of an installation shall ensure that:

(a) Its verification scheme is kept at an address in Great Britain;

(b) Records shall be kept for a period of six months after the scheme ceases to exist.

After regulation 15 there shall be inserted the following regulations:

15A-(1) Subject to paragraph (2), the operator of a fixed installation shall, at such time before completion of the design ensure that:

(a) A record is made of the safety-critical elements;

(b) Comment on the record by an independent and competent person is invited;

(c) A verification scheme is drawn up by or in consultation with such persons;

(d) A note is made of any reservation expressed by such person as to the contents of:

(i) the record; or
(ii) the scheme; and

(e) Such scheme is put into effect.

(2) The Operator shall not operate the installation unless the provisions of the above paragraph and of regulation 15B have been complied with.

15B. A verification scheme shall provide for the matters contained in Schedule 9.

15C. The operator, in the case of a fixed installation and the owner, in the case of a mobile installation, shall ensure that, as often as may be appropriate:

(a) The verification scheme is reviewed and, where necessary, revised or replaced by or in consultation with an independent and competent person; and

(b) A note is made of any reservations expressed by such persons in the course of drawing it up.

15D. The operator or owner of a fixed or mobile installation shall ensure that effect continues to be given to its verification scheme.

For a new fixed installation, a verification scheme is required to be put in hand before the completion of design work. For a mobile installation the verification system must take effect when it enters UK waters. However the adequacy of the verification scheme, including matters relating to design, construction and the installation's history must be verified before the vessel enters UK waters.

A detail assessment by the independent and competent person of the methodology used to identify the safety-critical elements is not required. This person is expected to use professional judgement, expertise and experience to reach a view on whether the list is appropriate for the installation in question.

Defence

15E.-(1) In any proceedings for an offence for a contravention of any of the provisions of regulations 15A to 15D it shall be a defence for the person charged to prove:

(a) that the commission of the offence was due to the act or default of another person not being one of his employees; and

(b) that he took all reasonable precautions, and exercised all due diligence, to avoid the commission of the offence.

SCHEDULE 9   MATTERS TO BE PROVIDED FOR IN A VERIFICATION SCHEME

1. The duty holder is required to provide particulars of how the independent and competent person(s) will be selected to carry out the work under the scheme, including its review to be explained in the scheme. The information required relates to matters such as the degree of technical expertise, knowledge and experience which will be needed to carry out tasks within the verification scheme, and the appropriate levels of independence.

2. An effective communication system to convey information to person(s) carrying out verification work under the scheme is required so that they can perform their tasks. It is important that duty holders ensure that those responsible for managing their verification scheme have a thorough understanding of the implications of all facets of the different stages of the installation's life cycle and how they fit together.

3. The nature and frequency of examination and testing.

4. Arrangements for review and revision of the scheme.

5. The arrangements for the making and preservation of records showing:

   (a) the examination and testing carried out;
   (b) the findings;
   (c) remedial action recommended; and
   (d) remedial action performed.

6. Arrangements for communicating the output of the verification process to the appropriate levels in the duty holder's organisation. An appropriate level would be that with sufficient authority to ensure that the action required in the light of verification findings is taken.

3. PREVENTION OF FIRE AND EXPLOSION, AND EMERGENCY RESPONSE REGULATIONS 1995 — SI 743

Introduction — The PFEER regulations apply to fixed and floating production installations and to accommodation vessels in support of such operations. They also apply to drill ships and jack-ups involved in exploration, production or accommodation support. The PFEER regulations do not apply to diving support vessels. The definition of installations is taken from the *Management and Administration Regulations* (MAR), SI 1995/738.

Regulation 1   Citation and Commencement
These Regulation may be cited as the Offshore Installations (Prevention of Fire and Explosion, and Emergency Response) Regulation 1995 and shall come into force on the 20th June 1995.

Regulation 2   Interpretation
Major accident means:

(a) a fire, explosion or other release of a dangerous substance involving death or serious personal injury to persons on the installation or engaged in an activity on or in connection with it;

(b) any event involving major damage to the structure of the installation or plant affixed thereto or any loss in the stability of the installation;

(c) the collision of a helicopter with the installation;

(d) the failure of life support systems for diving operations in connection with the installation, the detachment of a diving bell used for such operations or the trapping of a diver in a diving bell or other sub-sea chamber used for such operation; or

(e) any event arising form a work activity involving death or serious personal injury to five or more persons on the installation or engaged in an activity in connection with it.

Regulation 3   Application
This regulation applies to installations located in the territorial waters of Great Britain and the UK Continental Shelf.

Regulation 4   General Duty
(1)   The duty holder shall take appropriate measures with a view to:

(a) protecting persons on the installation from fire and explosion; and

(b) securing effective emergency response.

The purpose of the regulations overall is to promote a risk-based, systematic approach to managing fire and explosion hazards and emergency response. The general principles of risk control and health and safety management set out in the *Approved Code Of Practice* on MHSWR are relevant to this process and should be followed. They are summarised as follows:

Principles for preventative and protective measures

- It is always best if possible to avoid a risk altogether;
- combat risks at source, rather than by palliative measures:
- wherever possible adapt work to the individual;
- take advantage of technological and technical progress to improve work methods and make them safer;
- risk prevention measures should form part of a coherent policy and approach to reduce risks progressively that cannot be prevented or avoided altogether;
- give priority to measures which protect the whole workplace;
- make sure workers understand what they need to do;
- create an active health and safety culture which affects the organisation as a whole.

Principles for health and safety management

- **Planning.** Adopting a systematic approach which identifies priorities and sets objectives. Whenever possible, risks are eliminated by the careful selection and design of facilities, equipment and processes or minimised by the use of physical control measures;
- **Organisation.** Putting in place the necessary structure with the aim of ensuring that there is a progressive improvement in health and safety performance;
- **Control.** Ensuring that the decisions for ensuring and promoting health and safety are being implemented as planned;
- **Monitoring and Review.** Like quality, progressive improvement in health and safety can only be achieved through the constant development of policies, approaches to implementation and techniques or risk control.

These principles should be applied for fire and explosion hazard management and emergency response by:

- developing an approach to fire and explosion hazard management based on hazard identification and risk assessment, involving a balanced and integrated approach to designing out hazards, preventing hazardous events occurring, detecting that they have occurred, are occurring, controlling their escalation, and mitigating their consequences;
- a systematic approach to every response by considering the various stages of an emergency and how each of those stages can be most effectively managed, to reduce risks to health and safety.

Regulation 5   Assessment

The duty holder must perform an assessment to identify which events could give rise to a major accident involving fire or explosion, or the need to evacuate the installation. The assessment must establish appropriate *performance standards* to be met by the protective measures introduced. All requirements needed to comply with the rest of the regulations will be determined by the results of the assessment.

Regulation 6 Preparation for Emergencies

This regulation reflects the importance of adequate preparation for emergencies, including training and deals mainly with the organisation of personnel. One person, the offshore installation manager (OIM) should be given responsibility for taking overall charge in an emergency and the roles and responsibilities of those in the command structure should be clearly defined and displayed on an emergency duties plan.

Appropriate instruction should be given to all persons on the installation on or before arrival at the installation

Regulation 7 Equipment for Helicopter Emergencies

The duty holder must ensure that equipment is provided that can be used to deal with an accident involving a helicopter, including crash rescue equipment. These arrangements are in addition to those provided for dealing with a fire or explosion which are covered under regulations 9, 12 and 13.

Regulation 8 Emergency Response Plan

This regulation requires the duty holder to prepare an emergency response plan, which documents the organisation and arrangements for dealing with an emergency on the installation. The plan should contain procedures which cover all reasonably foreseeable emergencies, that is, it should set out who does what, when, where, how and to what effect.

The plan should cover both the offshore and onshore arrangements and take account of the views of those who are likely to become involved, such as HM Coastguard and the pipeline owners and operators. The plan should be exercised and tested with sufficient frequency and depth so that it can be relied upon to work effectively in an emergency, and should be monitored and reviewed in the light of these exercises and tests.

Regulation 9 Prevention of Fire and Explosion

This regulation requires the duty holder to take appropriate measures to prevent fire and explosion. Good design is the most effective means of preventing fire and explosion and this involves addressing issues such as the choice of process and production methods to minimise the risk of fire and explosion by optimising plant layout.

Measures to control activities which might lead to a release or ignition hazard should include adequate procedures and arrangements for the control of hot work and the use of electrical equipment not designed for use in areas of flammable and explosive atmospheres. In effect, a *permit to work system.*

Regulation 10 Detection of Incidents

Detection measures need to be provided for the full range of reasonably foreseeable events which require emergency response with a view to enabling information regarding such incidents to be conveyed forthwith to places from which control action can be instigated. These measures should cover:

(i) Fire detection

(ii) Flammable or toxic gas detection

(ii) Detection of leakages of flammable liquids

(iii) Man overboard

(iv) Monitoring of ships and supply boat operations

(v) Monitoring of helicopter flights

(vi) Monitoring of diving operations

Detection measures for major accident hazards should be based on the findings of the assessment and should be automatic in operation, where this is reasonably practicable, as should the transmission of information from detection systems to the point at which control action is instigated. Where detection and the relay of information cannot be done automatically, adequate arrangements should be in place to detect incidents and to instigate control action.

Regulation 11   Communication
The duty holder shall make appropriate arrangements to give an audible warning (and visual where necessary) of an emergency, to all persons on the installation and for the purpose of emergency response, for communication between persons not on the installation but engaged in activities in connection with it and beyond it such as diving operations. These arrangements are to remain effective in an emergency.

All arrangements should be based on the findings of the assessment required by regulation 5 for major accident hazards and should take into account the findings of the risk assessment required by regulation 3 of MHSWR for non-major accident hazards.

The alarm and communications systems provided should be capable of transmitting clear information to personnel wherever they are likely to be on the installation. Where the method of raising the alarm does not automatically initiate a general visual or acoustic alarm, there should be clear procedures for instituting the appropriate alarm and conveying information to the relevant personnel. This might include HM Coastguard, other installations, shore-based personnel and rescue and recovery services.

Duty holders should ensure that there are arrangements in place to alert personnel to an incident on the installation and that everyone is aware of the meaning of the different alarm signals, this should be included in installation induction and refresher training. There should be:

(i) Red flashing lights to provide warning against a release of toxic gas.
(ii) Yellow flashing lights for all other warning signals.
(iii) Acoustic signals provided for evacuation:
   a) continuous signal of variable frequency;
   b) continuous signal of constant frequency for toxic gas;
   c) intermittent signal constant frequency for all other cases.

Regulation 12   Control of Emergencies
Duty holders must have appropriate control measures to limit escalation of an emergency and this includes measures to monitor the extent of an emergency for the purpose of exercising managerial command and control.

Measures should be provided for the timely and effective shut-down of systems which could exacerbate an emergency and for the initiation of emergency shut-down systems and plant used to limit the spread of fire. These systems should be operable from a safe location such as the control point or TR which should be staffed at all times. These measures will include plant, equipment and workplace procedures and examples are:

(i) Emergency response procedures.
(ii) ESD systems.
(iii) Vents and drains.
(iv) Automatic isolation valves.
(v) Blowdown and flare systems.
(vi) Portable fire-fighting equipment.

There also specific requirements in respect of emergency shut-down valves in the Pipeline Safety Regulations (PSR), SI 1996/825.

Regulation 13 Mitigation of Fire and Explosion
Based on the findings of the assessment, the duty holder must put in place measures that will mitigate the effects of fire and explosion and the measures are likely to include:

- (i) Deluge systems.
- (ii) Fixed extinguishing systems.
- (iii) Fire-resistant coatings.
- (iv) Manual response equipment and procedures.
- (v) Ventilation control systems.
- (vi) Fire and blast walls.

The measures taken should provide for automatic systems unless manual operation can be justified in the assessment.

In determining what portable equipment is to be provided duty holders should take into account the numbers and location of personnel on the installation, the practicality of its effective use, the hazards and the availability of other systems.

Regulation 14 Muster Areas etc.
The regulation requires the duty holder to make provision for clearly defined muster areas where personnel on the installation can assemble safely while the emergency is assessed and control action taken. It also requires procedures to be in place for accounting for people and provision for personnel to access safely means for leaving the installation if necessary.

Muster points should remain useable and egress routes passable based on the scenarios identified in the assessment and the emergency response plan and appropriate communications facilities should be provided.

Adequate emergency lighting should include, where appropriate, floor level illumination and direction of escape indicators.

Personnel should be provided with appropriate information about the location, their muster station.

Regulation 15 Arrangements for Evacuation
The duty holder shall ensure that such arrangements are made as will ensure, so far as is reasonably practicable, the safe evacuation of all persons to a place of safety. Evacuation in this context means by helicopter, direct sea transfer or by lifeboat, that is, without getting wet. It does not include life rafts or methods which involve escape directly into the sea. These are considered to be a last resort and are covered in regulation 16.

There are a number of means of evacuation and the preferred means should be the normal means of getting persons to and from the installation, that is by helicopter. Alternative means must also be provided for when the normal means fail and in most cases, lifeboats (TEMPSC — totally enclosed motor propelled safety craft) provide the alternative means. There should be sufficient places provided by TEMPSC for 150% of the persons on board unless an alternative standard is justified.

Regulation 16 Means of Escape

Escape, means leaving in installation in an emergency when the evacuation arrangements have failed due to a catastrophic incident when a planned and orderly evacuation cannot be achieved. Means are to be provided for descent to the sea such as ladders and stairways, davit launched life-rafts, chute systems, and individually controlled descent devices; and items in which personnel can float on reaching the sea such as throw-over life-rafts. Preference is to be given to devices which provide some protection from the elements and avoid the need to enter the sea directly. Entering the sea directly is described as a last resort and should be avoided if at all possible.

Regulation 17 Arrangements for Recovery and Rescue

This regulation deals with arrangements which must be set in place for the recovery of persons following their evacuation or escape from the installation and for their subsequent safety. Basically it involves the recovery of individuals who have fallen into the sea or who have ditched into the sea following a helicopter incident.

There are many circumstances for which only a suitable vessel standing by will provide effective arrangements and in these circumstances such a vessel will need to be provided.

Regulation 18 Suitability of Personal Protective Equipment for Use in an Emergency

This regulation deals with the requirement to supply personal protective equipment and to ensure that there is prepared and operated a written scheme for the systematic examination and testing by a competent person, of the equipment.

Personal protective equipment must be designed to increase the chances of survival of an individual in an emergency by enabling them to reach muster areas, the temporary refuge, evacuation or escape points. The equipment should provide protection against the effects of fire, heat, smoke, fumes or toxic gas and in the event of immersion in the sea. A set of protective equipment should be located in the accommodation for every individual and should include a smoke hood, portable light source and heat proof gloves, survival suit and life-jacket. Additional equipment should be provided at appropriate locations on the installation.

There should also be appropriate personal protective equipment for use by those with specific emergency duties such as for helicopter duties and fire-fighting. The provision of personal protective equipment for use in major accident hazards should be based on the findings of the assessment required by regulation 5, and for non-major accident hazards on the assessment required by regulation 3 of MHSWR.

A *written scheme for the systematic examination* of the equipment is to be provided to ensure that the equipment is effectively maintained. The written scheme should specify which equipment is to be covered and the nature and frequency of examinations, and tests where appropriate. This should take into account what the equipment may be needed for and the *performance standards* set for availability and reliability. Examinations should also verify the location of equipment against plans.

The regulations do not cover clothes or equipment required for normal working activities.

Regulation 19 Suitability and Condition of Plant

All plant and equipment installed to comply with the PFEER regulations 7 to 19 must be suitable by design, construction or adaptation for the actual work it is provided to do. Performance standards determined in the assessment required by regulation 5, should be the basis for assessing that plant required to deal with major accident hazards is suitable for its purpose. They may be described in terms of (FARSI) functionality, availability, reliability survivability and on how they interact with other systems. They should be measurable and auditable.

Written schemes are to be provided for the *systematic examination* of the equipment to ensure that it is effectively maintained and the examinations are to be carried out by a *competent and independent person(s)*. Examinations are also to be carried out before the equipment is first used on the installation and following modifications or repair.

The written schemes replace the examination of life-saving appliances and fire-fighting equipment required under previous law; the scope of the scheme is similar, but now the duty holder is responsible for preparing the scheme. The written schemes are in addition to any schemes required as part of the routine maintenance and testing programmes.

The written scheme is designed to provide a careful and critical scrutiny to access the suitability of the plant for its purpose. It may be set out in a single document, or a series of documents or recorded electronically, but information about the scheme should be readily accessible.

Regulation 20 Life Saving Appliances
All survival craft, life-rafts, life-buoys, life-jackets and plant are to be suitably marked and equipped.

Regulation 21 Information Regarding Plant
Personnel are to be advised of the location of manual fire-fighting equipment, life saving appliances and personal protective equipment, and hazardous areas. This may be by plans, signs etc..

4. DESIGN AND CONSTRUCTION, ETC REGULATIONS 1996 — SI 913.

Introduction — The objectives of the Design and Construction Regulations (DCR) are to replace the certification regime established by the *Offshore Installations (Construction and Survey) Regulations* 1974 (SI 289). They came into force on the 30th June 1996 and apply to both fixed and mobile installations. In doing so, the Regulations also implement the relevant health and safety aspects of the *Extractive Industries (Borehole) Directive* (EID)(92/91/EEC), and support and complement the *Offshore Installations (Safety Case) Regulations 1992* (SI 1992/2885)(SCR), particularly the requirement to provide a *verification scheme* for the *safety-critical elements* of an installation.

The Regulations contain requirements for ensuring that offshore oil and gas installations, and oil and gas wells are designed, constructed and kept in a sound structural state, and other requirements affecting them, for purposes of health and safety. Part II of the Regulations deal with the integrity of installations and the provisions in it impose duties on the duty holder relating to:

- the maintenance of its integrity at all times;
- its design;
- work to it;
- the way it is operated;
- arrangements for maintaining its integrity;
- the reporting of danger to it;
- decommissioning and dismantlement.

Part III contains requirements relating to the helicopter landing area and Part IV imposes requirements affecting oil and gas wells and covers such areas as well design, well control, well examination and information, instruction, training and supervision of operatives.

Part V deals mainly with extremely important changes to the Safety Case Regulations, in particular the requirement to provide a verification scheme for safety-critical elements of an installation.

PART 1 — INTERPRETATION AND GENERAL

Regulation 1   Citation and Commencement
These Regulations may be cited as the Offshore Installations and Wells (Design and Construction, etc,) Regulations 1996 and shall come into force on the 30 June 1996.

Regulation 2   Interpretation
The definition of an installation in DCR differs from the definition given in the MAR and PSR regulations. Any well connected to the installation, and any pipeline or equipment connected to the pipeline (within or without 500 metres of the main structure) of the installation, is excluded from the DCR. This is to avoid duplication with Part IV of DCR — Wells, which covers wells separately and explicitly, and PSR which covers pipelines and associated equipment up to and including the 500 metres adjacent to the installation.

Regulation 3   Application
The Regulations apply to all installations (fixed and mobile) located in the relevant waters of the UK and to all installations covered by inter-governmental treaties.

PART II — INTEGRITY OF INSTALLATIONS

Regulation 4 – General Duty
The duty holder shall insure that an installation possesses such integrity as is reasonably practical. Integrity means, *structural soundness and strength, stability and, in the case of a floating installation buoyancy in so far as they are relevant to the health and safety of persons.*

Regulation 5 – Design of an Installation
The duty holder shall ensure that the designs to which an installation are constructed are such that:

- it can with stand such forces acting on it as is reasonably practical;
- its layout and configuration, including those of its plant, will not prejudice its integrity;
- fabrication, transportation, construction, commissioning, operation, modification, maintenance and repair of the installation may proceed without prejudicing its integrity;
- it may be decommissioned and dismantled safely; and
- in the event of reasonably foreseeable damage to the installation it will retain sufficient integrity to enable action to be taken to safeguard the health and safety of persons on or near it.

The phrase, *capable of withstanding reasonably foreseeable forces*, has been used to ensure that the design takes appropriate account of forces imposed by environmental conditions, the weight of equipment that may be placed on it, the activities in connection with it such as vessel or aircraft impact and accidents, fire and explosion.

The greatest scope for risk reduction is at the design stage and the layout and configuration of the plant should be optimised to assist in reducing the risks to the installation's integrity.

Regulation 6 – Work to an Installation

The duty holder shall ensure that the work of fabrication, construction, commissioning, modification, maintenance and repair of an installation and activity in preparation for its positioning are carried out in such a way that, so far as is reasonably practicable, its integrity is secured. The duty holder may contribute to meeting the requirements of this regulation through contractual agreements with various parties involved in the processes.

Regulation 7 – Operation of an Installation

The duty holder must ensure that the installation is operated in a manner which does not prejudice its integrity. This regulation refers primarily to the meteorological and environmental conditions and is most relevant to mobile installations.

Regulation 8 – Maintenance of Integrity

This regulation requires the duty holder to make suitable arrangements to insure that the integrity of an installation is maintained throughout its life-cycle. Such arrangements include planned maintenance, inspection of structures and periodic assessment of the installation and the frequency, scope and method of inspection should be sufficient to provide assurance, that the integrity of the installation is being maintained.

Regulation 9 – Reporting of Danger to an Installation

The duty holder shall ensure that within 10 days after the appearance of evidence of a significant threat to the integrity of an installation, a report is made to the HSE. Examples of what may be reportable are the identification of defects in the structure which could be a threat to the integrity of the installation.

Regulation 10 – Decommissioning and Dismantlement

The duty holder shall ensure that an installation is decommissioned and dismantled safely.

PART III — FURTHER REQUIREMENTS RELATING TO INSTALLATIONS

Regulation 11 – Helicopter Landing Area

The duty holder shall ensure that the helicopter landing area is adequate for its purpose and meets the requirements of relevant aviation legislation. Further information on offshore helicopter landing areas may be obtained from the Civil Aviation Authority (CAA) guidance notes CAP 437.

Regulation 12 – Additional Requirements

The duty holder shall ensure that the additional requirements set out in the Extractive Industries (Boreholes) Directive (EID) 92/91/EEC are complied with by 3rd November 1999. The Schedule deals primarily with the health and safety aspects likely to be encountered on an installation such as the temperature requirements for work and rest places, ventilation, lighting, sanitation arrangements and accommodation. The requirements are broadly equivalent to the onshore Workplace (Health, Safety, and Welfare) Regulations 1992 (SI 1992/3004). Sleeping quarters should contain no more than two persons.

PART IV — WELLS

Introduction – The well provisions of the DCR Regulations apply to wells drilled onshore and offshore and the Regulations replace SI 1980/1759, the *Offshore Installations (Well Control) Regulations* 1980, and SI 1991/308, the *Offshore Installations (Well Control)(Amendment) Regulations* 1991, and regulation 4 and Schedule 15 of SI 1976/1019, the *Offshore Installations (Operational Safety, Health and Welfare) Regulations* 1976.

It is likely that wells on, or attached to, an installation will be safety-critical to the safety of that installation and so should be included in the verification scheme.

Regulation 1   Citation and Commencement
These Regulations may be cited as the Offshore Installations and Wells (Design and Construction, etc.,) Regulations 1996 and shall come into force on the 30 June 1996.

Regulation 2   Interpretation
The Regulations apply to all wells in the offshore sector, including wells used for injection and monitoring, whether they are connected to a fixed installation, mobile installation or ship, or stand-alone on the sea bed.

The well is defined in terms of its pressure containment boundary and any equipment that is vital to controlling the pressure within the well is therefore covered. This equipment includes blowout preventers and Christmas trees. The well may be considered to end at:

- above the top blowout preventer in the BOP stack and outside the choke and kill valves;
- downstream of the swab and production wing valves of a Christmas tree;
- at the top of the stuffing box of a wireline BOP.

Regulation 13   General Duty
The well-operator shall ensure that a well is designed, modified, commissioned, constructed, equipped, operated, maintained, suspended and abandoned so that as far is reasonably practicable, there can be no unplanned escape of fluids form the well and that the risks to the health and safety of persons from it or anything in it are as low as is reasonably practicable. The focus overall is therefore on the safe physical condition of the well rather than the actual operation being carried out in the well.

Regulation 14   Assessment of Conditions Below Ground
Before the design of a well is commenced the well-operator shall assess the hazards associated with the geological strata and formations, and fluids within them.

Regulation 15   Design with a View to Suspension and Abandonment
The well-operator shall ensure that a well is so designed and constructed that, so far as is reasonably practicable it can be suspended or abandoned in a safe manner.

Regulation 16   Materials
The well-operator must ensure that every part of a well is composed of materials which are suitable. This requirement applies not only to such items as cement, casing or other well tubulars, but also the well-head equipment, e.g. drilling spools, casing heads, tubing heads and the well control equipment.

Regulation 17   Well Control
The well-operator shall ensure that suitable well control equipment is provided. Well control equipment includes equipment whose primary purpose is to prevent, control or divert the flow of fluids from the well. As such, well control equipment includes blowout preventers, downhole preventers, Christmas trees, wireline lubricators and stuffing boxes, rotating heads, tubing injection heads, circulating heads, internal blowout preventers and kelly cocks, choke and kill lines, choke manifolds and diverters. Plugs or other isolating devices installed in a borehole to prevent the well from flowing are also included.

In terms of a well-operator discharging his duty for ensuring the provision of well control equipment this will normally be achieved by review of the contractor's arrangements.. If necessary, the well-operator should check that the contractor has suitable policies, procedures and management controls to ensure that suitable equipment is supplied.

Regulation 18   Arrangements For Examination
Before the design of a well is commenced the well-operator shall put into effect arrangements in writing for an examination scheme for the well and its equipment, which will be carried out by independent and competent persons.

The objective of the examination is to obtain assurance that the well is designed and constructed properly, and maintained adequately. The examination must demonstrate that the pressure boundary of the well and the associated pressure containment equipment is suitable for this purpose. However, it is not anticipated that examination schemes will necessarily rely on the physical examination of wells, but more on documentary evidence of well safety.

When a well is connected to a fixed or a mobile installation, there may be an overlap between the well examination scheme and the verification scheme for the installation. Where equipment is deemed to be safety-critical it would need to be included in the installation verification scheme. Where a well is not attached to an installation, a separate well examination scheme will be required.

Regulation 19   Provision of Drilling Information
The well-operator must report to the HSE on a regular basis when carrying out drilling, workovers or any other operation involving substantial risk of the unplanned escape of fluids from a well. The report is required to include a significant amount of operational data such as the setting of casings and the depths achieved.

Regulation 20   Co-operation
Every person to which the Regulations apply shall co-operate with the operator and owner of an installation.

Regulation 21   Information, Instruction, Training and Supervision
The well-operator must ensure that persons involved in the drilling, well intervention or workover operations are suitably trained.

PART V — MISCELLANEOUS

Regulation 22 – Defence

Regulation 23 – Certificates of exemption
The HSE may exempt by a certificate in writing any person or installation from the provisions of the Regulations, provided that they are satisfied that the health and safety of persons who are likely to be affected by the exemption will not be prejudiced in consequence of it.

5. MANAGEMENT AND ADMINISTRATION REGULATIONS 1995 — SI 738

Introduction — The MAR regulations primarily deal with the more routine health and safety arrangements which relate to offshore installations. The Regulations apply to fixed and mobile offshore installations, but not to crane barges or sub-sea installations.

These Regulations cover:

- (a) the notification to the HSE of changes of owner of an installation;

- (b) installation managers powers;

- (c) the keeping of records of persons onboard;

- (d) permit to work systems;

- (e) various operational matters such as providing information to workers on the location of the relevant HSE offices;

- (f) the provision of health surveillance;

- (g) the provision of food and water supplies;

- (h) co-operation

- (i) arrangements for helideck operations;

- (j) amendments to the Safety Representatives and Safety Committees Regulations SI 1989/971;

- (k) consequential amendment to other regulations such as the Pipeline Works (First Aid) Regulations, SI 1989/1671.

The Regulations add more detailed requirements necessary for the management of offshore operations.

Regulation 1   Citation and Commencement
These Regulations may be cited as the *Offshore Installations and Pipeline Works (Management and Administration) Regulation 1995* (SI 1995/738) and came into force on the 20th June 1995, except regulation 23(2) which came into force on 20th June 1997.

Regulation 2   Interpretation
A list of relevant definitions are provided.

Regulation 3   Meaning of Offshore Installation
An *Offshore Installation* is an installation standing in, or stationed in relevant waters, and used for the exploitation, or exploration and storage of mineral reserves, or is used mainly for the provision of accommodation for persons who work on or from one of the previously defined structures.

The definition of an offshore installation in these Regulations replaces the definition previously used in the *Mineral Workings Act* and the *Safety Case Regulations*. It is worth noting that vessels providing accommodation for workers on offshore installations such as flotels are classed as installations whilst heavy lift vessels are not, even if they are providing accommodation as a secondary service to a heavy lift capability. Well service vessels, survey vessels and shuttle tankers are not considered to be installations.

Things which are part of offshore installations

Sub-sea wells are classed as part of an installation if they are connected by a pipeline or cable, and any pipeline, apparatus or work within 500 metres of the main structure of the offshore installation to which it is attached. Sub-sea installations are not subject to these Regulations, or the PFEER and SCR if they are free standing and not connected to a fixed or floating installation.

Regulation 4   Application

The Regulations apply to all installations (fixed and mobile) located in the relevant waters and to all installations covered by inter-governmental treaties. Relevant waters are defined as tidal waters and parts of the sea in or adjacent to Great Britain up to the seaward limits of territorial waters and parts of the Continental Shelf designated by the Act of 1964.

Regulation 5   Notification Concerning Offshore Installations

The duty holder shall notify the HSE prior to the arrival or departure of a mobile offshore installation to or from relevant waters. Also, prior to the change of duty holder of an offshore installation.

Regulation 6   Managers

The duty holder shall appoint an installation manager (OIM) to ensure that the installation is at all times under the charge of a competent person.

The installation manager will be in overall charge of the installation and responsible for the day-to-day management, and in charge of the health, safety and welfare of persons on or about the installation, including ensuring the maintenance of order and discipline. The OIM will also take charge of the installation in an emergency.

The OIM will be competent within the meaning of regulation 6 of the Health and Safety at Work Act *competent* meaning one who has *sufficient training and experience or knowledge and other qualities*. Competence assessment involves taking account of the persons, skills, judgement, experience, qualifications and relevant knowledge and training in combination, all in relation to the tasks to be undertaken. To ensure that an individual remains competent his/her performance should be monitored and refresher training and demonstrations of sustained competence be provided.

Regulation 7   Restraint and Putting Ashore

The OIM has legal powers to put people under restraint and send them ashore, when necessary, to ensure the safety of the offshore installation and the health and safety of those on board.

Regulation 8   Co-operation

Safety requires co-operation between everyone who has a contribution to make in ensuring health and safety on the installation and this includes operators, owners, concession owners, employers, managers and people in charge of visiting vessels or aircraft.

Duty holders cannot evade their responsibilities by seeking to pass their responsibilities to others.

Regulation 9   Records

To ensure that in an emergency, everyone can be accounted for, a record must be maintained on the installation of all the individuals that are on board, a record often referred to as Personnel on Board (POB). This record will identify each individuals name, and the name and address of his employer, a more complete record of each individuals particulars is to be maintained at an onshore address. The records must be amended whenever anyone arrives or departs from the installation.

Regulation 10   Permits to Work

The duty holder must establish a system of written permits to work to be used in circumstances where the nature and scale of the risk arising from the work to be carried out demand such a stringent system of control. These permits must be issued by a competent person who has sufficient knowledge to decide on conditions and precautions necessary to ensure that safety is not jeopardised.

The HSE provide a guidance booklet on permit to work systems.

Regulation 11   Instructions

Written instructions should be provided and all relevant individuals made aware of these instructions for activities considered to have an element of risk in there execution. Examples of such activities are isolation procedures, over-side working, transfer procedures, helideck operations and inspection/ maintenance procedures.

Regulation 12   Communication

Effective communications must be provided which are appropriate for health and safety purposes. Communications must be provided between the installation and the shore, vessels, aircraft (including helicopters) and other installations.

Regulation 13   Helicopters

A competent person, the helicopter landing officer (HLO), must be appointed who will be responsible for the day-to-day management of the offshore installation helideck, in control of the associated helideck operations and the helideck crew. Equipment should be provided for safe helideck operations, including chocks and tie-down straps/ropes, refuelling equipment and a suitable power source for starting helicopters may also be required in some instances.

Procedures must be established for safe helideck operations and an HLO should accompany any flight to an unmanned offshore installation to carry out the HLO's function immediately after the helicopter has landed and up to the point of its subsequent take-off.

Regulation 14   Operational Information

Arrangements should be provided to measure and record such meteorological and oceanographic information as is necessary for securing the safe operation on the installation and the persons on or near it. Environmental conditions may affect the safety of activities such as the loading or unloading of ships and helicopter operations. Operational parameters will have been established as part of the Safety Case.

Regulation 15   Information to Persons

The address and telephone number of the office of the HSE Executive for the onshore installation must be posted where it can be seen by all individuals on the installation, and on vessels supplying a service to the installation such as pipelaying or well servicing operations. This is to ensure that the workforce can contact the HSE inspectors if required.

Regulation 16   Health Surveillance

Health and safety surveillance must be provided which is appropriate to the health and safety risks likely to be encountered on board the installation. This surveillance should commence before workers are exposed to any hazardous substances. The regulations are not concerned with screening workers to identify any pre-existing health conditions or to meet fitness requirements for offshore work.

Regulation 17 Drinking Water
An adequate supply of clean drinking water must be provided and marked as such on each installation. Supplies should be tested regularly.

Regulation 18 Provisions
The duty holder shall ensure that all provisions for the consumption of persons on the installation are fit for human consumption, palatable and of good quality.

Regulation 19 Identification Of The Offshore Installation
Each offshore installation, where physically possible, must display a name or other marking which will allow any vessel or helicopter approaching the installation to be able to identify it without confusion or ambiguity. The HSE publish guidance on the subject under the cover of *Standard Marking Schedule for Offshore Installations*.

Regulation 20 Certificates Of Exemption
The HSE may issue a certificate exempting any person or offshore installation from these regulations provided that the health and safety of people who are likely to affected are not prejudiced if an exemption is issued. The HSE may not grant exemptions from any of the provisions of relevant EC health and safety directives.

Regulation 21 Application of The Employer's Liability (Compulsory Insurance) Act 1969

Regulation 22 Repeals and Modifications of the 1971 Act

Regulation 23 Revocation and Modification of Instruments.

6. THE PIPELINES SAFETY REGULATIONS 1996 – SI 825

Introduction – The Pipeline Safety Regulations replace earlier prescriptive legislation and introduce a goal-setting, risk-based approach to managing pipeline safety. The statutory instrument is accompanied by a publication, *Guidance On Regulations*.

Relationship With Other Regulations
The Pipeline Safety Regulations cover pipelines and associated equipment up to, and including the 500 metres adjacent to the installation. Pipelines are not included in the DCR regulations but fall partly within the scope of the offshore safety case regime, *protection of people from risks associated with the pipeline*, and certain parts such as the risers and ESDVs are likely to be classed as safety-critical and thus will require verification.

Regulations 1 Citation and Commencement
These regulations may be cited as the The Pipeline Safety Regulations and revoke the Emergency Pipelines Regulations, SI 1029 in their entirety, and a number of the regulations within the other pipeline statutory instruments.

Regulation 2 Interpretation
Emergency shut-down valve means a valve which is capable of adequately blocking the flow of fluid within the pipeline. Minor leakage may be tolerated providing it does not represent a threat to safety. The rate of leakage should be based on the installation's ability to control safely the hazards produced by such a leak.

Regulation 3   Meaning of Pipeline

The Regulations cover the sub-sea pipeline and where pig traps are included, they are classed as part of the pipeline. The limit of the pipeline on the offshore installation is the inboard flange of the emergency shut-down valve (ESDV) or the inboard flanges of the primary shut off valves on the pig receiver, where fitted, which connect the pipeline to the installation. The limit of the pipeline at the onshore reception facility is similarly defined.

Regulation 4   Application

This regulation defines the scope of the requirements as onshore pipelines in Great Britain, and all pipelines in territorial waters and the UK Continental Shelf.

Regulation 5   Design of a Pipeline

The operator shall ensure that no fluid is conveyed in a pipeline unless it has been so designed that, so far as is reasonably practicable, it can withstand:

（a）   forces arising from its operation;

（b）   the fluids that may be conveyed in it; and

（c）   the external forces and the chemical processes to which it may be subjected.

The external forces and the chemical processes to which the pipeline will be subjected will need to be identified and evaluated and this will include terrain, sub-terrain or sea bed conditions. Account should also be taken of foreseeable mechanical and thermal stresses and strains to which the pipeline may be subjected during its operation.

In general British Standards provide a sound basis for the design of pipelines. Other national or international standards are likely to be acceptable provided that they provide an equivalent level of safety.

Regulation 6   Safety Systems

Safety systems cover a means of protection such as emergency shut-down valves and shut-off valves. Safety systems are not meant to cover all control or measuring devices. However, safety systems do include control or monitoring equipment, such as flow detectors and pressure monitors, which have to function properly in order to protect the pipeline or to secure its safe operation.

Regulation 7   Access for Examination and Maintenance

The design of the pipeline should take due account of the need to facilitate examination and maintenance. Consideration should be given at the design stage to any requirement to provide suitable and safe access and operation for in-service inspections, such as pigging.

Regulation 8   Materials

The operator shall ensure that no fluid is conveyed in a pipeline unless it is composed of materials which are suitable. Any changes to the fluid conveyed or the operating conditions of the pipeline, including an extension of the pipeline design life, will warrant a reassessment of the pipeline material to ensure it is capable of conveying the fluid safely.

Regulation 9   Construction and Installation

The operator shall ensure that no fluid is conveyed in a pipeline (save for the purpose of testing it) unless it has been so constructed and installed that, so far as is reasonably practical, it is sound and fit for the purpose for which it has been designed.

Regulation 10 Work on a Pipeline
The operator shall ensure that modification, maintenance or other work on a pipeline is carried out in such a way that its soundness and fitness for the purpose for which it has been designed will not be prejudiced.

Regulation 11 Operation of a Pipeline
The operator shall ensure that:

(a) no fluid is conveyed in a pipeline unless the safe operating limits of the pipeline have been established; and

(b) a pipeline is not operated beyond its safe operating limits, save for the purpose of testing it.

Regulation 12 Arrangements for Incidents and Emergencies
Adequate arrangements must be in place to cover the eventuality of an accident occurring or damage being discovered which requires immediate action to be taken, particularly where pipelines are inter connected.

The Prevention of Fire Explosion and Emergency Response Regulations (PFEER) requires the installation owner to draw up an emergency response plan to cover emergencies which may affect the pipeline.

Regulation 13 Maintenance
The operator shall ensure that a pipeline, and all attachments such as the pig trap and any safety systems, is maintained in an efficient state, and in good repair. This includes both how and when the pipeline should be surveyed and examined.

Regulation 14 Decommissioning
The operator shall ensure that a pipeline which has ceased to be used for the conveyance of any fluid is left in a safe condition. Where decommissioning is considered, advice must be sought from the Department of Trade and Industry whose formal approval must be sought prior to decommissioning. The pipelines are subject to the Petroleum Act of 1998 edit.

Regulation 15 Damage to Pipelines
No person shall cause such damage to a pipeline as may give rise to a danger to persons. This regulation applies to the operator of the pipeline and to third parties, for example when anchoring vessels within the field.

Regulation 16 Prevention of Damage to Pipelines
Consideration should be given to reducing the potential for damage to the pipeline by use of concrete coatings, trenching, burial, protection structures or mattresses etc.

The operator shall inform persons of its existence and whereabouts. As part of the offshore Pipeline Works Authorisation issued by the Department of Trade and Industry under the Petroleum and Submarine Pipelines Act 1975, information regarding the location of offshore pipelines is normally passed to the Hydrographer of the Royal Navy for inclusion on Admiralty charts.

Regulation 17 Co-operation
Where there are different operators for different parts of a pipeline, each operator shall co-operate with the other so far as is necessary to enable the operators to comply with the requirements of these regulations.

Regulation 18   Dangerous Fluids

These are identified in some detail in Schedule 2 but basically are fluids which are flammable, toxic, react violently with water, are oxidants or Acrylonitrile.

Regulation 19   Emergency Shut-Down Valves

Emergency shut-down valves (ESDVs) are required to be fitted to all risers of major accident hazard pipelines of 40 mm or more in diameter at offshore installations. (Schedule 3 sets out the requirements more fully and they are described more fully in the explanation of ESD systems in chapter three.)

Regulation 20   Notification Before Construction

The operator must notify the HSE of their intentions at least 6 months prior to the construction work on a pipeline commencing. This would normally be at the end of the conceptual stage of design where the proposals are sufficiently detailed to assess arrangements concerning the safe operation of the pipeline and its construction.

The HSE will be sympathetic to requests for shorter notification periods where good reasons can be demonstrated. This notification does not form part of the role the HSE undertakes as consultee on the route of the pipeline for planning purposes. Onshore pipeline routing will be dealt with by the Department of Trade and Industry.

Regulation 21   Notification In Other Cases

The HSE require a notification period of 14 days prior to a pipeline being brought into use for the first time, or after major modifications have been carried out. In exceptional circumstances a shorter notification period will be approved.

Regulation 23   Major Accident Prevention Document

This regulation deals with the operator's methods for controlling the design, construction and installation, operation, maintenance and final decommissioning. Before the design of a pipeline is completed, a document relating to the pipeline must be prepared which contains sufficient particulars to demonstrate that:
    (a)   all hazards with the potential to cause a major accident have been identified;
    (b)   the risks arising form those hazards have been evaluated;
    (c)   the safety management system is adequate; and
    (d)   adequate arrangements of audit and the making of reports.

Major Accident Prevention Document — MAPD

Major accident means death or serious injury involving a dangerous fluid. The major accident prevention document (MAPD) is a management tool the intention of which is to ensure that the operator has assessed the risks from major accidents and has introduced an appropriate safety management system to control those risks. The MAPD can be made up of a number of documents and a single MAPD can cover a number of pipelines.

Safety Management System — SMS

The SMS should cover the organisation and arrangements for preventing, controlling and mitigating the consequences of major accidents. Extensive guidance is given on the requirements of the SMS and the requirements for audit in the *Guidance on Regulations*.

Offshore Engineering

The remaining regulations cover administration requirements and onshore requirements:

Regulation 24   Emergency Procedures
Regulation 25   Emergency Plans In Case Of Major Accidents
Regulation 26   Charge by a Local Authority for a Plan
Regulation 27   Transitional Provision
Regulation 28   Defence
Regulation 29   Certificates of Exemption
Regulation 30   Repeal of Provisions of the Pipelines Act 1962
Regulation 31   Revocation and Modifications of Instruments

7. THE PROVISION AND USE OF WORK EQUIPMENT REGULATIONS 1992 — SI 2932

Introduction — The Regulations apply to onshore and offshore activities covered by the 1989 Order on, or associated with, oil and gas installations, including mobile installations, diving support vessels, heavy lift barges and pipe-lay barges. The regulations cover an extremely wide range of equipment from overhead projectors to the process plant in its entirety.

Regulation 1   Citation

These Regulations may be cited as the Provision and Use of Work Equipment Regulations (PUWER) 1992. The Regulations came into force on 1st January 1993 for new equipment with existing equipment having until the 1st January 1997 to comply.

Regulation 2   Interpretation

Use in relation to work equipment means any activity involving work equipment and includes starting, stopping, programming, setting, transporting, repairing, modifying, maintaining, servicing and cleaning, and related expressions shall be construed accordingly.

Work equipment means any machinery, appliance, apparatus or tool and any assembly of components which, in order to achieve a common end, are arranged and controlled so that they function as a whole.

Regulation 3   Disapplication of These Regulations

These Regulations shall not apply to or in relation to the master or crew of a sea-going ship.

Regulation 4   Applications of Requirements Under These Regulations

Employers have a general duty under Section 2 of the Health and Safety at Work etc. Act 1974 to provide and maintain, so far as is reasonably practicable, machinery, equipment and other plant that is safe. The Regulations places duties on all employers providing work equipment to ensure that it is suitable, properly maintained etc.

Application to the Offshore Industry

These matters will also be considered in the context of the *Offshore Installation (Safety Case) Regulations* 1992 which will place a duty on owners or operators to demonstrate in their Safety Case that their management system is adequate to ensure that relevant statutory provisions in respect of the installation and connected activities will be complied with.

Regulation 5 Suitability of Work Equipment
Every employer shall ensure that work equipment is so constructed or adapted as to be suitable for the purpose for which it is used or provided.

This Regulation lies at the heart of this set of Regulations. It addresses the safety of work equipment from its initial integrity, the place where, and the purpose for which it will be used. For example, is the equipment suitable for use in a wet environment, or in a flammable atmosphere?

Regulation 6 Maintenance
Every employer shall ensure that all equipment is maintained in an efficient state, in efficient working order and in good repair. The extent of maintenance required will vary tremendously depending on the equipment and a substantial, integrated programme will be required for a complex process plant.

Equipment may need to be checked frequently to ensure that safety-related features are functioning correctly. A fault in a safety critical system could remain undetected unless maintenance procedures provide adequate inspection or testing. The frequency at which equipment needs to be checked is dependent on the equipment itself and the risk involved.

Routine maintenance — This includes periodic lubrication, inspection and testing, based on the recommendations of the equipment manufacturer, and should take into account any legal requirements.

Planned preventative maintenance — When inadequate maintenance could cause the equipment, guards or other protection devices to fail in a dangerous way, a formal system of planned preventative maintenance may be necessary.

There is no requirement for a maintenance log to be kept but if there is a log, it must be kept up to date.

Regulation 9 Training
Every employer shall ensure that all persons who use work equipment have received adequate training for the purposes of health and safety. An employer's obligation to train extends not only to those who use work equipment but also to those supervising or managing them. Statements of competence may be embodied in qualifications accredited by the National Council for Vocational Qualifications (NCVQ) and the Scottish Vocational Education Council (SCOTVEC).

Regulation 10 Conformity with Community Requirements
Every employer shall ensure that any item of work equipment provided for use in the premises or undertaking of the employer complies with any enactment (whether in an Act or Instrument) which implements in Great Britain any of the relevant Community Directives which are applicable to that item of work equipment.

There are legal requirements covering all those involved in the manufacturing and supply chain of work equipment which are designed to ensure that new work equipment is safe and complies with current legislation.

Article 100A Directives

The aim of this group of Directives is to achieve the free movement of goods in the Community Single Market by eliminating differing national controls and harmonising essential technical requirements.

The Directives set out *essential safety requirements* that must be met before products may be sold in the Community and products which comply with the Directives must be given free circulation within the Community.

Suppliers must ensure that their products when placed on the market comply with the legal requirements implementing the Directives applicable to their product. Compliance can be claimed by the manufacturer by affixing a mark — the "CE Mark" to the equipment.

The situation is complicated by the fact that not all work equipment is covered by a product Directive, nor are product Directives retrospective. If an employer provides second-hand equipment for the first time in the workplace it does not need to be modified to meet the essential safety requirements of the relevant product Directives but it must comply immediately with the regulations 11 to 24 of PUWER. (Second-hand equipment imported form outside the European Community has to comply immediately with essential safety requirements of the relevant product Directives.)

There is a movement within the EC to harmonize standards and this is being lead by the CEN and CENELEC. When developed these standards will be transposed in the UK by the British Standards Institution and will bear a common number (i.e. EN XXXX will be BS EN XXXX here). Once these Directives are fully in force only products which conform and bear the "CE Mark" may be placed on the market in the UK.

Regulation 11 Dangerous Parts of Machinery

Employers must take effective measures to prevent contact with dangerous parts of machinery. The term "dangerous part of machinery," is well established in health and safety law.

A risk assessment carried out under regulation 3 of the MHSWR should identify hazards presented by machinery. The risk assessment should not just deal with the machine in its normal operating mode, but must also cover activities such as setting, maintenance, cleaning or repair.

The Regulation sets out various requirements for guards and protection devices. These are largely common sense, and in large part are detailed in relevant national and international standards.

Regulation 12 Protection Against Specified Hazards

Every employer shall take measures to ensure that the exposure of a person using work equipment to any risk is either prevented, or, where that is not reasonably practicable, adequately controlled.

This Regulation covers risks arising from certain listed hazards during the use of equipment such as:

(a) material falling from equipment such as scaffold boards;

(b) overheating or fire due to friction in electric motors, ignition by welding torch, thermostat failing, cooling system failure or similar;

(c) explosion of the equipment due to pressure build-up, perhaps due to the failure of a pressure-relief device or the unexpected blockage or sealing off of pipework.

A risk assessment carried out under regulation 3 of the Management of Health and Safety at Work Regulations 1992 should identify these hazards, and assess the risks associated with them.

Regulation 13   High or Very Low Temperature
All parts of work equipment and substances produced which are at a high or very low temperature shall have protection where appropriate so as to prevent injury to any person by burn, scald or sear. The risk from contact with hot surfaces should be reduced by engineering methods, i.e. reduction of surface temperature, insulation, shielding, barricading and guarding.

Regulation 14   Controls for Starting or Making a Significant Change in Operating Conditions.
The controls provided should be designed and positioned so as to prevent, so far as possible, inadvertent or accidental operation. Buttons and levers should be of appropriate design, for example including a shrouding or locking facility.

Regulation 15   Stop Controls
Every employer shall ensure that, where appropriate, such equipment is provided with one or more readily accessible control, the operation of which will bring the work equipment to a safe condition in a safe manner.

Regulation 16   Emergency Stop Controls
An emergency stop control may be required in addition to the other safeguards to prevent risk when some irregular event occurs. The location of emergency stop controls should be determined by the risk assessment.

Regulation 17   Controls
Every employer shall ensure that all controls for work equipment shall be clearly visible and identifiable, including appropriate marking where necessary.

Regulation 18   Control Systems
Every employer shall ensure, so far as is reasonably practicable, that all control systems of work equipment are safe. Failure of any part of the control system or its power supply should lead to a fail-safe condition and not impede the operation of the stop or emergency stop controls.

Regulation 19   Isolation From Sources of Energy
Every employer shall ensure that, where appropriate, work equipment is provided with suitable means to isolate it from all its sources of energy. Isolation means establishing a break in the energy supply in a secure manner, i.e. by ensuring that inadvertent reconnection is not possible. The possibilities and risks of reconnection should be identified as part of the risk assessment, which should then establish how security can be achieved. For some equipment an isolating switch or valve may have to be locked in the *off* or *closed* position to avoid unsafe reconnection.

Regulation 20   Stability
All work equipment should be stabilised by clamping or otherwise where necessary for safety.

Regulation 21   Lighting
Every employer shall ensure that suitable and sufficient lighting, which takes account of the operations to be carried out, is provided at any place where a person uses work equipment.

Regulation 22 Maintenance Operations
Regulation 6 requires that equipment is maintained. Regulation 22 requires that equipment is constructed or adapted in a way that takes into account the risk associated with carrying out maintenance work, such as routine and planned preventative maintenance, as described in the guidance to regulation 6. Compliance with this Regulation will help to ensure that when maintenance work is carried out, it is possible to do it safely without risk to health, as required by Section 2 of the HSW Act.

Regulation 23 Markings
Every employer shall ensure that work equipment is marked in a clearly visible manner with any marking appropriate for reasons of health and safety.

Regulation 24 Warnings
Warnings or warning devices should be provided where risks to health or safety remain after other hardware measures have been taken. They may be incorporated into the permit-to-work systems and can reinforce measures of information, instruction and training. A warning is normally in the form of a notice or similar. Examples are positive instruction such as *Hard Hats Must Be Worn*. A warning device is an active unit giving a signal; the signal may typically be visible or audible, and is often connected into equipment so that it is active only when a hazard exists.

Warnings must be easily perceived and understood, and unambiguous.

Regulation 25 Exemption Certificates
The Secretary of State for Defence may, in the interests of national security, exempt any of the home forces, or any visiting force from these Regulations.

Regulation 26 Extension Outside Great Britain
The Regulations apply to offshore activities covered by the 1989 Order on, or associated with, oil and gas installations, including mobile installations, diving support vessels, heavy lift barges and pipe-lay barges.

Regulations 27 Repeals, Saving and Revocations
Refer to regulations.

Part 5. ASSOCIATED INFORMATION

1.0 HEALTH AND SAFETY EXECUTIVE OFFSHORE SAFETY DIVISION (OSD)

The *Offshore Safety Division* (OSD) was created in 1990 as a department within the *Health and Safety Executive* (HSE). In 1991, all duties pertaining to the occupational health, safety and welfare of offshore installations were transferred from the *Petroleum Engineering Division* (PED) of the *Department of Energy*, to the Offshore Safety Division of the Health and Safety Executive.

The Offshore Safety Division is first and foremost a regulator whose core purpose is to ensure that risks to people from work activities in the offshore industry are properly controlled. Their objectives are influenced by political, economic, social and technical issues and their duties range from direct enforcement to contributing to research.

Two of the main tasks performed by the OSD are the assessment of safety cases and the inspection of offshore installations. The inspections are aimed primarily at a review of operating practises to ensure that the duty holder is complying with the procedures outlined in their safety case.

The OSD are represented at all the geographic locations of offshore activity in the UK and are organised into the following groups:

OD 1 Aberdeen

- Inspection Teams
- Well Operations Specialists
- Inspection Management Support
- Planning and Finance

OD 2 Aberdeen

- Inspection Teams
- Occupational and Environmental Health Topic Team
- Safety Case Management Support
- Pipeline Liaison

OD 3 Aberdeen

- Inspection Team
- Diving Safety Specialists
- Legal and Operational Strategy Cross Business Support and Aberdeen Business Support Information Management

OD 4 Norwich and Bootle

- Inspection Teams
- Safety Management Specialists
- Corporate Audit Divisional Safety Health and Safety Incident Response OSD Projection in East Anglia
- Research Budget
- Personnel/Training/IT
- Heavy Lifting

OD 5 London

- Integrity of Structures and Marine Systems Verification-marine, aviation and workplace safety

OD 6 Bootle

- Inspection
- Fire and Explosion
- Process Safety
- Electrical Engineering and Control
- Mechanical Engineering
- Quantified Risk Assessments
- PFEER Assessment
- PFEER Summaries — EER
- Communication Command Control
- Human and Organisational Factors Research
- Human Factors Engineering
- Isle of Man
- OPITO
- Cross Business Support for Bootle

1.1   HSE CORRESPONDENCE

Official correspondence emanating from the Offshore Safety Division of the Health and Safety Executive may appear in any one of three guises, depending on the nature and urgency of the subject matter.

Distribution will be dependant on the content of the correspondence but will normally include the operators of offshore installations and those companies who subscribe to the HSE publications department. Each item of correspondence is uniquely numbered and retains the format originally introduced by the Department of Energy.

The three levels of communication are:

i)   **OPERATIONS NOTICE**

An Operations Notice provides notification of a change to existing legislation, frequently an addition to, or modification of a Statutory Instrument. Adherence with the provisions of the Notice are mandatory.

ii)   **SAFETY NOTICE**

The Safety Notice takes the form of a letter designed to highlight a potentially dangerous situation and recommends preventative action. Failure of a component such as a particular type of burner assembly or piece of lifting equipment are typical examples. The Notices should be retained on file by the installation and adopted as a guidance note.

iii)   **DIVING INFORMATION SHEET**

The diving information sheets represent the equivalent of a Safety Notice for the attention of personnel involved in diving operations and replaces the Diving Safety Memorandum (DSM).

2.0 UK OFFSHORE OPERATORS ASSOCIATION (UKOOA)

UKOOA was officially established in 1973, having previously existed (from 1964) as an informal organisation. Every oil company engaged in exploration or production on the *UK Continental Shelf* (UKCS) is a member of UKOOA and it provides a forum for discussion on matters relating to technology, the environment and safety.

The Association operates purely in an advisory capacity and works closely with the Government to ensure that legislation obtains the desired results and reflects the requirements of the industry.

As a service to the industry, UKOOA provide a range of specialist guidance documents which relate to various aspects of the offshore industry such as, *Guidelines For The Management Of Emergency Response For Offshore Installations,* and a wide range of well illustrated literature which is of more general interest. They can be contacted at:

UK Offshore Operators Association (UKOOA)
30 Buckingham Gate
London
SW1E 6NN

Telephone: 0171-802-2400
Facsimile: 0171-802-2401
email: info@ukooa.co.uk
web site: oilandgas.org.uk

Chapter Three

SAFETY SYSTEMS

PART 1.   SAFETY CASE

PART 2.   FIRE FIGHTING EQUIPMENT

PART 3.   LIFE SAVING APPLIANCES

PART 4.   NAVIGATIONAL AIDS

PART 5.   HAZARDOUS AREAS

PART 6.   EMERGENCY SYSTEMS

THE CONOCO/CHEVRON BRITANNIA LIFEBOATS.
The accommodation module doubles as a safe haven or temporary refuge and provides direct access to the helideck, the primary means of evacuation, and the lifeboats, the secondary means of evacuation. The lifeboats are of the free fall type and are launched from their cradles rather than being lowered to the water on wire rope falls.
(Picture reproduced with kind permission of Conoco (U.K.) Limited and Chevron U.K. Limited.)

SAFETY SYSTEMS

Part 1. THE SAFETY CASE

1.0 STATUTORY OPERATIONS MANUAL

Under the now defunct *Construction and Survey Regulations* (SI 289), every offshore installation had to be provided with a *Statutory Operations Manual*, a document which contained a summary of all the information necessary to enable the *Offshore Installation Manager* (OIM) to operate and maintain the installation in a safe condition. The Department of Energy *Guidance Notes on the Design, Construction and Certification of Offshore Installations* outlined the requirements for the content of the operations manual and the manual had to be approved by the Certifying Authority prior to issue.

The operations manual enabled visitors to the platform to quickly familiarise themselves with the installation layout and design conditions, and being such a practical document, it was adopted by numerous operators world-wide, regardless of local regulations.

2.0 THE SAFETY CASE

In the UK, the requirement to provide an operations manual has been superseded by the Safety Case Regulations which require a *safety case* to be prepared for each installation. The safety case shares many similarities with the operations manual and contains similar material. However, most importantly it also provides an account of the duty holders *safety management system* and their approach to the use of *quantitative risk assessments* as a means of ensuring that all hazards have been adequately addressed.

The requirements for the preparation of safety cases are outlined in the *Safety Case — Guidance on Regulations*. A safety case must be prepared, and submitted to the *Health and Safety Executive* (HSE) for assessment at the design stage, and 6 months prior to admitting hydrocarbons onto the installation. The safety case is considered to be an essential ingredient in the safe operation of the installation and as such should be reviewed and updated at regular intervals, at least every 3 years, to incorporate any changes which may have taken place in working practises, advances in technology, or modifications to the installation. Safety cases are also to be prepared for combined operations involving other installations and for the abandonment phase of the installation.

The Safety Case Regulations are dealt with in more detail in chapter two, but the information which is to be included within the safety case is listed below.

1. **GENERAL INFORMATION**

    The name and address of the operator of the installation.

2. **STRUCTURE — LAYOUT AND NOTATION**

    A description, with scale diagrams of the main and secondary structures and plant, and the connections to pipelines and any wells connected to the installation.

3. **PLANS AND ENVIRONMENTAL CRITERIA**

    A scale plan of the location of the installation and of anything connected to it with particulars of the meteorological and oceanographic conditions the installation may be subjected to and the properties of the sea-bed and subsoil at its location.

PLATFORM SAFETY NOTICE

PUBLIC ADDRESS AND ALARM INDICATIONS			
Function	Status Lights	Tone	Action
Installation Safe	Green	"Bell" followed by announcement	• Work as normal
General Alarm • fire detector • gas detector • manual initiation	Yellow	Intermittent single tone	• stop work • make area safe • muster
Prepare to abandon installation	Yellow	Intermittent single tone	• Prepare lifeboats • await instructions

Think *Safety* at all times
Study the location of safety equipment
Know your muster station

IF IN DOUBT - ASK

Fire
1. Raise alarm by activating break glass unit
2. Announce **FIRE** over public address system
3. Try to extinguish a small fire after raising alarm
4. Retreat to muster area

Man Overboard
1. Throw man a lifebouy
2. Keep man in sight at all times
3. Raise alarm by public address system

Survival Equipment Lifejackets, immersion suits, smoke hoods, gloves and torches are located:
- in cabins
- at muster points
- in boxes as shown on Life Saving Plan

Muster Points
- Accommodation Locker Room
- Lifeboat station

Lifeboats
- AP Platform - West Side
- AD Platform - East Side

STATION BILL

SAFETY SYSTEMS

4. OPERATIONAL PARAMETERS

Particulars of the types of operation, and activities in connection with and operation, which the installation is capable of performing.

5. ACCOMMODATION ARRANGEMENTS

The maximum number of persons expected on the installation and for whom accommodation is to be provided.

6. WELL CONTROL ARRANGEMENTS

Particulars of the plant and arrangements for the control of operations on a well, including how to control well pressures, how to prevent the uncontrolled release of hazardous substances and how to minimise the effects of damage to sub-sea equipment by drilling equipment.

7. PIPELINES

A description of any pipeline with the potential to cause a major accident which includes its dimensions and layout, maximum operating conditions and safety devices.

8. FIRE AND GAS DETECTION

Particulars of plant and arrangements for the detection of toxic or flammable gas, and for the detection, prevention and mitigation of fires.

9. TEMPORARY REFUGE

A description of the arrangements made to protect persons from the hazards of explosion, fire, heat, smoke, toxic gas or fumes following an accident and for enabling personnel to be evacuated. This description should include details of the temporary refuge (including how incidents can be monitored from within), escape routes and evacuation arrangements.

10. PERFORMANCE STANDARDS

A statement of performance standards which have been established in relation to the arrangements required for compliance with paragraph 9 which should include a description of how long a period of time the temporary refuge is designed to provide protection for.

11. QUANTITATIVE RISK ASSESSMENTS

A demonstration by reference to the results of a quantitative risk assessment that the measures taken in relation to the hazards referred to in paragraph 9 will reduce risks to health and safety to the lowest reasonably practical level.

12. SPECIFICATIONS AND DESIGN CODES

Particulars of the main requirements in the specification for the design of the installation and its plant, including any limits for safe operation.

Offshore Engineering

13. DESIGN CRITERIA

Sufficient particulars to demonstrate that the design of the installation, its plant and pipelines are such that the risks from a major accident are at a level that is as low as is reasonably practical.

14. MODIFICATIONS AND REPAIRS

Particulars concerning any remedial work to be carried out to the installation or plant and the time in which it must be carried out.

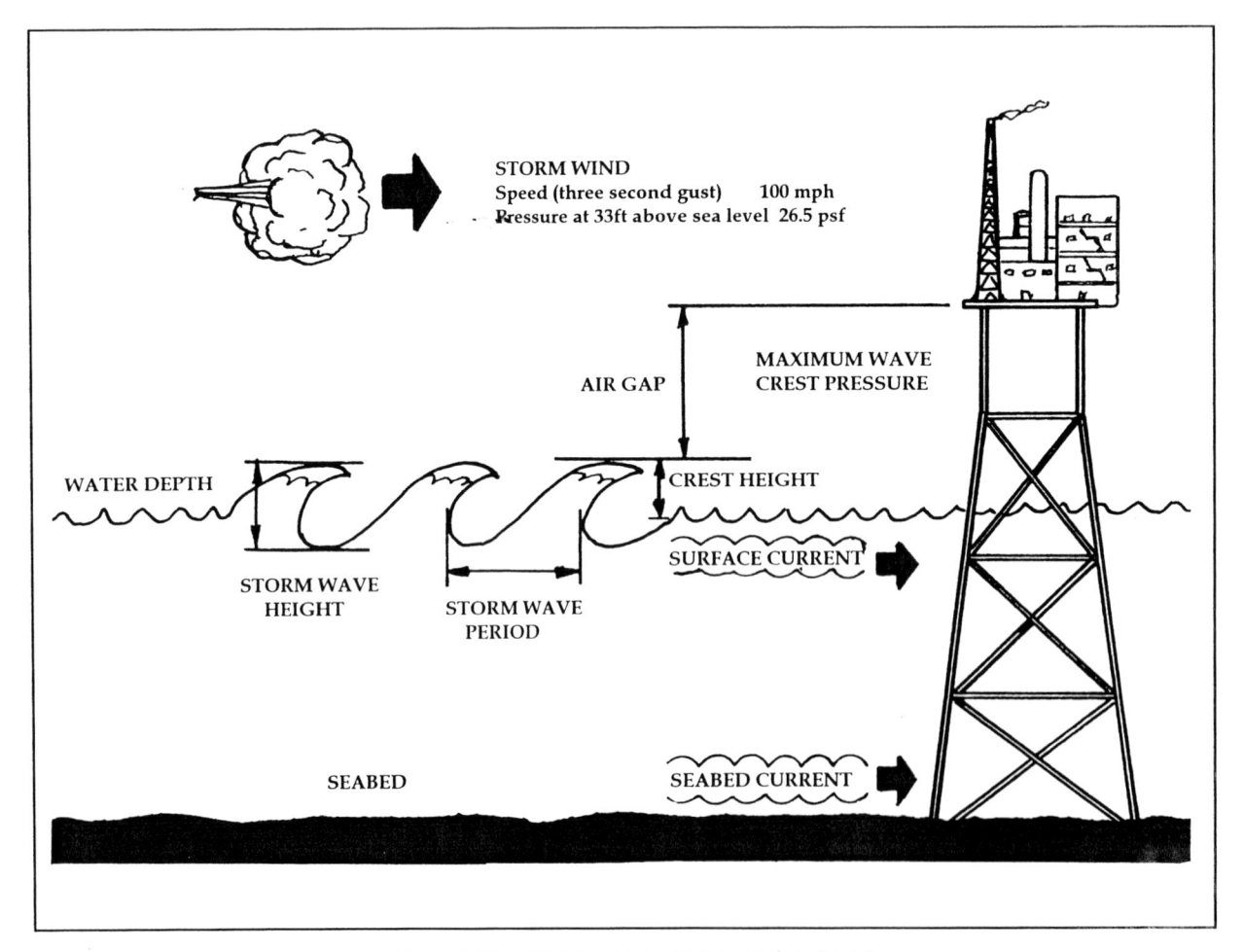

ENVIRONMENTAL CONDITIONS

Storm wind: Speed (three second gust) 100 mph; Pressure at 33ft above sea level 26.5 psf

Part 2. FIRE-FIGHTING EQUIPMENT

INTRODUCTION

Legislation requires operators to provide a combination of active and passive fire-fighting protection measures to provide protection for personnel against the hazards of fire and the release of gas on offshore installations. These measures will include fire and gas detection systems, remote control safety devices, fire alarms, fire mains, water deluge systems, automatic sprinkler systems, fixed and portable fire-fighting equipment, and the protection of strategic locations with passive fire protection materials.

1.0 HISTORY-UK CONTINENTAL SHELF

Up until 1995, the requirements for active fire-fighting equipment on offshore installations located in the territorial waters of the United Kingdom were governed by Statutory Instrument No. 1978/611, *Offshore Installations (Fire-fighting Equipment)*. In 1995, SI 611 was revoked and replaced by the *Offshore Installations (Prevention of Fire and Explosion, and Emergency Response) Regulations, SI 1995/743*, the Regulations being referred to as *PFEER*. The *Health and Safety Executive* (HSE) publish an *Approved Code of Practise and Guidance* to assist with the interpretation of the Regulations.

The intention of PFEER is to provide a less prescriptive form of legislation and to encourage a more integrated approach to the hazards associated with fires and explosions. The Regulations do not require any significant changes to be made to the type of equipment previously installed under SI 611, but, rather than dictating exactly what equipment must be provided and what precautions must be taken as was previously the case, PFEER requires the duty holder to carry out an *Assessment*, essentially a *fire and explosion analysis* (FEA) of the installation.

The object of the assessment is to identify events which could give rise to a major accident resulting in a fire or explosion and to evaluate the likelihood and consequences of such an incident occurring. It is the results of the analysis which will ultimately dictate what equipment should be installed and what precautions should be taken to prevent such an occurrence arising and should one occur, recommend steps which will detect, control and mitigate the effects.

2.0 EXAMINATION AND TESTING

Having decided what equipment and operating procedures are required, the duty holder is obliged to prepare and operate a *written scheme* for *the systematic examination* of the arrangements by a *competent and independent person*. The scheme replaces the bi-annual examination of fire-fighting equipment and life-saving appliances, previously required under SI 611 and SI 486, surveys carried out initially by the *Department of Transport's* (DoT) marine surveyors, and subsequently by the *Certifying Authorities*.

A summary of the main features of the assessment and the written scheme of examination is to be included in the safety case and they are to be updated at regular intervals to incorporate changes which may have taken place in working practises, advances in technology, or modifications to the installation.

The PFEER regulations and the requirements relating to written schemes of examination are more fully described in chapter two.

2.1 EQUIPMENT APPROVAL

Under the old legislation, each and every piece of fire-fighting equipment had to be of a type approved by the Department of Transport (DoT) and the requirements under the new regulations are basically the same but it is now the *Maritime and Coastguard Agency* (MCA), rather than the DoT who are responsible for the approval of equipment, and the MCA have subsequently authorised the major certifying authorities to carry out approvals on their behalf. Equipment manufactured outside the UK is acceptable providing it complies with a recognised national standard and is approved by a National Administration, a National Standards Organisation or an equivalent body.

2.2 GUIDANCE NOTES AND STANDARDS

At the present moment in time, the industry is going through a period of change and there are a number of initiatives under discussion. Consequently, operators are referencing a range of documents and guidance notes in addition to the PFEER guidance previously mentioned, and are likely to continue to for the foreseeable future, at least until a common understanding is agreed.

2.2.1 Department of Energy Guidance

During the last 20 years, the under listed documents provided guidance to the operators of UK registered installations and they are still widely referred to, even though the regulations they were written to provide guidance for have recently been revoked. Also, numerous National Authorities worldwide adopted, and still use the guidance.

i) **Offshore Installations: Guidance on Fire-fighting Equipment**

Whilst the SI 611 Regulations have been superseded, the guidance notes contain excellent advice on the installation and testing of fire-fighting equipment. In fact, the HSE have stated that the *guidance still contains some sound technical advice and that it will be for the duty holder to justify the appropriateness and suitability of any standards used.*

ii) **Offshore Installations, Guidance Notes on the Design, Construction and Certification**

Like the guidance notes on fire-fighting equipment, these guidance notes have been discontinued but are still referred to extensively, particularly for advice on passive fire fighting protection measures. Their relevance is acknowledged by the HSE in a similar manner to the SI 611 guidance notes.

2.2.2 International Convention for The Safety of Life at Sea (SOLAS)

Whilst the SOLAS regulations are aimed primarily at the marine industry, most of the requirements pertaining to life saving equipment and fire-fighting arrangements on offshore installations are based on the requirements of these regulations.

2.2.3 Petroleum and Natural Gas Industries – Offshore Production Installations – Control and Mitigation of Fires and Explosions – Requirements and Guidelines

This standard, published by the *International Standards Organisation* (ISO) is currently in draft form having been issued for comment. How these requirements and guidelines will fit in with the newly introduced PFEER regulations, remains to be seen.

A brief description of the essential component parts of active and passive fire protection systems will now be given. Where performance standards are quoted such as for fire hoses and nozzles, these relate to what has historically been accepted by the industry and they are not mandatory requirements. Reference to the regulations and specific design specifications pertaining to a particular installation should be made to ensure the requirements of both the duty holder and administration are met in each case.

3.0. ACTIVE FIRE PROTECTION

The Regulations require the provision of fire and gas detection systems, remote control safety devices, fire extinguishers and fireman's equipment on all installations (manned or un-manned) and fire alarms, fire mains, hydrants and hoses, water deluge systems or monitors, and automatic sprinkler systems on all installations that are normally manned.

3.1　FIRE AND GAS DETECTION

Automatic fire detection systems must be fitted throughout the installation and the equipment used must be of a self monitoring design so that if a fault develops a warning alarm will be initiated. Similarly, gas detection must also be provided on all parts of the installation where flammable gas may accidentally accumulate.

Manually activated alarm points must be provided at strategic positions in the accommodation and working areas.

Detectors sensitive to smoke, heat, ultra violet light and gas may be used individually or in combination depending on the hazard most likely to occur in any particular location.

The main control room is usually dedicated for use as the fire control station because the operation of production equipment and communications can most easily be co-ordinated at that point. The main control room is also normally manned 24 hours per day, at least on the larger installations (where the control room is not normally manned a repeater alarm must sound in a manned location or cabin).

It is a requirement that should a detector sense a hazard or a manual alarm point be activated, then an audible and visual alarm will be initiated at the hazard and in the main control room. There is no requirement for the alarm to sound throughout the installation or for the automatic operation of the fire-fighting equipment. The location of the hazard must be identified in the main control station and the operations staff can then decide what action is to be taken.

3.2　FIRE-FIGHTING EQUIPMENT

The active fire-fighting system centres around a ring main which is pressurised by at least two fire pumps as shown on the sketch. The system is normally maintained at a constant pressure of approximately 10 bars (150 psi) by a sea water service pump and pressure tank. The fire pumps may be manually activated from strategic locations such as the main control room, helideck and process areas, or automatically by a significant drop in ring main pressure as will occur when a fire hydrant is opened or the deluge system activated.

i)　**FIRE PUMPS**

The fire pumps consist of a centrifugal deepwell pump that is normally powered by a self-contained diesel engine via a gearbox. Electric submersible pumps are occasionally used, again, self-contained units being preferred with a dedicated diesel engine and generator to provide the necessary electrical power.

The number of fire pumps required will be determined from the fire and explosion analysis but normally, at least two independently powered fire pumps will be found on an offshore installation. The number of pumps installed should reflect the possibility of the unavailability of equipment due to breakdown or maintenance requirements. They should be located remotely from each other in areas of minimal risk and be capable of operating for a clearly defined period of time, normally at least 12 hours unattended.

An engine protection system will be provided but it should not be capable of automatically shutting the engine down under fault conditions. The protection system should initiate alarms under fault conditions which must then be addressed by platform staff and suitable corrective action taken. The only exception to this rule is the ingestion of gas at the air inlet manifold which will shut the engine down prior to an overspeed situation developing.

Each pump should be capable of supplying adequate water to operate the largest section of deluge equipment in addition to maintaining a pressure of 3.5 kg/cm^2 at two fire hydrants which are fitted with nozzles of 19 mm diameter.

ii)   **FIRE MAIN**

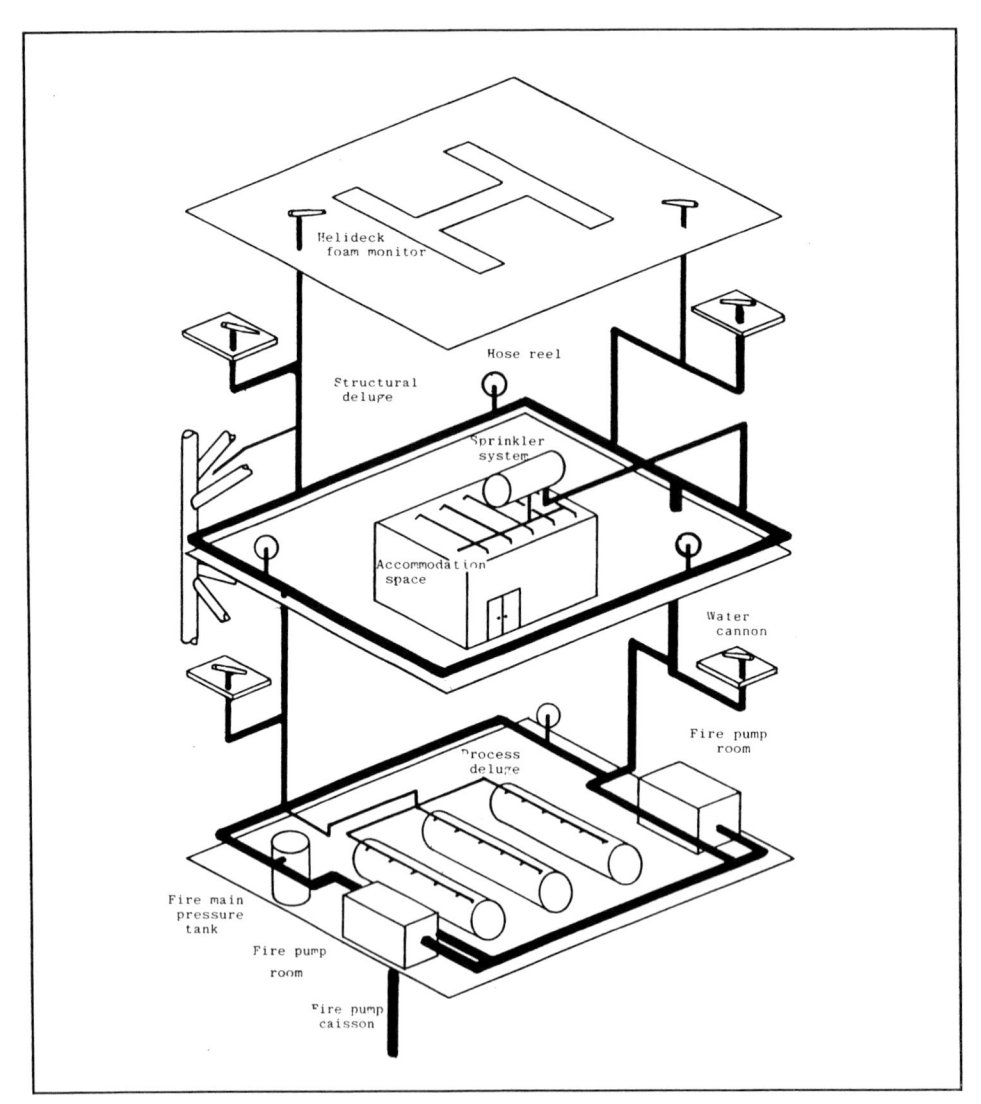

FIRE MAIN

SAFETY SYSTEMS

The fire main should be constructed of corrosion resistant materials such as galvanised mild steel, stainless steel or cunifer (copper/nickel alloy) and must be protected against the effects of both over pressure, and freezing.

Wherever possible, the fire main should be routed clear of hazardous areas with full use being made of any protective shielding afforded by structural members. Isolation valves should be installed so that a damaged section of fire main will not adversely affect the operation of the remaining system.

In some parts of the world *glass reinforced plastic* (GRP) fire mains have been installed on offshore installations but at the time of writing, they are not approved for use in the UK due to their failure to meet the material fire test requirements. At the present moment in time, they are only capable of passing fire tests under flowing conditions, failure occurring when the flow of water ceases. However, with the dramatic improvement in material properties, it is only a matter of time before full approval is obtained.

iii) HOSE REELS

The fire hydrants and hose reels provide a means of back-up to the fixed fire-fighting systems and permit boundary cooling to be applied to spaces in which an outbreak of fire has occurred. Hydrants should be located in readily accessible positions, particularly in way of access points and a sufficient number of hose reels stations must be provided so that water can be brought to bear on a fire in any location from at least two directions, at least one of the required jets must be supplied from a single standard length of hose.

The standard fire hose is regarded as being of 64 mm bore, 18 metres in length and made from unlined canvas. However, smaller diameter hoses are permitted such as the 45 mm bore lined varieties, provided they have a similar throughput to the larger hoses. Non-collapsible hoses of 25 mm diameter may also be used but only where deluge cover is provided as the primary means of fire protection for the space concerned. The 25 mm hoses may also be used in accommodation spaces, but again, only as an addition to the larger hoses.

All hydrants, nozzles and hose couplings should be of a standard type throughout the installation. The nozzles should be of robust construction and capable of providing a water jet of 12 metres length and a minimum discharge rate of 26 m3/hour whilst a pressure of not less than 3.5 bar is being maintained in the fire main. The nozzles should also be capable of providing a fine water spray curtain of approximately 4.1 metres diameter at a distance of 2.1 metres.

iv) DELUGE

The deluge system pipework is similar in design to a sprinkler system, but, on a much larger scale with the exception that *frangible bulbs* are not fitted, the pipework remaining open and dry until the deluge control valves on the fire ring main are activated. The outlet nozzles are of the *open type* and are fitted with deflector plates designed to create an even water droplet spray pattern.

The primary function of the deluge system is to minimise the effect of heat arising from a major fire involving hydrocarbons. The deluge water is intended to cool the process plant and reduce the escalation of a fire whilst safety systems, designed to isolate or vent the supply of hydrocarbons, take effect and any ignited hydrocarbons are consumed by the fire.

The deluge system covers the wellheads, gas process pressure vessels and pipework, process plant and storage vessels in their entirety and may be extended to cover the main structural steelwork to prevent it from weakening if exposed to fire. The system may be manual or automatic in operation with the preference being for automatically activated systems.

The water coverage provided by the deluge system will be determined by the size of the potential hazards within the area to be protected. A minimum water coverage figure of 12.2 litres per minute over the *reference area* is specified for all areas whilst this increases to 20 litres per minute where there are high pressure hydrocarbon inventories present (and for structures when required) whilst the figure rises to 400 litres a minute for wellhead areas.

v) WATER MONITORS

A monitor may be described as a permanently fixed device for directing water or foam in jet or spray form onto a fire, in effect a water cannon.

The monitors are essentially designed for the application of water in open deck areas and should be capable of providing the desired cover irrespective of weather conditions. In theory, it is acceptable to use water monitors in place of a deluge system to protect process areas but in practice the monitors are normally fitted to supplement the deluge systems.

Water monitors are also used to protect the helideck on normally manned installations. They must be capable of discharging a sufficient quantity of low expansion foam to completely cover the helideck, even in adverse weather conditions. A tool kit containing equipment to assist in releasing the occupants from helicopter wreckage will also be found in the immediate vicinity of the water monitors.

vi) SPRINKLER SYSTEM

Every accommodation space must be protected by an automatic sprinkler system, an accommodation space being defined as, *any room used for eating, sleeping, cooking or recreation, or as an office, sick bay, laundry room or locker room, any corridor giving access to any of these rooms, and any store-room in the vicinity of any of these rooms.*

The sprinkler system must be supplied from a permanently pressurised water source located remotely from the accommodation which is capable of operating for a period of at least 4 hours. Sprinkler systems are normally charged with fresh water via a pressure tank with changeover arrangements provided to admit sea water from the fire main once activated. However, statically charging the system with sea water is acceptable provided that it can be demonstrated that the efficiency of the system will not be impaired by corrosion, marine growth or sediment.

The sprinkler system frangible bulbs are designed to burst at between $68°C$ and $70°C$ in the accommodation spaces whilst in the galley, higher settings may be used, up to $30°C$ above the maximum deck head temperature. The heads should provide a water supply of 5 litres per square metre per minute in spray form. In cooking areas where cooking oils and fats are used, deflector plates should be positioned to prevent water impingement.

Operation of the sprinkler system should initiate an audible and visual alarm in the main control station and indicate the location of the hazard. The automatic activation of a general fire alarm is not required, further action being left to the discretion of the control room staff who can manually activate further alarms or alert personnel by means of the public address system, if required.

Arrangements are to be provided to enable the sprinkler system to be tested without detracting from the operational efficiency or disrupting the routine operation of the installation.

vii) WATER SPRAY

Water spray or *fog systems* share many similarities with sprinkler systems being simple in design, having no moving parts and thus requiring minimal maintenance. Unlike a sprinkler system, water spray systems can be used to protect machinery spaces and are suitable for fighting both oil and hydrocarbon fires.

In most cases the systems comprise of dedicated air (or Nitrogen) and water storage vessels which come complete with associated controls. The principle of operation is that stored air or nitrogen pressure, is used as both the propellant for the water supply, and for the atomisation of the water at the nozzles. The extinguishing mechanism is one of heat removal by the water, the microscopic water droplets or fog turning to steam which starves the fire of oxygen to a level at which combustion cannot be sustained.

Water fog systems are normally supplied in modular form as dedicated stand alone units for protecting specific areas. They are designed as a *one-shot* system and consume very small volumes of water with no further back-up being required. They operate at low pressures (3–5 bar) so the water can be distributed to the nozzles via small bore tubing or flexible hoses (which remain dry until activated) making installation particularly easy. Water spray systems are rapidly gaining in popularity, not least because they are seen as the most environmentally acceptable medium.

3.3 FIXED FIRE-FIGHTING SYSTEMS

Fire-fighting equipment which is installed to protect a specific compartment or area is generally described as a fixed fire-fighting system. An inert gas is predominately used as the extinguishant although dry powder systems are used extensively. Sprinkler systems and water fog systems could also be described as fixed fire-fighting systems but have been dealt with in the previous section because water is used as the extinguishant and this is more in keeping with the *wet and dry* philosophy of the mediums concerned.

3.3.1 Essential Features

It is a requirement that certain locations such as *control rooms* and *machinery spaces* are protected by fixed fire extinguishing systems. The favoured medium for the extinguishant has for many years been Halon gas although Carbon Dioxide (CO_2) and dry powder systems are also used extensively.

There are a number of requirements which must be addressed when using gaseous extinguishants such as Halon and CO_2.

The ventilation systems must be shut down and visual and audible alarms activated in the affected area at least 15 seconds before the release of the gas and status lights must be fitted outside all protected spaces to indicate whether the system is in the automatic, manual, or the discharged condition.

Initiation of the fixed fire-fighting equipment may be automatic but manual operation is often preferred for normally manned locations. Where systems are arranged for automatic release, warning notices should be provided at compartment entrance points. The systems must also be capable of being activated manually and full operating instructions should be provided at the control point.

Any electrical circuits essential to the operation of a fixed fire-fighting system should be run in heat-resistant cable, e.g. mineral insulated cable (MIC).

3.3.2 Storage

The gas storage cylinders associated with fixed fire-fighting systems are to be located outside the space to be protected in a well ventilated, gas tight storage room, preferably accessed from the open air. Pressure relief devices fitted to CO_2 cylinders, tanks, or manifolds should be so arranged that when operated there will be no danger to personnel from the resultant discharge of the gas. Halon cylinders may be stored within the space to be protected subject to enhanced safety measures. Continuous monitoring of the levels within the gas cylinders is to be provided and automatic over-pressure protection which in the event of the container being exposed to a fire will release the gas and thus protect the space.

At a period agreed by the authorities but at least after 20 years, the gas storage cylinders should be hydraulically tested to confirm their soundness. After the 20 year test, 10% of the cylinders should be tested every five years.

3.3.3 Fire-Fighting Mediums

The various mediums may briefly be described as follows:

i) **HALON**

Halon is a lighter than air gas which relies on a chemical reaction to break the combustion chain and extinguish the fire. It is virtually instantaneous in operation, leaves no mess and is easily dispersed once the area has been made safe. Halon can be stored at low pressures (25 bar) and requires only a 5–7% concentration to be effective which makes the equipment light and compact.

On existing installations Halon may be found providing protection to control rooms, plant rooms, machinery spaces and any locations housing delicate equipment. It is also widely used for gas turbine enclosures where its operation is invariably automatic.

Halon is the ideal fire-fighting gas, but, because of the effect it is reported to have on the thinning of the ozone layer its use is to be phased out by the year 2000 by all countries who are signatories to the Montreal Protocol. From July 1992 the use of Halons on new installations has been restricted to those locations designated as *essential*, that is normally manned spaces such as main control rooms.

ii) **CARBON DIOXIDE (CO_2)**

Carbon Dioxide was used almost universally prior to the introduction of Halon and has made somewhat of a resurgence as Halon is being phased out. Whilst carbon dioxide is termed a greenhouse gas and is thus as environmentally unacceptable as Halon, it is produced as a bi-product of chemical manufacturing processes and would normally be vented to atmosphere anyway.

Carbon Dioxide is particularly suitable in locations such as vent stacks where the heavier than air smothering effect which characterises the gas can be put to the most effective use. The disadvantages with CO_2 are that it is a high pressure gas (60 bar) and a 35% concentration is required to be effective so the equipment is both heavy, and occupies more space than equivalent mediums. It is also an asphyxiant and great care must therefore be taken in both the storage arrangements and use in manned locations.

SAFETY SYSTEMS

Air to be supplied from a safe area.
Gas detection is desirable but not mandatory
(gas not expected in a safe area).

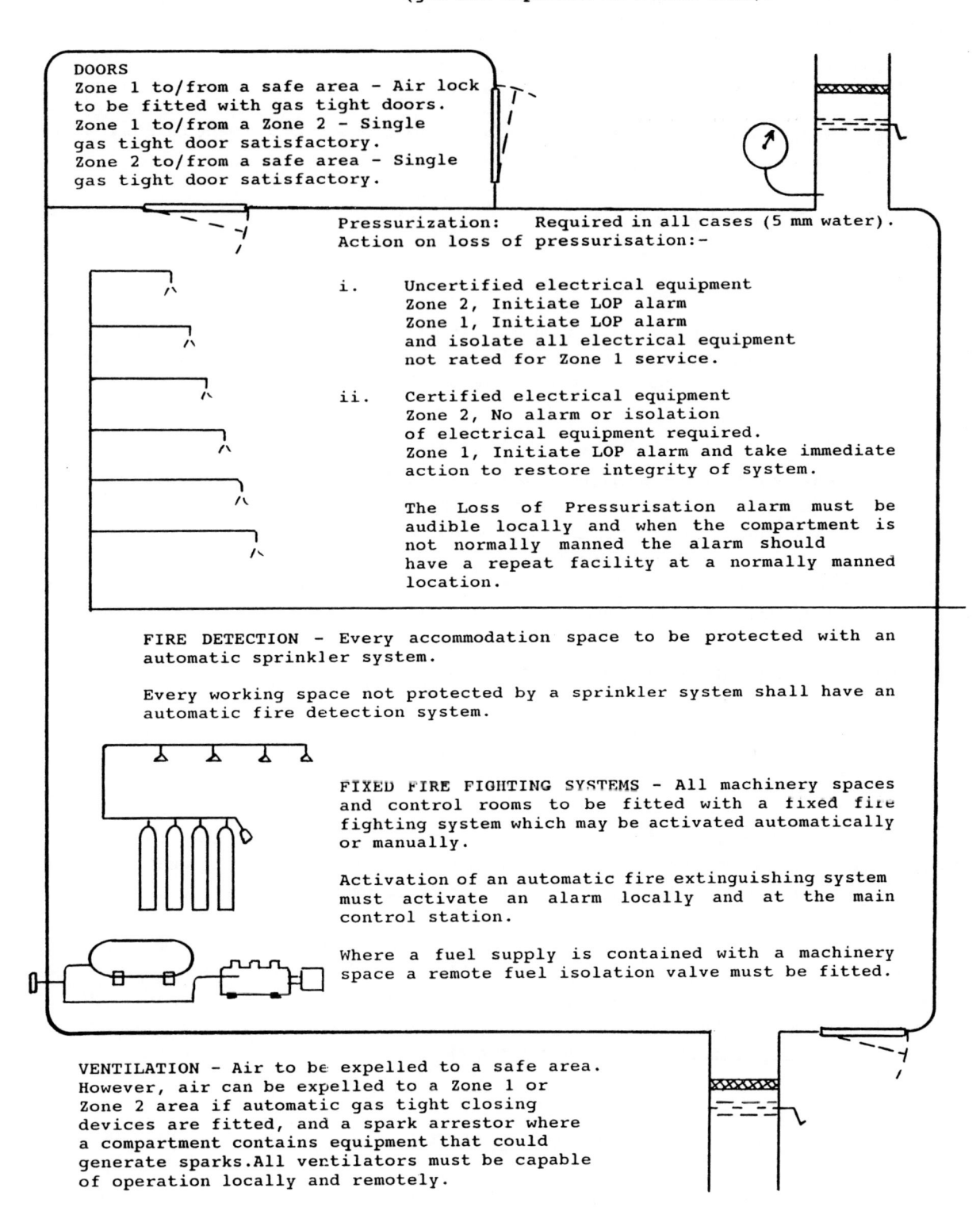

DOORS
Zone 1 to/from a safe area - Air lock
to be fitted with gas tight doors.
Zone 1 to/from a Zone 2 - Single
gas tight door satisfactory.
Zone 2 to/from a safe area - Single
gas tight door satisfactory.

Pressurization: Required in all cases (5 mm water).
Action on loss of pressurisation:-

i. Uncertified electrical equipment
   Zone 2, Initiate LOP alarm
   Zone 1, Initiate LOP alarm
   and isolate all electrical equipment
   not rated for Zone 1 service.

ii. Certified electrical equipment
    Zone 2, No alarm or isolation
    of electrical equipment required.
    Zone 1, Initiate LOP alarm and take immediate
    action to restore integrity of system.

The Loss of Pressurisation alarm must be
audible locally and when the compartment is
not normally manned the alarm should
have a repeat facility at a normally manned
location.

FIRE DETECTION - Every accommodation space to be protected with an
automatic sprinkler system.

Every working space not protected by a sprinkler system shall have an
automatic fire detection system.

FIXED FIRE FIGHTING SYSTEMS - All machinery spaces
and control rooms to be fitted with a fixed fire
fighting system which may be activated automatically
or manually.

Activation of an automatic fire extinguishing system
must activate an alarm locally and at the main
control station.

Where a fuel supply is contained with a machinery
space a remote fuel isolation valve must be fitted.

VENTILATION - Air to be expelled to a safe area.
However, air can be expelled to a Zone 1 or
Zone 2 area if automatic gas tight closing
devices are fitted, and a spark arrestor where
a compartment contains equipment that could
generate sparks. All ventilators must be capable
of operation locally and remotely.

FIRE AND GAS PROTECTION REQUIREMENTS

iii) DRY POWDER

Like Halon, dry powder relies on a chemical reaction to break the combustion chain and whilst it is very effective it leaves a tremendous mess. It is also extremely effective in fighting vent stack fires and may be used as an alternative to CO_2.

iv) ALTERNATIVE GASES

Chemical manufactures have produced alternatives to Halon and whilst some have been approved by statutory bodies, they are not at present commercially available.

Two such gases are Inergen and Argonite.

a) INERGEN

Inergen is a mixture of Nitrogen, Argon and CO_2 (50/42/8%), which has many of the benefits of Halon and can be used as a straight replacement for Halon with relatively few modifications. The main disadvantage with Inergen is the volume required, approximately 50% concentration by volume so a considerable number of storage cylinders are required.

b) ARGONITE

Argonite is a mixture of naturally occurring Nitrogen and Argon and like Inergen has many of the benefits of Halon, including the use of similar equipment. When Argonite is injected into a room it reduces oxygen to levels that will not sustain combustion, yet will permit essential safety personnel to operate. It is of a similar molecular weight to air and lingers longer than Halon so the protected spaces do not need to be hermetically sealed, a considerable cost saving advantage.

3.4 PORTABLE FIRE-FIGHTING EQUIPMENT

A considerable number of portable fire-fighting extinguishers are located throughout offshore installations. Extinguishers using water, foam, dry powder and CO_2 will be installed at vantage points as specified on the installation fire-fighting plan.

Foam concentrate in hand held extinguishers should be replaced at intervals not exceeding 10 years and in the case of extinguishers of 45 litres capacity or greater, the charge should be withdrawn at least once in four years and the internal condition of the extinguisher verified.

Portable CO_2 extinguishers should be recharged if the weight of gas falls to less than 90% of the stated capacity. If when the cylinders are emptied, four years have elapsed since the date of the last hydraulic test, the cylinders should be tested to the bursting pressure of the safety disc before being recharged. Portable fire-extinguishers other than CO_2 should be tested to 21 bar every four years whilst dry powder extinguishers should be tested to 1.5 × working pressure or 21 bar, whichever is the greater, every four years.

3.5 REMOTE STOPS AND CLOSING DEVICES

In order to minimise the spread of fire and maximise the efficiency of the fixed fire-fighting mediums the following mandatory requirements apply:

STAINLESS STEEL
FORGED FLANGES
Size Range ½" to 48" NPS

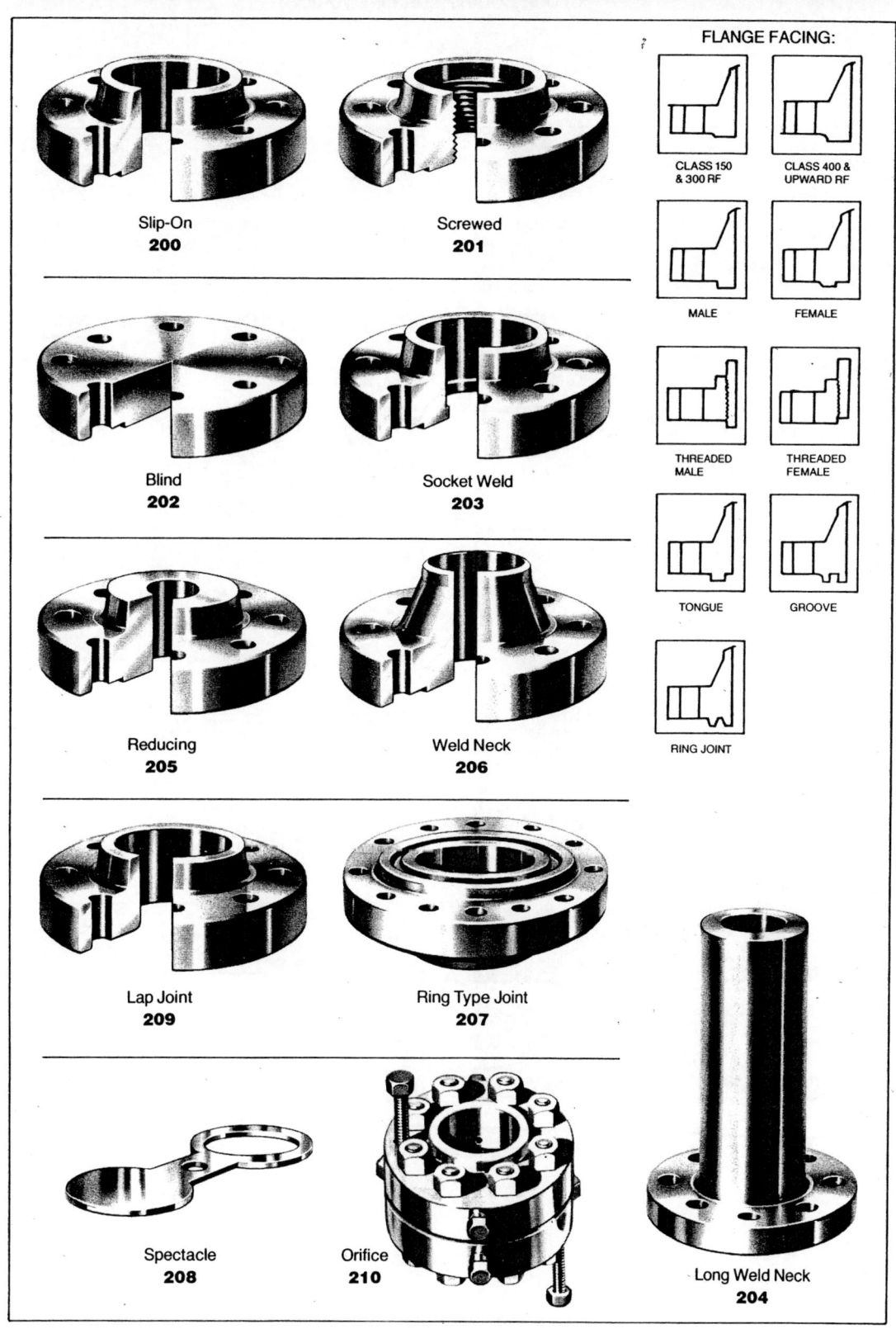

FLANGE FACING:

CLASS 150 & 300 RF	CLASS 400 & UPWARD RF
MALE	FEMALE
THREADED MALE	THREADED FEMALE
TONGUE	GROOVE
RING JOINT	

- Slip-On **200**
- Screwed **201**
- Blind **202**
- Socket Weld **203**
- Reducing **205**
- Weld Neck **206**
- Lap Joint **209**
- Ring Type Joint **207**
- Spectacle **208**
- Orifice **210**
- Long Weld Neck **204**

9 – Flange types. (Courtesy of RGB Stainless Limited).

COLOUR CODING FOR PIPEWORK

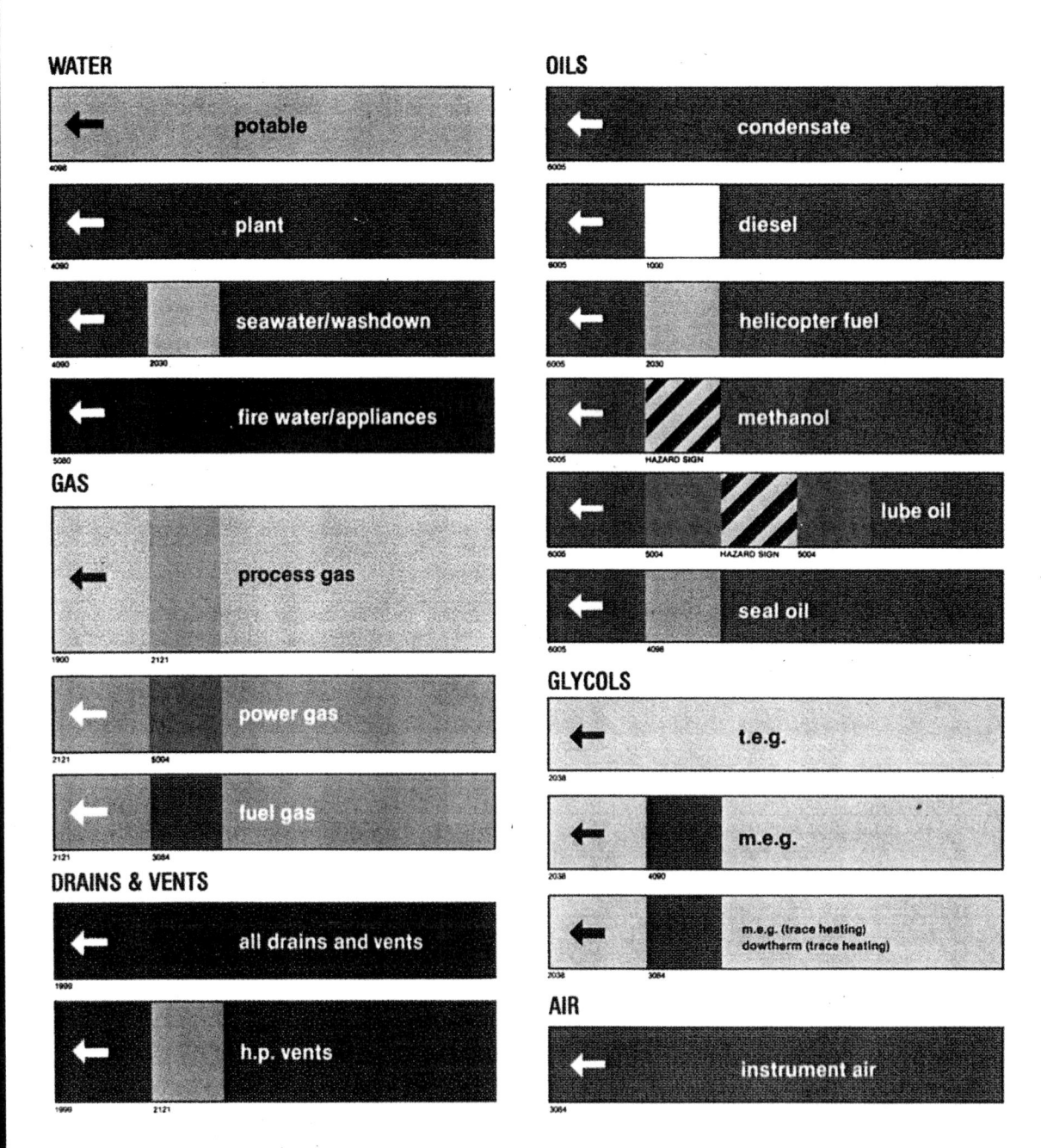

WATER
- potable
- plant
- seawater/washdown
- fire water/appliances

GAS
- process gas
- power gas
- fuel gas

DRAINS & VENTS
- all drains and vents
- h.p. vents

OILS
- condensate
- diesel
- helicopter fuel
- methanol
- lube oil
- seal oil

GLYCOLS
- t.e.g.
- m.e.g.
- m.e.g. (trace heating) / dowtherm (trace heating)

AIR
- instrument air

NUMBERS REFER TO HEMPELS COLOUR CODE

10 – Piping systems, typical colour coding plan.

11 – The equipment ...
State of the art diving support vessel the Deepwater 1 involved in survey work during the installation of a jacket.
Inset — the rugged splendour of an offshore supply boat equipped for diving support duties, the Smit Manta.

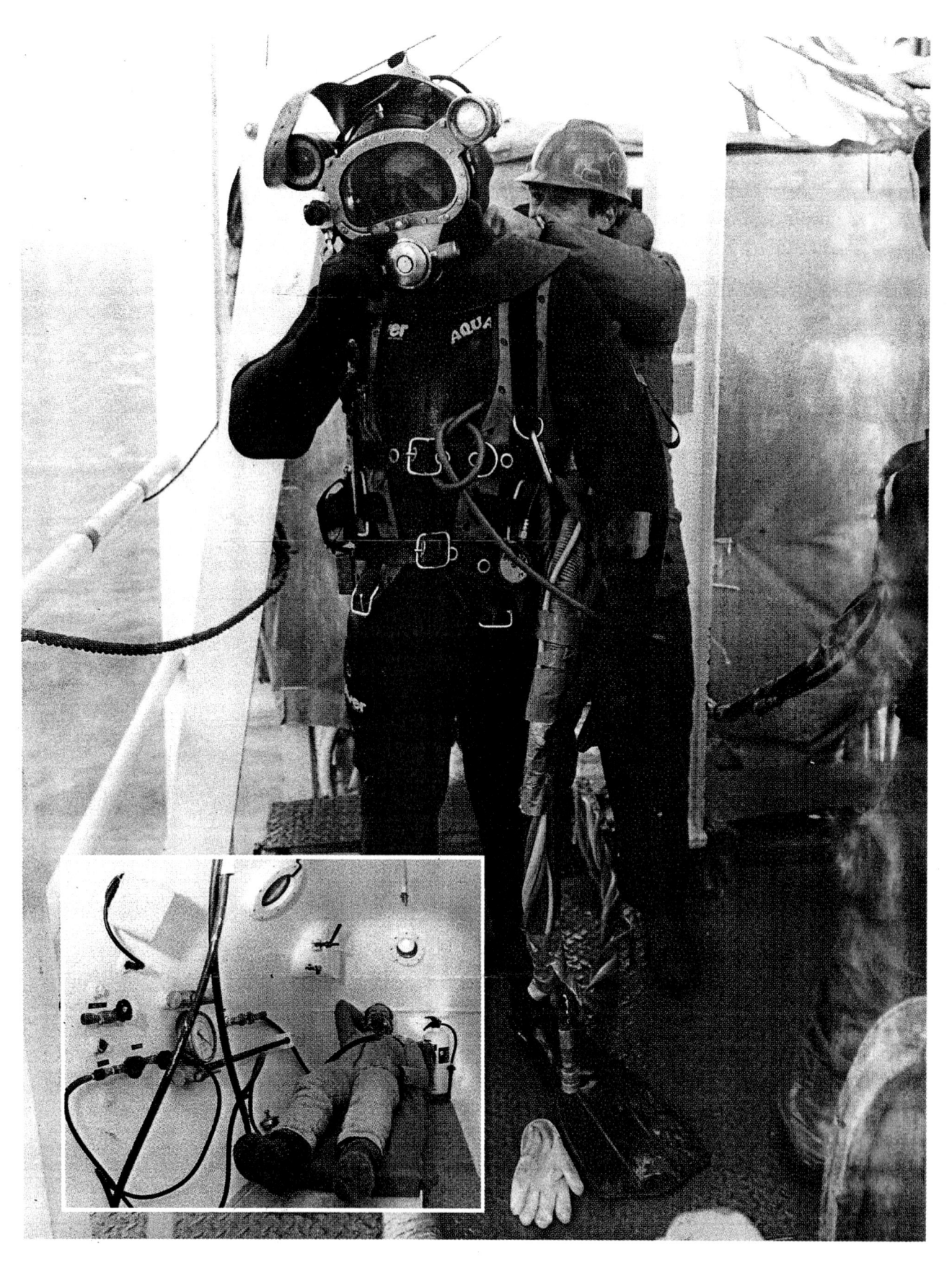

12 – The men ...
Typical North Sea air diving dry suit and associated attire.
Inset — the spartan interior of an air diving deck decompression chamber.
The diver can be seen inhaling oxygen to accelerate the decompression process.

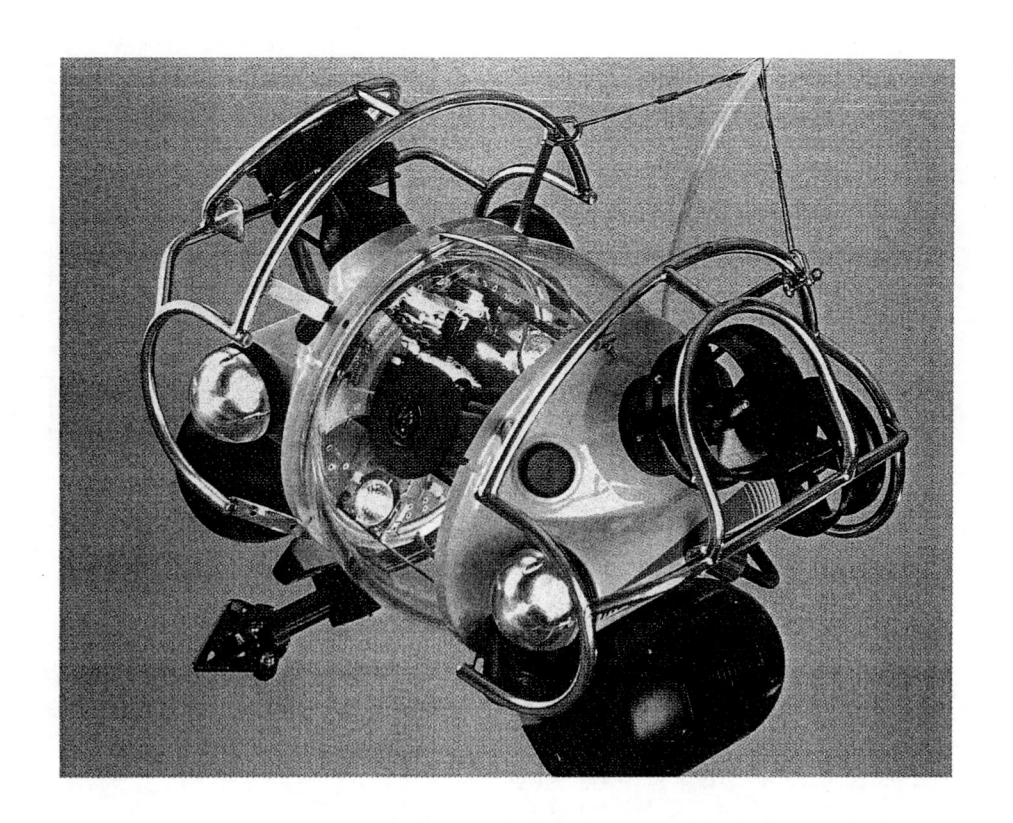

13 – The competition ...
The Offshore Hyball, a lightweight Remotely Operated Vehicle (ROV)
equipped with video monitoring inspection equipment.
(Courtesy of Hydrovision Limited).

The Trojan Remotely Operated Vehicle (ROV).
The propeller thruster units and articulated manipulators are clearly visible.
(Courtesy of Slingsby Engineering Limited).

Installation of the mud-mat using the drill string.

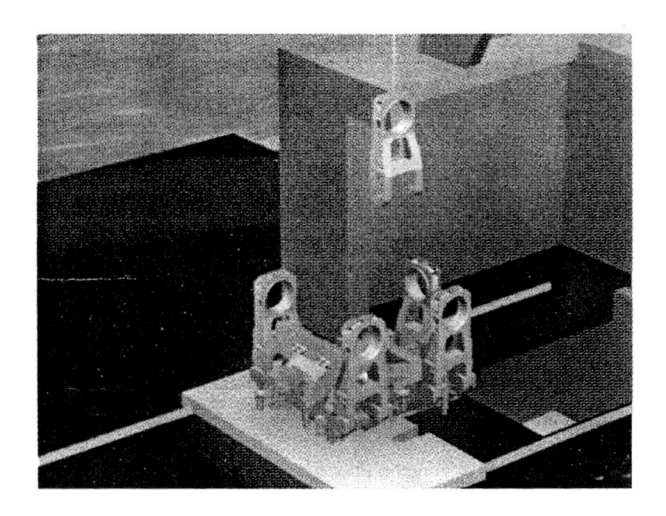

Assembly of the HOST on board the drill ship.

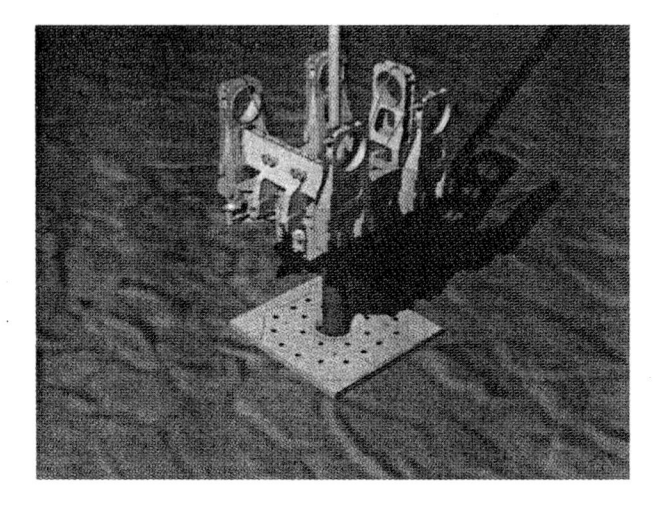

The partially assembled template is lowered through the moon pool and landed on to the mud mat.

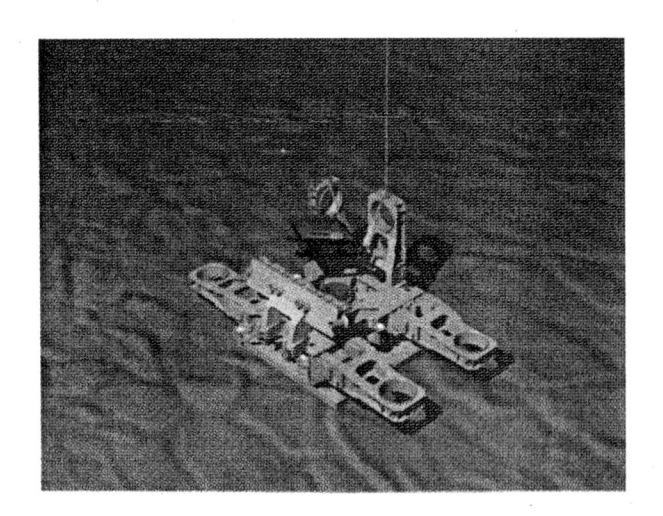

Deployment of the wing assemblies using an ROV.

Drill the first 36 inch conductor.

14 – The HOST GL Diverless Wellhead System.
Installation Sequence.

Install the 36 inch conductor and Permanent Guide Base (PGB).

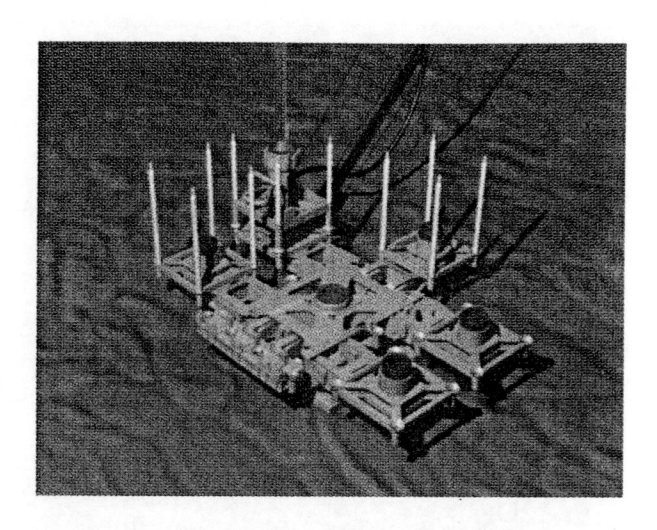

Drilling 26 inch intermediate casing and installation of the casing string string.

Install the sub-sea BOP stack, drill, case and complete the first well.

Install the first Christmas tree.

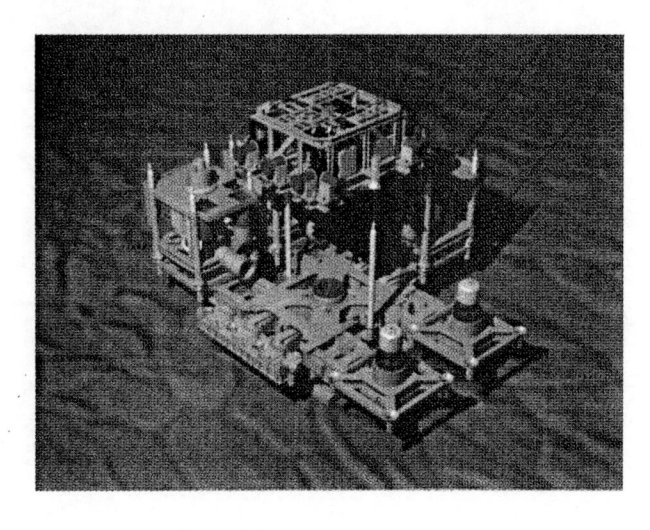

Install the manifold module.

Pull in the flowlines and umbilical control cables.

15 – The completed assembly, five wells installed and connected completely diverless.
(Suitable for installation in water depths up to 1,500 metres.)
Picture reproduced with permission of Konsberg Offshore as.

16 – The task ...
The installation of a drilling template. The template is being installed on the end of a drill string from a jack-up.
The template has slots for 9 wells and there are two pile guides.
Inset – sub-sea wellhead protection frame loaded on to the deck of an offshore supply boat for transportation to its final resting place.

i) VENTILATION SYSTEMS

Remote stops for ventilation fans and remote closing devices for ventilator flaps are to be provided. Modern installations may employ a centralised ventilation control system which can be operated from the main control room. This facilitates the isolation of specific areas which can be returned to service as and when required to assist in the dispersion of smoke.

ii) FUEL SYSTEMS

Compartments containing gas or diesel powered machinery must be provided with fuel isolation valves located externally to the machinery space.

4.0 PASSIVE FIRE PROTECTION

The requirements for passive fire protection are outlined in the *Department of Energy Guidance Notes on the Design, Construction and Certification of Offshore Installations* (withdrawn) and are generally in accordance with the recommendation of the *International Convention For The Safety Of Life At Sea* (SOLAS).

Passive fire protection involves the application of non-combustible heat resisting materials to insulate strategic locations against the effects of fire and smoke. The materials used are generally *mineral wools, fibre* partitioning boards, *cementitious and intumescent* coatings and all must be certified as having had their effectiveness proved by independent prototype testing within a controlled furnace atmosphere. On satisfactory completion of testing the materials are issued with a fire test certificate which is generally valid for a period of five years. It is most important that passive fire protection materials are installed strictly in accordance with the approved arrangements if the desired protection is to be achieved. The manufacturers of these materials will readily provide literature to show how their products must be installed, and copies of the fire test certification.

4.1 FIRE DIVISIONS — DESIGNATIONS

Various designations are used to specify the structural fire protection requirements and A, B and H-Class divisions will be terms frequently encountered. A time interval is normally included within the fire rating so that if, for example, an A-60 barrier is specified then this will maintain the integrity required of an A-Class division for a period of 60 minutes.

The most commonly encountered divisions may be briefly described as follows:

i) A-Class Divisions

These are constructed from steel, or an equivalent material suitably stiffened so as to maintain their structural integrity, and be capable of preventing the passage of smoke and flame to the end of the 60 minute *standard fire test*.

The standard or *cellulosic* fire test as defined by ISO 834 involves heating one side of the insulated material and its support structure to a furnace temperature in excess of $925°C$. The requirements are that the average temperature on the unexposed side of the division shall not increase by more than $139°C$ above the initial temperature, nor shall the temperature at any one point, including any joint, rise by more than $180°C$ above the initial temperature within the duration of the test, normally 60 minutes. Barriers which pass this test are classified as A-60.

Offshore Engineering

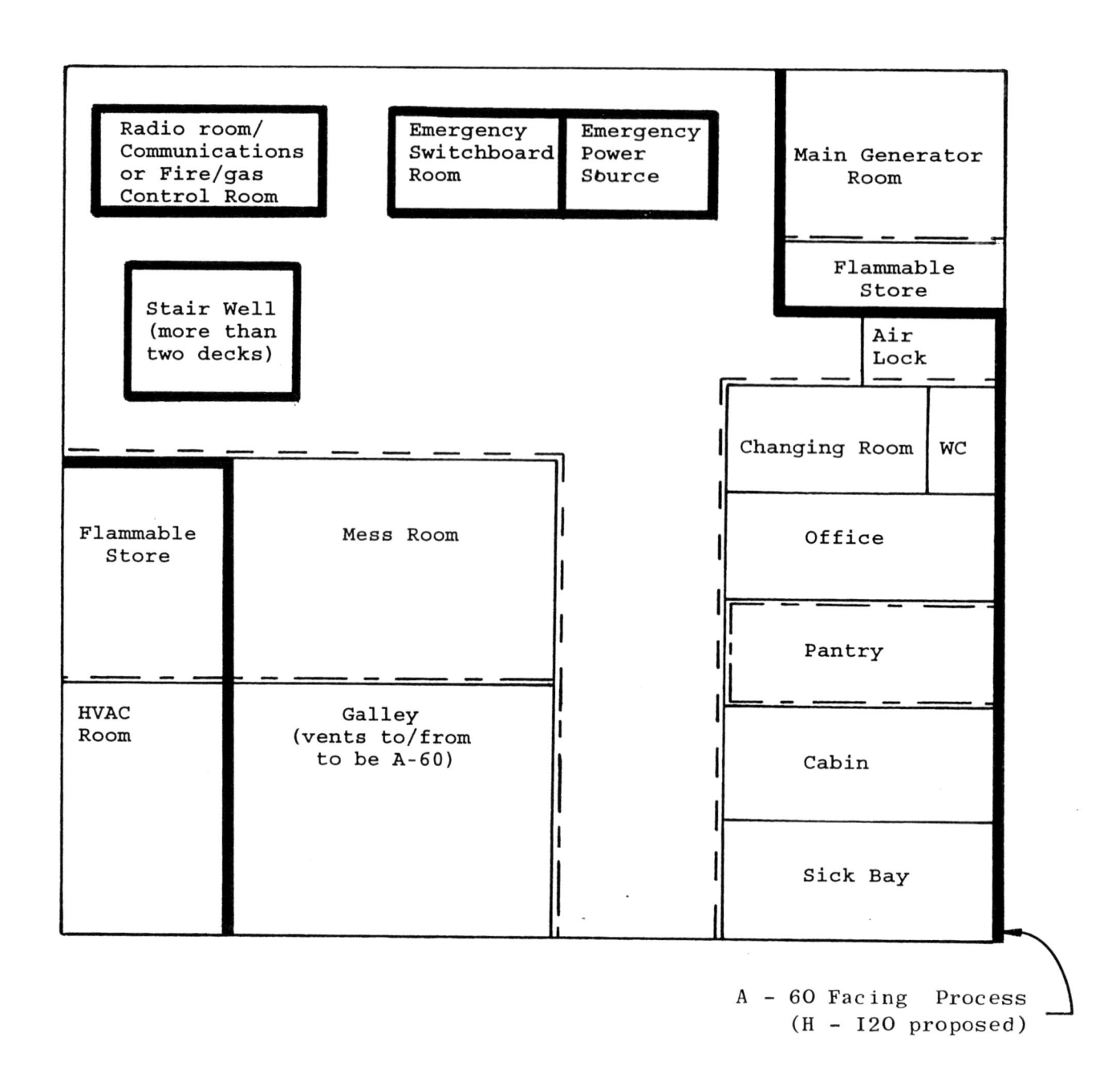

| Radio room/ Communications or Fire/gas Control Room | Emergency Switchboard Room | Emergency Power Source | Main Generator Room |

Stair Well (more than two decks)

Flammable Store

Air Lock

Changing Room | WC

| Flammable Store | Mess Room | | Office |

| HVAC Room | Galley (vents to/from to be A-60) | | Pantry |

Cabin

Sick Bay

A - 60 Facing Process
(H - 120 proposed)

▬▬▬ A-60

⸺ ⸺ B-15

═ ═ ═ Non combustible (60 mins.)

PASSIVE FIRE PROTECTION

As previously stated, an A-60 barrier must remain intact for a period of 60 minutes and this most arduous of specifications is used to protect locations such as the main control room, emergency sources of power, communication systems, and all accommodation boundaries that face a hazardous area on offshore installations. A-0 designated barriers will also be encountered and these will withstand the standard fire test, and thus prevent the passage of heat and smoke for a period of 60 minutes, but, they are not insulated. The A-0 barriers are used to classify items such as galley serving hatch shutters.

Typical A-60 arrangements are shown in the sketches, a steel bulkhead providing structural integrity whilst a dense mineral wool provides the thermal insulation.

ii) B-Class Divisions

These divisions are constructed from non-combustible materials (not necessarily steel because they are not required to be load bearing) and for testing are exposed to a standard fire test in a similar manner to A-Class divisions. They must be capable of preventing the passage of flame (but not smoke) towards the end of the first 30 minutes of the fire test but the accredited notation is typically only 15 minutes e.g. B-15. The temperature rise restrictions on the unexposed face are the same as for A-Class divisions.

The B-15 fire rated fibre board partitioning panels are used extensively for the construction of bulkheads, ceilings and corridors in the vicinity of non-critical areas such as accommodation rooms located in safe areas.

iii) H-Class Divisions

These are constructed in a similar manner to A-Class divisions but additional layers of mineral wool or improved *ceramic fibre* based insulations are applied to withstand the more rigorous demands of a *hydrocarbon fire test*. The furnace temperature specified for a hydrocarbon fire test is in excess of $1100°C$, the temperature rise restrictions on the unexposed face being the same as for A and B-Class divisions. The tests are carried out for periods ranging from 30 to 120 minutes, the latter giving an H-120 rated division.

The H-120 specification is a relatively new requirement and is recommended for use on accommodation boundaries that face locations where a hydrocarbon fire could occur such as wellheads and process areas. It may also be specified for the protection of sub-sea riser pipes in the vicinity of the *emergency shutdown valve* (ESDV).

iv) J-Class Protection

In addition to the standard and hydrocarbon fire tests, there is a third fire test which is used to determine the suitability of materials to resist *jet fires*. The test attempts to simulate the effects of a fracture in a pipeline riser or flow line which is characterised by a particularly fierce and intense fire with greater heat flux and higher levels of turbulence of the flame, and with the added danger of erosion of the insulating material.

Work is in progress to establish an International Standard for jet fires with input from oil companies, manufacturers of fire protection materials and certifying authorities but at the present moment in time fire tests are carried out to provisional standards, or guidance notes. In the UK, the Health and Safety Executive (HSE) have published guidance in the form of an *Offshore Technology Report*-OTO 95 634 — *Jet Fire Resistance Tests of Passive Fire Protection Materials* and tests are currently being carried out, and materials approved for jet fire protection against this document.

TYPICAL PASSIVE FIRE PROTECTION ARRANGEMENTS

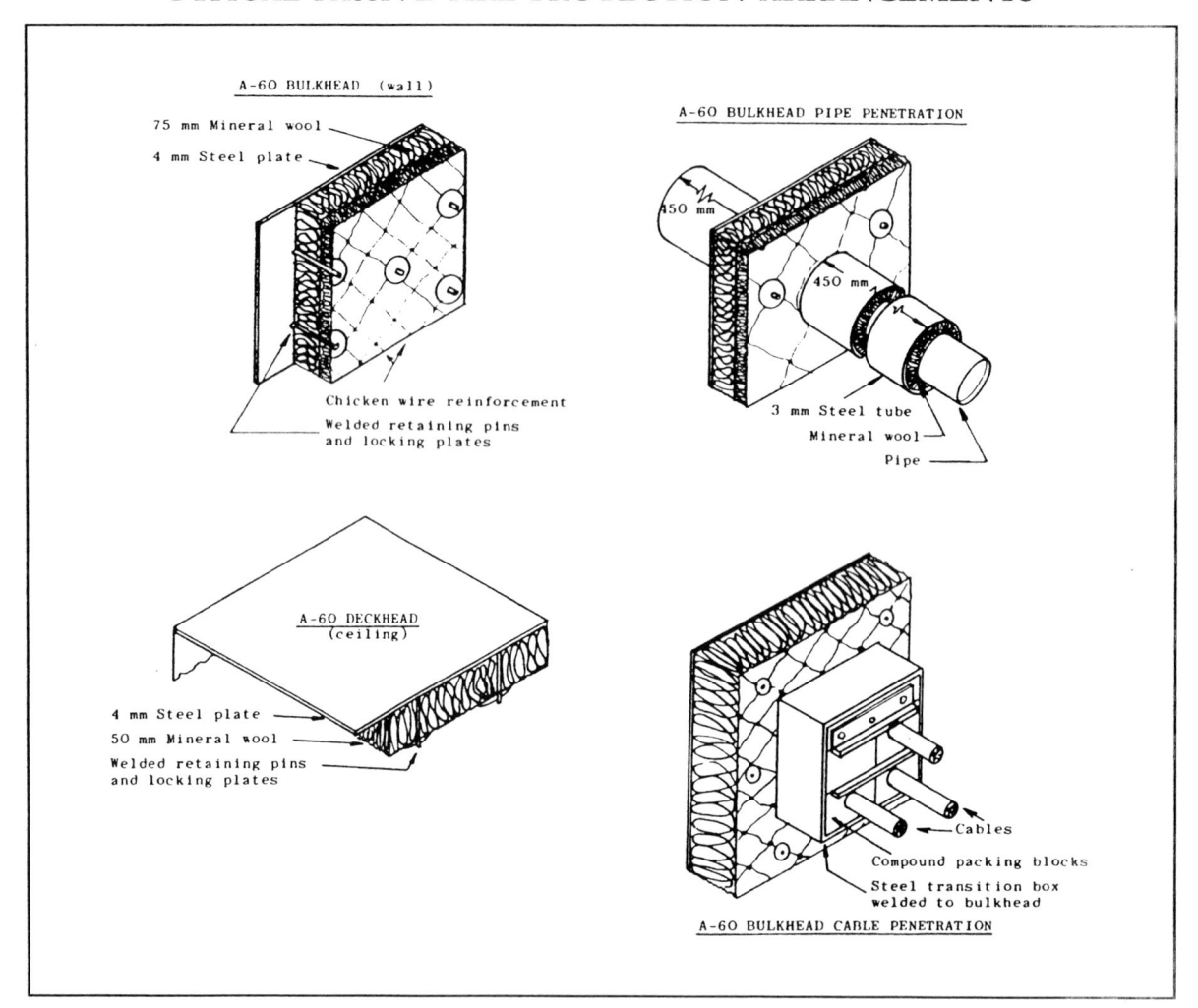

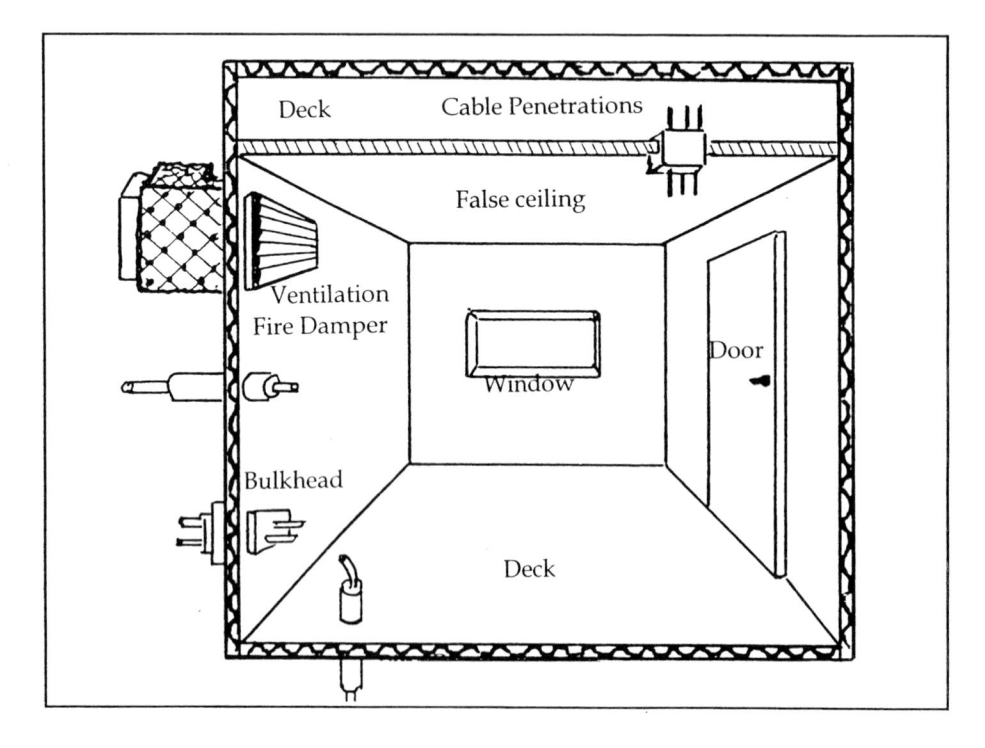

A *J* rating is used to identify jet fire tested materials and the period of effectiveness ranges from 30 to 120 minutes, the basic requirement being to keep the core temperature of the protected material to below 400°C. However, jet fire conditions and ratings are considerably more complex than the A-Class and H-Class barriers and each case must be considered separately and account taken of the particular hazards determined for the intended application.

4.2 FIRE DIVISIONS — PENETRATIONS

All items which penetrate fire rated divisions are to be certified to the same standards as the division itself. That is to say, all doors, windows, ventilators and ducting, pipe and cable transits must be approved as H-120, A-60, or B-15, or to whatever fire rating the design drawings specify.

All passive fire protection materials and fittings must be installed strictly in accordance with manufacturers recommendations in order to achieve the desired protection and to comply with the fire test approval certificates. This is particularly important where doors are concerned because they are approved as a complete assembly which includes the frame, hinges and locks. Electrical cables are another area where careful consideration is required because they can be supplied as either fire proof, fire retarding or low flame spread.

4.3 DRAUGHT STOPS

Draught stops are to be fitted in void spaces behind ceilings, bulkheads and stair wells both to prevent the passage of smoke and fire, and to prevent cold air from feeding a fire. They should be installed at intervals of 14 metres and may take the form of a fixed partition, or a curtain made from mineral wool.

4.4 FURNISHINGS

In addition to structural fire protection requirements it is recommended that all furnishings and internal decor be constructed to comply with the requirements of the various parts of BS 476 which defines standards for the resistance, in-combustibility and non-flammability of building material and structures. This includes primary deck coverings such as self levelling compounds which are used extensively on ships and offshore installations and should be of an approved type.

4.5 NEW DEVELOPMENTS

As a general rule, all materials used in the construction of an offshore installation, and this includes that major equipment items, shall be composed of materials which are consistent with their function, and these shall be incombustible. Consequently, any items constructed from new materials such as GRP for the replacement of traditional materials, may only be used if they have been proved as being incombustible by fire testing prototype samples. This includes such items as fuel tanks, fire mains and deck gratings.

Offshore Engineering

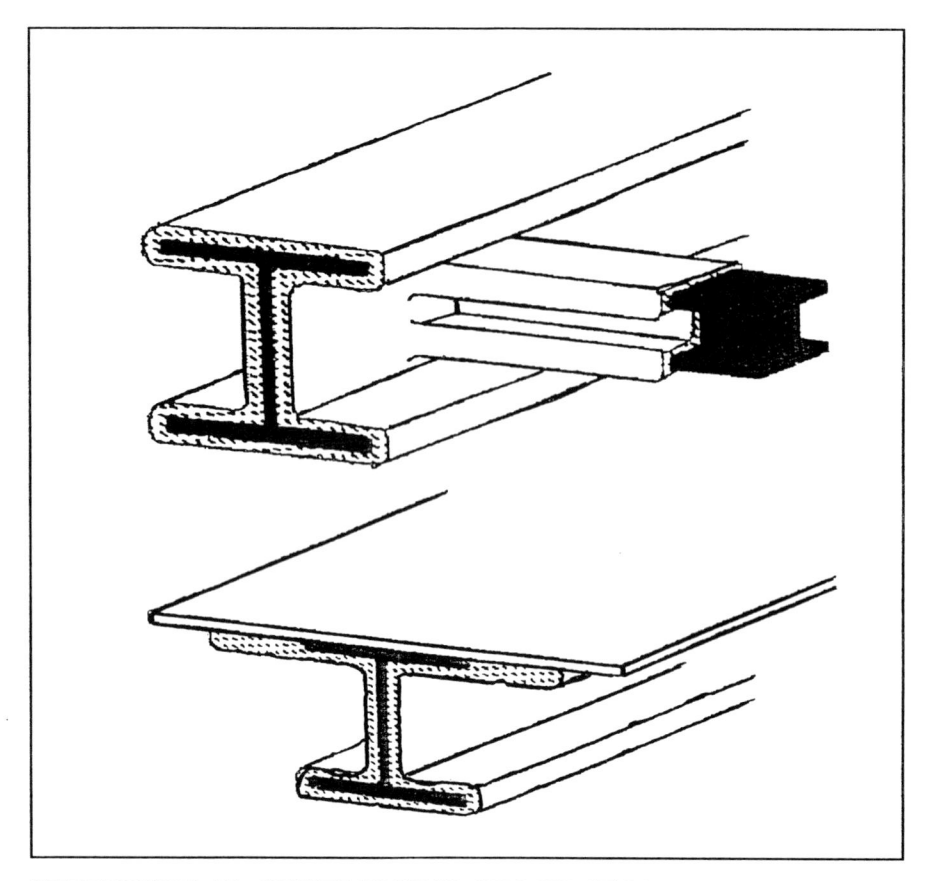

STRUCTURAL STEELWORK CEMENTITIOUS COATINGS

REFERENCE STANDARDS

1. Department of Energy Guidance Notes on the Design, Construction and Certification of Offshore Installations (withdrawn)

2. Department of Energy Offshore Installations: Guidance on fire-fighting equipment

3. Offshore Installations (Prevention of Fire and Explosion, and Emergency Response) Regulations, SI 1995/743, (PFEER) and Approved Code of Practise and Guidance

4. SOLAS (Safety Of Life At Sea) rules.

5. BS 476 Fire tests on buildings and structures

6. Offshore Technology Report-OTO 95 634 — Jet Fire Resistance Tests of Passive Fire Protection Materials (Health and Safety Executive)

Part 3. LIFE-SAVING APPLIANCES

INTRODUCTION

This section deals with the legislation covering the requirements for life-saving equipment on offshore installations located in the territorial waters of the UK Continental Shelf. The basic equipment is also described, the equipment being the same for installations the world over.

1.0 HISTORY-UK CONTINENTAL SHELF

Up until 1995, the requirements for life-saving appliances and emergency procedures on offshore installations located in the territorial waters of the United Kingdom were governed by Statutory Instrument No. 1977/486, *Offshore Installations — Life-Saving Appliances*, and the *Emergency Procedures Regulations* 1976 (SI 1976/1542). In 1995, these regulations were revoked and replaced by the *Offshore Installations (Prevention of Fire and Explosion, and Emergency Response) Regulations, SI 1995/743*, the Regulations being referred to as *PFEER*. The *Health and Safety Executive* (HSE) publish an *Approved Code of Practise and Guidance* to assist with the interpretation of the Regulations.

1.1 SI 743 PFEER-PREVENTION OF FIRE AND EXPLOSION, AND EMERGENCY RESPONSE REGULATIONS

The intention of the new Regulations is to provide a less prescriptive form of legislation which will encourage a more integrated approach to the selection of life-saving appliances and the formulation of emergency procedures. The Regulations do not require any significant changes to be made to the type of equipment previously installed under SI 486, but, rather than dictating exactly what equipment must be provided and what precautions must be taken as was previously the case, PFEER requires the duty holder to carry out an *Assessment* of the installation.

The object of the assessment is to identify events which could give rise to a major accident involving fire or explosion, the need for emergency response, or the need to evacuate the installation. It is the results of the assessment which will ultimately dictate what equipment should be installed and what precautions should be taken. In addition to identifying what equipment and procedures are required, it is a requirement that *performance standards* are set for the equipment and procedures to ensure that they will fulfil their objectives. Having established suitable performance standards and requirements, equipment will have to be selected and procedures prepared which will meet these requirements.

1.2 APPLICATION OF SOLAS — INTERNATIONAL CONVENTION FOR THE SAFETY OF LIFE AT SEA

As most items of live saving equipment are manufactured primarily for the marine industry, they comply with the requirements or performance criteria specified by SOLAS. To comply with PFEER, the duty holders will have to review the SOLAS requirements for a particular item to determine whether or not it meets the performance requirements identified as a result of their assessment for a particular installation. It is thought that in most cases they will, as offshore rescue and assistance is normally only a matter of hours away, whilst at sea, rescue can take several days. If the performance standards specified by SOLAS meet or exceed those required by the duty holder, the equipment may be used.

1.3 SURVEY AND EXAMINATION

Having installed suitable equipment and formulated suitable operating procedures, the duty holder is obliged to prepare and operate a *written scheme* for the *systematic examination* of these arrangements by a *competent and independent person*. The scheme is similar to, but, replaces the bi-annual examination of fire-fighting equipment and life-saving appliances previously required under SI 486, surveys initially carried out by the marine surveyors of the *Department of Transport* (DoT), and subsequently by the *certifying authorities*.

A summary of the main features of the assessment and the written scheme of examination is to be included in the safety case and they are to be updated at regular intervals to incorporate changes which may have taken place in working practises, advances in technology, or modifications to the installation.

A more complete account of the PFEER Regulations can be found in chapter two.

1.4 GUIDANCE NOTES AND STANDARDS

At the present moment in time, the industry is going through a period of change and there are a number of initiatives under discussion. Consequently, operators are referencing a range of documents and guidance notes in addition to the PFEER guidance previously mentioned.

i) **Offshore Installations: Guidance On Life-Saving Appliances**

Whilst the SI 486 Regulations have been superseded, the guidance notes provided by the HSE contain excellent advice on the installation and testing of life-saving appliances. In fact, the HSE have stated that the *guidance still contains some sound technical advice and that it will be for the duty holder to justify the appropriateness and suitability of any standards used*.

ii) **International Convention For The Safety Of Life At Sea (SOLAS)**

The SOLAS regulations are aimed primarily at the marine industry, but most of the requirements pertaining to life-saving equipment on offshore installations are based on the requirements of these regulations. They provide advice on the design, manufacture, certification, testing, commissioning and maintenance of life-saving appliances and the document is widely referred to.

2.0 GENERAL REQUIREMENTS

INTRODUCTION

Life-saving appliances are normally provided in accordance with standards laid down by the national administration of the country concerned, such as, the *American Coastguard* in the USA and the UK *Maritime and Coastguard Agency* (MCA). These requirements are normally based on *The International Life-Saving Appliance Code*, a mandatory marine code under the terms of the *International Convention For The Safety Of Life At Sea 1974* (SOLAS), a code developed by the *International Maritime Organization*. Whilst SOLAS is mandatory only for ships operating under the national flag of signatory nations, the majority of oil and gas producing countries are signatories to SOLAS and use the Code as the basis for specifying the minimum requirements for life-saving appliances on offshore installations.

SAFETY SYSTEMS

PLAN ON CELLAR DECK

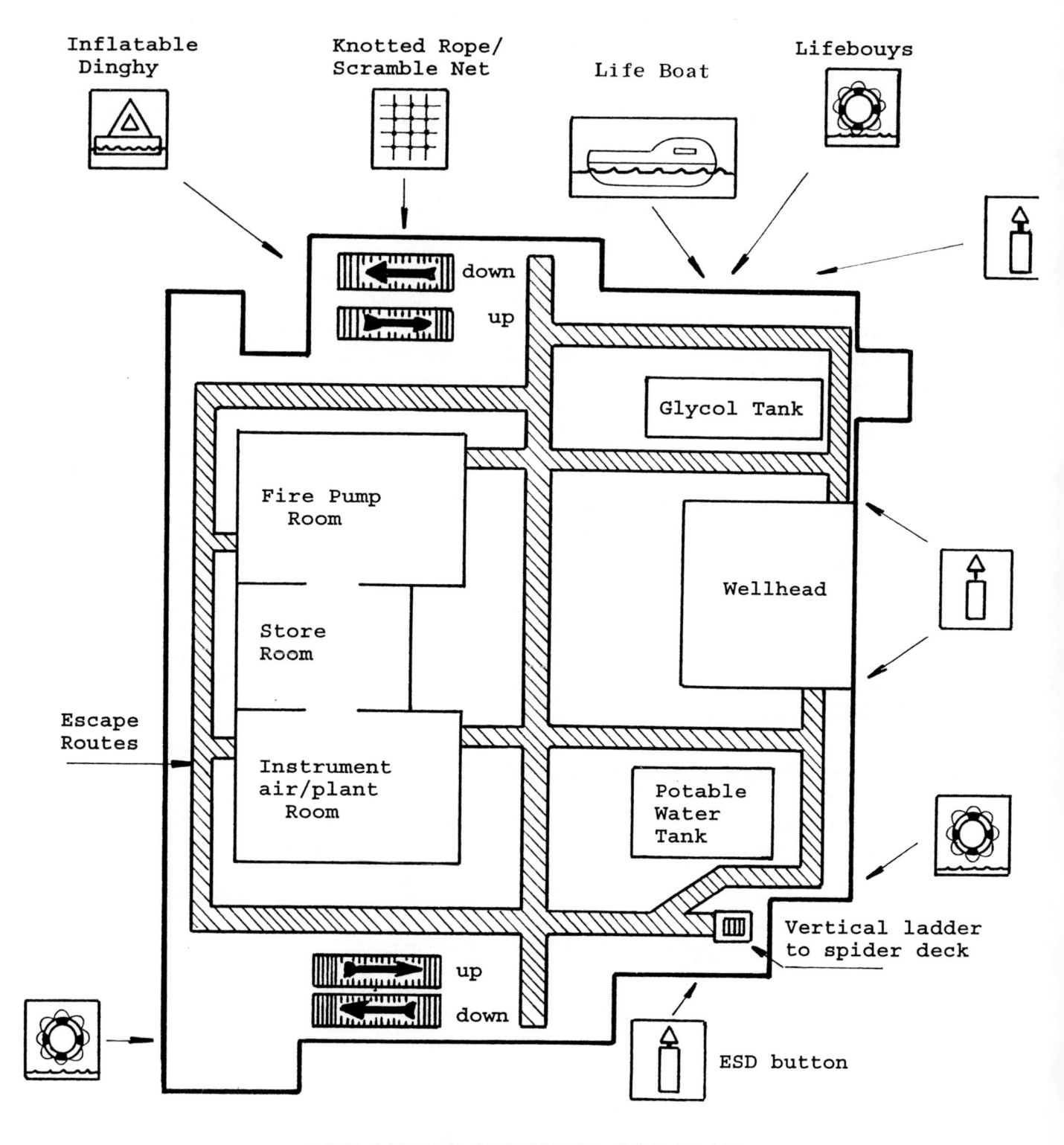

LIFE SAVING APPLIANCE (LSA) PLAN

2.1 APPROVAL OF EQUIPMENT

The International Life-Saving Appliance Code specifies requirements for the manufacture and testing of life-saving equipment and generally speaking, all equipment must be manufactured in accordance with a type approval certificate issued by the National Administration or flag state.

Type approval certificates are issued following confirmation by the national administration that the equipment provides safety standards at least equivalent to the SOLAS recommendations and that successful prototype tests have been completed. National administrations such as the *American Coastguard* and the UK *Maritime and Coastguard Agency* (MCA), previously the Department of Transport (DoT), frequently choose to delegate the approval process to organisations such as the major ship *classification societies*.

2.2 MANUFACTURE AND CERTIFICATION

Large items of equipment such as lifeboats and davits will be built under the survey of representatives of the local administration, or one of the ship classification societies, whilst smaller items may be supplied simply with a type approval certificate issued by the manufacturer. Some of the equipment which is not specifically for marine use such as torches, gloves and smoke hoods may be manufactured in accordance with a national standard and be accompanied with a *certificate of conformity* issued by the manufacturer against appropriate company standards. It may also be supplied against a CE number which shows that it has been manufactured in accordance with the relevant European regulations.

3.0 LIFE-SAVING EQUIPMENT

3.1 LIFE-SAVING APPLIANCE PLAN

All life-saving appliances should be installed in accordance with an approved Life-saving Appliance Plan, copies of which can should be provided in the *Statutory Operations Manual* or the *Safety Case*. The Plan should be clearly displayed in the accommodation and at an appropriate working space such as the main control room. The Plan should also show escape routes and muster areas and normally, the locations of fire-fighting equipment.

3.2 ALARM AND PUBLIC ADDRESS SYSTEM

A general alarm must be provided and a public address system which permits aural communication, a talk back facility, which is clearly audible at all locations on the installation.

Where the alarm and public address systems are located in noisy surroundings a visual form of indication is to be provided. Both systems must be supplied with two separate sources of electrical power, one of which is designed to function in an emergency situation.

SAFETY SYSTEMS

3.3 TEMPORARY REFUGE AND MEANS OF ESCAPE

A temporary refuge or safe haven is to be provided where personnel can congregate should an emergency situation develop involving fire and explosion, or a release of gas. Clear lines of access should be provided for personnel entering the refuge and for egress to the evacuation stations. Life support systems are to be provided which will protect the occupants for a period of time determined from the results of the *fire and explosion* (FEA) and *evacuation, escape and rescue analyses* (EERA). The temporary refuge must provide means to monitor the emergency situation and facilities for communication with the emergency services.

There must be suitable means provided for personnel to reach the helideck and the sea by at least two routes. Escape to the sea may be effected using fixed ladders, knotted ropes or scramble nets. Personnel descent devices, a form of abseiling equipment, may also be installed, often referred to as *donuts*.

All escape routes are to be clearly signposted, kept clear of obstructions and provided with adequate emergency lighting arrangements.

3.4 LIFEBOATS

GENERAL REQUIREMENTS

Every normally manned installation must be provided with a totally enclosed motor propelled survival craft (TEMPSC), capable of containing all persons on board (POB).

Clear instructions are to be provided which will enable one person to launch and operate the lifeboat. The lifeboat must be self-righting, constructed of fire retarding materials and protected against heat by an external sprinkler system. There must be sufficient fuel on board to propel the boat for 12 hours and a compressed air supply which will enable the occupants to breath, and the engine to run for at least ten minutes. A radio which can transmit distress signals must also be included in the survival equipment.

3.4.1 Testing and Commissioning

Lifeboats and davits are to be tested in accordance with IMO *Assembly Resolution A.689(17), Testing of Life-Saving Appliances*. The requirements are quite onerous, particularly for new lifeboats and davits.

The lifeboat is to be load tested by installing test weights which represent $1.1 \times$ SWL (safe working load). The brakes should be capable of holding this load and they should be tested at some point whilst the load is lowered to the water. It is to be confirmed that the lifeboat can be released from the falls when fully waterborne in the 110% overload condition, and in the light condition. Whilst in the water, the action of the hydrostatic release mechanism can be confirmed (if fitted) and that there are 3 turns of wire remaining on the winch drums, allowances being made for the lowest expected tide.

On completion of load testing, the speed of lowering of the lifeboat should be confirmed as being not less than 0.45 metres per second (90 feet per minute) and not more than 0.9 metres per second (180 feet per minute).

3.4.2 Davits: Winches and Loose Gear

Lifeboat davits and winch structural members are to be designed with a safety factor of 4.5 and on completion of manufacture should be subjected to a static proof load test of 2.2 × SWL, the SWL for each davit being, half the weight of the lifeboat, fully loaded with equipment and weights representing the maximum number of people, based on a weight of 75 kilograms per person. Each winch should be capable of lowering and holding a test load of 1.5 × SWL.

Attachments which support the sheave blocks, the blocks themselves and the attachments of the rope terminations are to be tested to 2.5 × SWL. The breaking tensile strength of the wire rope falls, chains, and links should be at least 6 × SWL. Each lifeboat release mechanism must be statically proof tested to 2.5 × SWL by an approved test house and be permanently marked with:

- manufacturers name
- type and serial number
- date of manufacture
- SWL
- test certificate number

Once in service, the wire rope falls should be changed end for end at intervals of 30 months and should be replaced every five years unless they are made of stainless steel wire, in which case a longer interval may be agreed with the authorities.

3.4.3 In Service Examination

Once installed the lifeboats should be examined on a regular basis and this should include an in water test, preferably every three months but at least once a year. It is only by manoeuvring the lifeboat that the equipment can be fully tested and particular attention should be given to the lifeboat release mechanisms. Some are designed to release the boat from the wire rope falls only when it is *off-load*, that is when the boat is fully waterborne. Other systems are designed to release *on-load*, that is before the boat is fully buoyant. Some designs are launched from a cradle and *free-fall* into the sea. Only personnel fully trained in the operation of the release mechanisms should carry out these tests because there have been numerous fatalities over the years associated with the routine testing of lifeboats.

3.5 LIFE-RAFTS

The requirements for life-rafts depend on the capacity of the lifeboats. If one lifeboat can accommodate the entire POB then additional capacity equal to the POB must be provided by life-rafts. That is to say the combined capacity of lifeboats and life-rafts equals 2 × POB.

The alternative is to provide two lifeboats whose combined capacity equals 1.5 × POB in which case no life-rafts need be fitted. In practice inflatable life-rafts are positioned at strategic locations regardless of lifeboat capacity. These inflatable life-rafts should be serviced annually. Life-rafts are often referred to generically as RFDs, a trade name of one of the many manufacturers.

The lanyard lengths should be checked on life-rafts and the drop lengths of knotted ropes and scramble nets to ensure that they are adequate and will indeed reach the water.

All survival craft must contain a first aid kit, an adequate supply of drinking water and a waterproof electric hand lamp suitable for signalling purposes.

SAFETY SYSTEMS

3.6 LIFEJACKETS AND PERSONAL SURVIVAL EQUIPMENT

There must be located on each installation a quantity of lifejackets equal to 1.5 × the POB. One must be stored by each bed with the remainder being located at embarkation or muster points. Donning instructions must be provided wherever the jackets are stored. Each jacket should be capable of turning the wearer so that they float, face up within 5 seconds of entering the water, and subsequently of keeping them afloat for a period of 24 hours. The jackets are to be manufactured from low flammability, rot proof, high visibility materials and should be provided with a non-metallic whistle, and preferably a light. A suitable ring or loop should also be fitted to the jacket to facilitate rescue.

In the UK, legislation requires the provision of personal protective equipment designed to increase the chances of survival of an individual in an emergency. The equipment should provide protection against the effects of fire, heat, smoke, fumes or toxic gas and immersion in the sea. A set of protective equipment should be located in the accommodation for every individual and should include a smoke hood, portable light source and heat proof gloves. A survival suit should also be provided in addition to a lifejacket.

3.7 LIFEBUOYS

The number of lifebuoys required will depend on the number of persons on board but at least eight must be positioned at locations readily available to assist a person who has fallen into the sea. The lifebuoy must be fitted with a light which will operate automatically when immersed for a period of at least 45 minutes. The light should have a luminosity of 2 candelas. Lights need not be fitted to lifebuoys which are installed in addition to the minimum requirement of eight.

Lifeboats and lifebuoys must be highly visible and bear the name of the installation.

3.8 MAINTENANCE

In the UK, the maintenance of life-saving appliances is frequently delegated to specialist companies who are authorised by the equipment manufacturers to service their products. Similar practises are followed to those on ships with most items of equipment, including the inflatable life rafts being serviced annually.

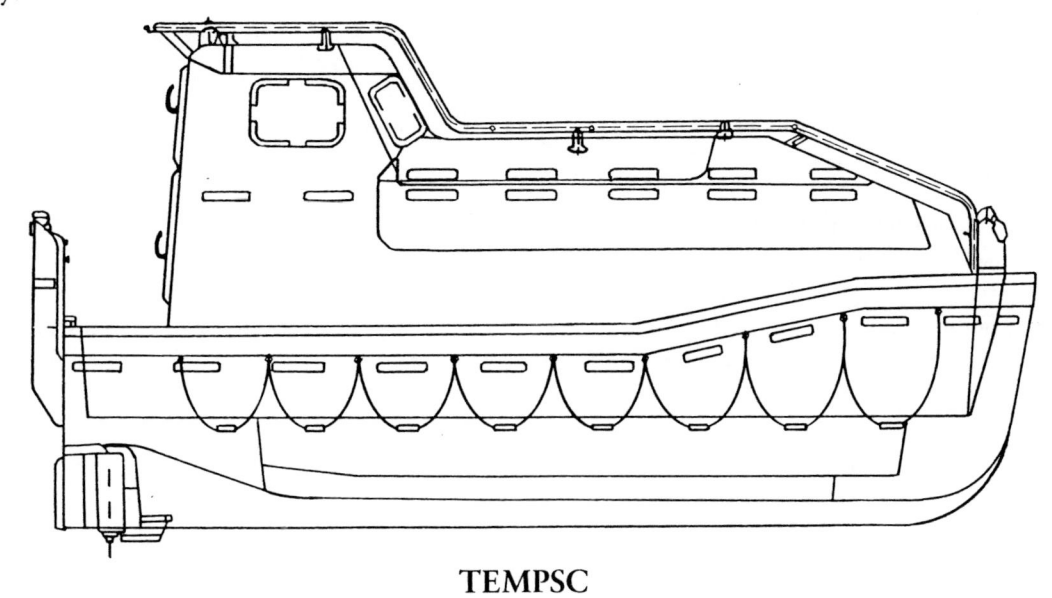

TEMPSC

SAFETY SYSTEMS

Part 4. NAVIGATIONAL AIDS – FIXED OFFSHORE INSTALLATIONS

As a means towards providing protection against shipping, every offshore installation located within the territorial waters of the UK must be fitted with visual and audible navigational aids in accordance with the requirements of a schedule provided by the *Department of Transport* under the *Coast Protection Act* 1949 as amended by Coast Protection Act (CPA) 1949, section 34 (as extended by the *Continental Shelf Act* (CSA) 1964 section 4(1).

1. VISUAL NAVIGATION AIDS

i) WHITE LIGHTS

It is a requirement that a flashing white light can be seen from whichever direction an installation is approached.

A light intensity of 12,000 candelas is specified which will give a range of approximately 15 miles. Should the bulb filament fail, then a secondary bulb with a range of 10 miles, should illuminate automatically and initiate a bulb failure alarm.

The lights which are normally mains powered are to be provided with an independent secondary source of power (normally batteries) capable of ensuring uninterrupted operation for a period of 4 days (96 hours). In the event of a mains power failure the change over to the secondary power supply should take place automatically and initiate an alarm at the main control station.

ii) RED LIGHTS

It is a requirement that the horizontal extremities of an installation be illuminated with flashing red subsidiary lights. A red light need not be fitted to an extremity where a white light already exists.

A light intensity of 1,200 candelas is required which will give a range of approximately 3 miles.

iii) LIGHTING — GENERAL

The sketches show typical arrangements. Two white and two red lights will normally suffice per installation (not per jacket).

The white lights must flash in unison. The red lights must also flash in unison but they need not flash in unison with the white lights.

The lights shall spell the letter "U" in Morse Code every 15 seconds. They should be mounted at a position between 12 and 30 metres (40 and 100 feet) above sea level in order that maximum visibility is obtained.

The lights shall be exhibited from 15 minutes before sunset until sunrise and at all times when meteorological visibility is less than 2 miles.

SAFETY SYSTEMS

MULTI JACKET INSTALLATION

KEY
- Red light
- White light
- Fog horn

SINGLE INSTALLATION

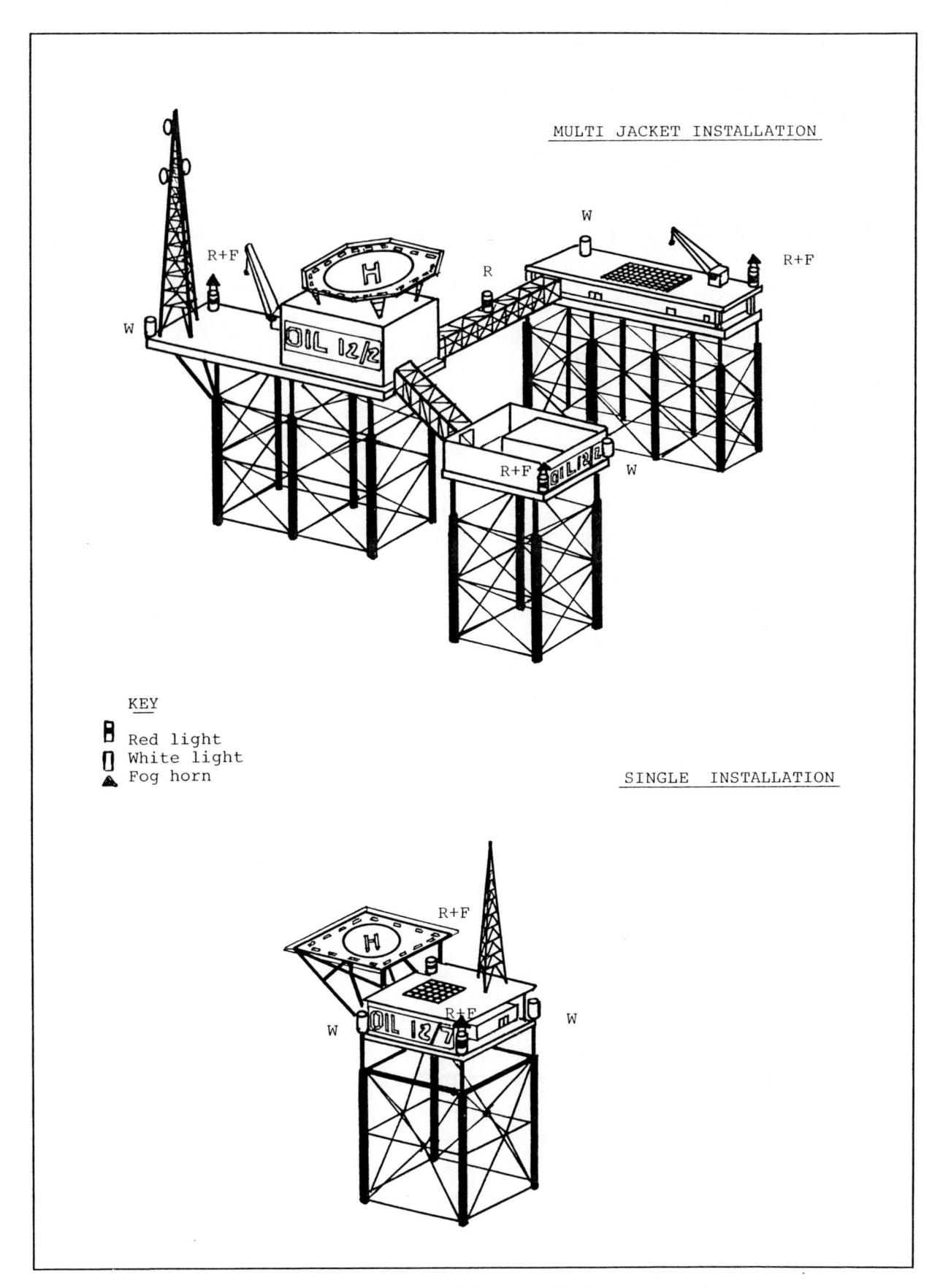

NAVIGATIONAL AIDS

2. AUDIBLE NAVIGATIONAL AIDS

Fog signals shall be activated whenever meteorological visibility falls below 4 miles and deactivated only after visibility improves to 5 miles.

The fog signal shall spell the letter "U" in Morse Code over a period of 30 seconds and have a range of at least 2 miles.

The primary fog signal which may be mains powered is to be provided with a secondary fog signal which will commence operation automatically on failure of the primary system and initiate an alarm at the main control station. The secondary system must have a range of at least half a mile.

The main and secondary fog signals shall each be capable of operating continuously at full power for at least four days (96 hours) from a power source independent of the mains supply (normally batteries).

The secondary power supplies for navigation aids are to be located within compartments structurally protected against fire to at least A-60 requirements.

3. NAVIGATION AID FAILURE

The Hydrographic Office must be informed of any failure of in installation's aids to navigation if:

- it is likely that it will not be possible to show main or subsidiary navigation lights all round when visibility falls below two miles;
- fog signals cannot be given;
- any element of the aids to navigation is operating on stand-by because of a failure of primary systems.

If a total failure of the navigational aids occurs the installation must be temporarily marked by one of two methods:

a) a temporary battery powered system of lights and fog signals must be installed which meets the requirements for light and fog secondary systems.

or

b) two Cardinal buoys are to be located on either side of the installation at one cables length from the installation.

The buoys must be provided with a light with an intensity of not less than 70 candelas and a fog horn consisting of a wave activated whistle or bell.

4. HELI-DECK ILLUMINATION

The requirements for the illumination of helidecks are outlined in the *International Standards and Recommended Practises for Aerodromes, Volume II Heliports* which is issued by the *International Civil Aviation Authority*. These requirements are also detailed in the document CAP 437, *Offshore Helicopter Landing Areas: A Guide to Criteria, Recommended Minimum Standards and Best Practise* issued by the *Civil Aviation Authority* in the UK.

SAFETY SYSTEMS

The perimeter of the safe landing area should be delineated by yellow lights of an intensity of 25 candelas. The lights, which are to be located at intervals no greater than 3 metres, with a minimum number of four on each side of the helideck, should be yellow in colour, although alternate blue and yellow lights have been used in the past. When it is intended for installations to be visited during the hours of darkness, the helideck should be floodlit in a manner which will not dazzle the helicopter pilot.

A white flashing helideck *wave-off* light or lights should also be provided to warn pilots when it is unsafe to land. The lights will normally be activated by a general alarm condition.

All heli-decks are subject to approval by the British Helicopter Advisory Board (BHAB) and should be re-examined at intervals of three years.

5. IDENTIFICATION PANELS

Identification panels are to be fitted which display a registered name or unique number by which the installation is registered. Figures should be 1 metre (40 inches) high on a yellow background and be so arranged that one panel is visible from any direction.

The panels are to be illuminated or constructed from retro-reflective materials.

The helideck must also be clearly identified with the installation name and number.

INFORMATION

Queries on navigation aids can be directed to:

Department of Transport,
Ports Division,
Zone 4/25,
Great Minster House,
76 Marsham Street,
London
SW1P 4DR

Offshore Engineering

Part 5. HAZARDOUS AREAS

Hazardous area classification was originally developed to provide guidelines for the selection of electrical equipment within onshore oil refineries. It has since been adapted to fulfil a similar role on offshore installations for both electrical and mechanical equipment. The hazardous and non-hazardous areas will be clearly marked on general arrangement drawings of the installation which can be found in the *Safety Case* or the *Statutory Operations Manual*.

The purpose of sub-dividing hazardous area into zones is to attempt to indicate the probability of a hazardous mixture of gas and air being present. The probability can then be matched to the probability of equipment becoming dangerous.

Definitions of the zones as provided by the present IEC (*International Electrotechnical Commission*) standards (Publication 79-10) are outlined below.

1. NON-HAZARDOUS AREAS

Non-hazardous or safe areas are those locations where explosive hydrocarbon mixtures are not expected to occur in sufficient quantities to present a hazard.

It should be noted that it is a requirement for all accommodation spaces to be located in a safe area.

2. HAZARDOUS AREAS

Hazardous areas are sub-divided into 3 categories:

i) Zone O — an area in which an explosive gas-air mixture is continuously present, or present for long periods.

It is unlikely that a Zone O area would be encountered offshore and should it exist then all electrical equipment should be excluded. Where this is impractical, intrinsically safe components may be used.

ii) Zone 1 — an area where an explosive gas-air mixture is likely to occur during normal operating conditions.

Such regions occur around flare booms and vent stacks. If a circle is drawn with its centre at the gas source and with a radius of 15 metres, then all regions within this circle will be classed as Zone 1. The zone will extend vertically upwards to a distance of 3 metres and vertically downwards to the sea level for gasses which are heavier than air. Whilst the primary constituent of natural gas is methane which is lighter than air, it often contains heavier gas fractions such as butane and heptane and liquid condensate is also often present. For this reason, the hazardous areas are normally extended downwards to sea level.

All electrical equipment within a Zone 1 area should be suitably certified (see reference tables). All electrical cables should be protected with a metallic sheath with the addition of a non-metallic impervious sheath.

SAFETY SYSTEMS

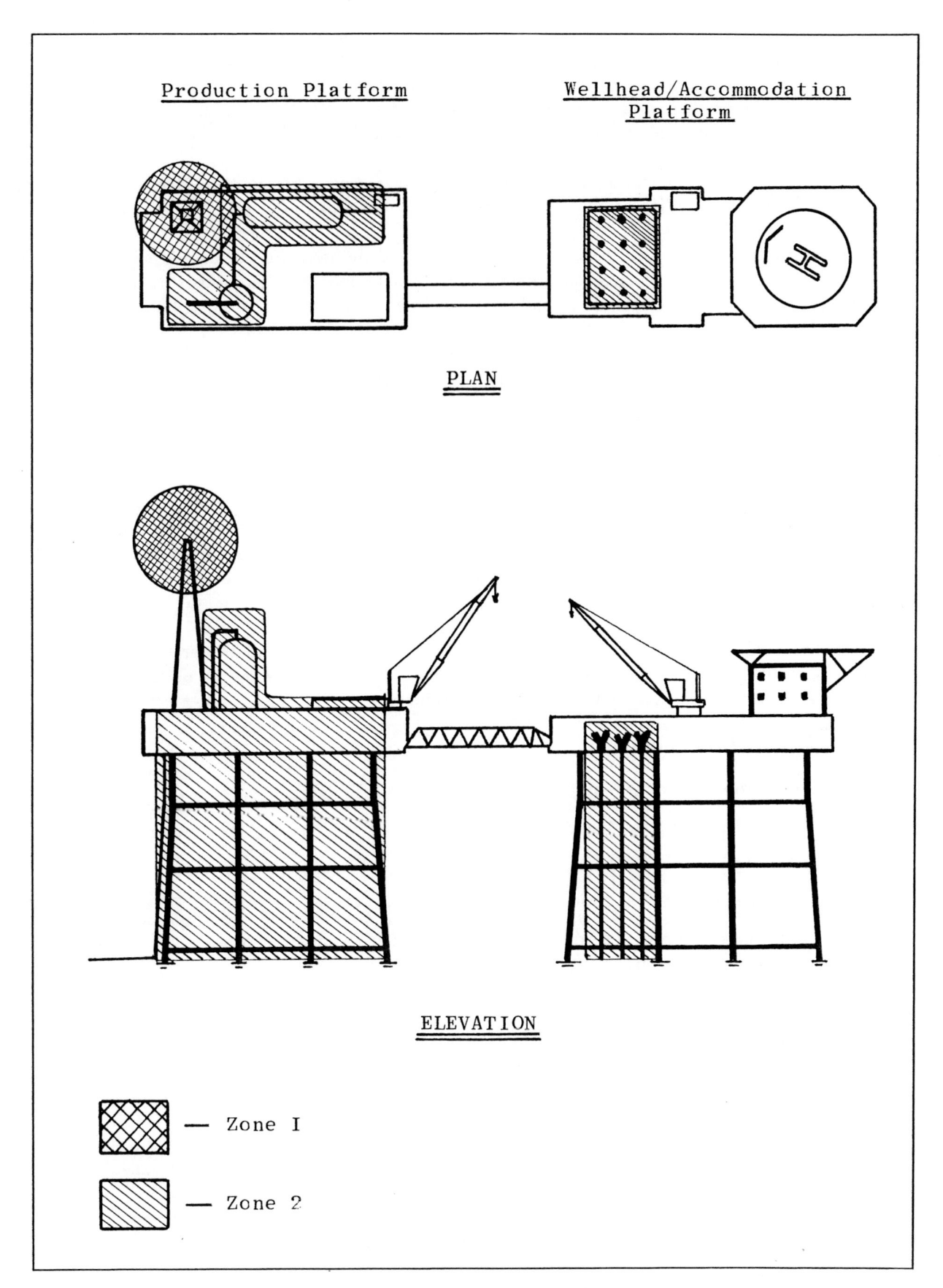

Production Platform   Wellhead/Accommodation Platform

PLAN

ELEVATION

⊠ — Zone 1

▨ — Zone 2

HAZARDOUS AREAS

131

iii) Zone 2 — an area where an explosive gas-air mixture is not likely to occur, and if it does, will only exist for a short period of time.

Gas processing areas, wellheads and parts of the drill floor are all normally classed as Zone 2 areas and all electrical and mechanical equipment should be suitably certified.

The hazardous area is normally considered to extend a horizontal distance of 7.5 metres from gas handling equipment and vertically downwards to the sea level.

Open deck areas of North Sea rigs are well ventilated by prevailing winds and natural draughts which prevent the build-up of large quantities of gas. All distances so far quoted for hazardous area boundaries are based on these open deck conditions. Enclosed areas must be considered separately because of the possibility of a build-up of gas occurring.

There are differences in the interpretation of area classification from country to country and from company to company. The American practise is still to divide hazardous areas into two divisions. *Division 1* is the more hazardous of the two divisions and embraces both Zone 1 and Zone 2 areas. The Zone 2 and *Division 2* areas are broadly equivalent.

Part 2 of the UK Code of Practise BS 5345 *Code of practise for the selection, installation and maintenance of electrical apparatus for use in potentially explosive atmospheres*, provides assistance in the classification of hazardous areas but the outdated Institute of Petroleum (IP) *Model Code of Safe Practice, Part 1 1965* is much more readily understood and is perhaps a better document for this purpose.

Some latitude exists as to the determination of hazardous area boundaries and in any plant, the decision as to the extent of each zone is made following a risk assessment by the plant operation team with assistance from, most practically, the chemical engineers.

Some of the factors to be considered are:

a) the probability of the presence of hazardous gas.

b) the quantity of hazardous vapour.

c) the degree of ventilation.

d) the nature of the gas (is it heavier than air etc.)

e) the consequence of an explosion. If an explosion could cause considerable loss of life the locations are frequently upgraded.

3. ENCLOSED COMPARTMENTS

3.1 ZONE 1 AREAS

It is permissible to use non-certified equipment in a Zone 1 area provided that the equipment is located within a positively pressurised (5 mm water gauge) compartment which is ventilated with air supplied from a safe area. Under these circumstances the compartment can be considered to be a safe area. An interlock must be fitted which will de-energise all uncertified electrical equipment and initiate an alarm in the main control room if a failure of the pressurisation occurs.

SAFETY SYSTEMS

All doors leading from an enclosed compartment located in a Zone 1 area to a non-hazardous area should be fitted with an airlock. If this is not practical, the HVAC (heating, ventilation and air-conditioning) system should be fitted with a standby fan which will start automatically on loss of pressurisation.

3.2 ZONE 2 AREAS

(a) An enclosed area located within a Zone 2 area in which it is possible for an explosive gas-air mixture to develop under abnormal conditions must be classed as Zone 1. In addition, the surrounding areas for a distance of 7.5 metres will also be considered to be Zone 2.

The enclosed area may be considered a safe area if it is suitably ventilated and pressurised (5 mm water gauge).

(b) An enclosed area must be classed as a Zone 1 area, even if it does not contain a source of hazard, unless suitable ventilation and pressurisation systems are fitted.

It is recommended that all access points which either:

i) lead from a Zone 1 area into a Zone 2 area, or

ii) lead from a Zone 2 area into a non-hazardous area,

should be constructed with a two door airlock. If this is impractical, gas tight self-closing doors should be fitted.

All doors leading from a safe area to a hazardous area should bear the legend *HAZARDOUS AREA* in letters approximately 50 mm (2 inches) in height.

4. HAZARDOUS AREA EQUIPMENT

4.1 SELECTION OF EQUIPMENT

All offshore installations operating in controlled waters designated under the *United Kingdom Continental Shelf Act* (CSA) 1964 must comply with all the relevant statutory requirements pertaining to electrical installations in hazardous areas, and the *Health and Safety at Work Act* (HSWA) 1974.

The selection of equipment for use in hazardous areas is governed by the hazardous area zone in which it will be located, and by the gasses that it could come into contact with.

Apparatus for use in hazardous areas should be selected in accordance with each of the following criteria:

a) classification of the area;

b) temperature classification;

c) apparatus group;

d) environmental conditions.

A typical equipment certificate number would be *Ex e IIB T4(IP 6)* and the build up of this designation will now be explained. The *Ex e* refers to the type of protection selected, the *IIB* to the suitability for particular gas groups, the *T4* to the surface temperature rating of the equipment and the *IP6* to the protection provided by the enclosure against the ingress of solids and water. These will now be more fully explained separately.

4.2 EQUIPMENT PROTECTION

The generic term for all methods of hazardous area protection used in Europe is *flame-proof* whilst the American practise is to use the term *explosion-proof*, they are to all intents and purposes the same.

Internationally, the Ex symbol, derived from the German *explosionsgeschutzten*, has been used for many years to prefix all protection techniques. More recently, compliance with one or more of the CENELEC (*European Committee for Electrotechnical Standardisation*) standard types of protection is denoted by the symbol EEx. In both cases the symbol is accompanied by one of nine lower case letters to indicate the type of protection provided, the definitions of each being provided in Table 1 and Table 2 of BS 5345: Part 1.

Of the nine protection categories, those which will be encountered most frequently are *Ex d* (flame proof), *Ex e* (increased safety) and *Ex i* (intrinsically safe) and they are shown in Table 1.

TABLE 1

Selection of Apparatus and Systems According to Hazardous Area Zone		
Zone	**Type of Protection**	**Marking**
0	Intrinsically-safe apparatus or system	Ex ia
	Special protection (specifically certified for use in Zone 0)	Ex s
1	Flammable enclosure	Ex d
	Intrinsically-safe apparatus or system	Ex ib
	Pressurization, continuous dilution and pressurized room	Ex p
	Increased safety	Ex e
	Special protection	Ex s
	Any equipment suitable for Zone 0	
2	Non-incendive	Ex n
	Oil-immersion	Ex o
	Sand filling	Ex q
	Any equipment suitable for Zone 0 and Zone 1	

SAFETY SYSTEMS

a) **Flameproof** — The principle on which flameproof *Ex d* equipment operates is containment of the explosion, should one occur, within the equipment thus preventing the propagation of the flame front to other locations. This involves constructing extremely robust equipment which tends to be both heavy and expensive and there is a limit to the size of what can be produced to contain an explosion, the limit for *Ex d* motors being 500 kW. It is the favoured method of protection in the UK whilst in Europe and the USA, *Ex e* equipment is preferred.

b) **Increased safety** — The increased safety, *Ex e* equipment is designed so that the tendency for arcing and sparking is reduced. Electric machines which use brushes are deemed to be sparking devices and hence are excluded from the *e* type concept. Preventing arcing and sparking in effect removes the potential ignition source and thus the hazard and this is achieved by improved design measures such as employing a higher specification of insulation and improved security of the electrical terminals.

c) **Intrinsically safe** — Intrinsically safe *Ex i* equipment is designed so that should a fault condition arise, an incendive spark will not occur. They are essentially lower power devices.

It should be noted that no *Ex* rated equipment is gas tight, or relies on being gas tight as a means of providing protection. The only equipment which relies on preventing the ingress of gas is pressurised equipment *Ex p*. The ingress of gas is prevented by maintaining the component or compartment at a positive pressure with air taken from a safe area and thus protected, un-certified electrical equipment can be used. Pressurised control rooms are used extensively with suitable alarms providing a warning and electrical power isolation occurring if pressurisation is lost.

As previously stated, a full summary and the definitions of the various concepts of protection can be found in BS 5345.

4.3 APPARATUS GROUP

There are two basic apparatus groups, the Roman numeral I being allocated to mining activity where the predominate risk is methane and coal dust, usually called *firedamp*.

All surface industry equipment, including that destined for use offshore is marked with the Roman numeral II in addition to being further sub-divided into gas groups IIA (propane), IIB (ethylene) and IIC (hydrogen). The gasses referred to in brackets are representative of typical gasses within a group as defined in BS 5345. An Ex e II rating without further reference to the letters A, B or C means that the equipment is suitable for all gas groups.

The gas groupings consider only the spark ignition and flame-propagation aspect of the explosion protection, that is the amount of energy or the size of the spark required to ignite a gas and for flame propagation to proceed. It should be noted that there is no correlation between the classification of the gasses for ignition energy and the classification for ignition temperature. For example, Hydrogen, which is classified as Group IIC (ignition energy $19\mu J$) has an ignition temperature of $560°C$ whilst acetaldehyde is classified Group IIA (ignition energy approximately $150\mu J$) and has an ignition temperature of $140°C$.

The BS 5345: Part 1: 1976 gives the temperature classification of the more common industrial gasses and is the most recent authoritative list.

TABLE 2 DATA ON TYPICAL FLAMMABLE PRODUCTS

COMPOUND	GAS GROUP	State At N.T.P.	Density (Air = 1)	Flammable Limits Lower %	Flammable Limits Upper %	Auto Ign °C	Temp Class
Methane	IIA	Gas	0.55	5.0	15.0	538	T1
Ethane	IIA	Gas	1.04	3.0	12.5	515	T1
Propane	IIA	Gas	1.56	2.1	9.5	466	T1
Butane	IIA	Gas	2.05	1.5	8.5	365	T2
Pentane	IIA	Liquid	2.48	1.4	8.0	285	T3
Hexane	IIA	Liquid	2.79	1.1	7.5	233	T3
Heptane	IIA	Liquid	3.46	1.1	6.7	215	T3
Kerosene	IIA	Liquid	4.50	0.7	5.0	210	T3
Hydrogen	IIC	Gas	0.07	4.1	74.2	560	T1
Ethylene	IIB	Gas	1.00	2.7	34.0	425	T2
Hydrogen sulphide	IIB	Gas	1.10	4.3	45.5	260	T3

4.4 TEMPERATURE CLASSIFICATION

Gas-air mixtures can be ignited by contact with hot surfaces and this must be taken into consideration when selecting equipment. The maximum surface temperature that can be generated by the apparatus must not exceed the ignition temperature of the gasses or vapours that could come into contact with it. Consequently, all electrical equipment used in hazardous atmospheres is classified according to the maximum surface temperature reached during prototype tests. The tests are carried out an ambient temperature of 40°C and if the equipment is to be used at higher temperatures than this, then it's temperature classification should be re-assessed.

TABLE 3

DATA ON TYPICAL FLAMMABLE PRODUCTS

Relationship Between T Class And Maximum Surface Temperature

T class	Temp °C
T 1	450
T 2	300
T 3	200
T 4	135
T 5	100
T 6	85

It is important to note that flashpoint and ignition temperature are not the same thing. Flashpoint is the temperature at which a liquid gives off sufficient vapour for ignition to occur when that vapour comes into contact with a naked flame. It is appreciably lower than the ignition temperature (often referred to as the auto ignition temperature) which is the temperature at which ignition will occur without an ignition source being present. For example, kerosene has a flashpoint of 38°C and an ignition temperature of 210°C whilst methane gas has an ignition temperature of 538°C.

It is also worth noting that some highly flammable gasses have relatively high ignition temperatures, *methane* (the main constituent of natural gas) being 538°C and *hydrogen* being 560°C. Consequently, T-1 rated equipment with a permissible surface temperature of up to 450°C may be used with these gasses. On the other hand, *hydrogen sulphide* (a constituent of sour gas) has an ignition temperature of 260°C so T-3 certified equipment must be used which has a maximum permissible surface temperature of 200°C.

The temperature classification awarded to simple electrical apparatus and components will generally be T-4 which means the maximum surface temperature will not exceed 135°C. Junction boxes and switches however may be certified with a T-6 designation because by their nature, they do not contain heat dissipating components, will not get hot and their maximum surface temperature will not exceed 85°C and they are thus suitable for virtually all gasses.

4.5 ENCLOSURE PROTECTION (IP)

Electrical apparatus should be constructed so as to provide a degree of protection against both electrical and mechanical failure. Protection should also be provided against the effects of the elements, the effects of heat from adjacent plant, and solvents.

A coded classification scheme has been developed to indicate the degree of protection afforded by an electrical enclosure against the entry of liquids and solid materials. The classification is supplementary to, and not an alternative to, the types of protection that are necessary to ensure protection in explosive gas-air mixtures and they should not be confused with explosion protection designations.

A two number coding system is used, the first numeral designating the degree of protection afforded against the ingress of solid materials, the second numeral designating the degree of protection provided against the ingress of liquids as shown in the Table 4.

TABLE 4

Ingress Protection (IP) Against Contact
And Ingress of Foreign Bodies and Water

0	No protection	0	No protection
1	Objects > 50 mm	1	Vertically dripping water
2	Objects > 12 mm	2	75-90° angled dripping water
3	Objects > 2.5 mm	3	Sprayed water
4	Objects > 1.0 mm	4	Splashed water
5	Dust protected	5	Water jets
6	Dust tight	6	Heavy seas
		7	Effect of immersion
		8	Indefinite immersion

The IP rating will not appear on all equipment as it is not relevant specifically to hazardous area equipment. It only appears on some equipment and then only if it exceeds the minimum requirement of IP 56. It will be found extensively on junction boxes.

Example of Apparatus Marking

e.g.–EEx d IIB T5,

    EEx d   Flameproof equipment (and thus suitable use in Zones 1 and 2)

    IIB   Surface/Offshore equipment complying with ethylene gas groups

    T5   Maximum surface temperature of 100°C (thus suitable for all gasses which have an ignition temperature of more than 100°C.)

4.6 CERTIFICATION AND MARKING

4.6.1 MARKING

It is a requirement that all electrical equipment installed within hazardous areas should be certified as suitable for the location and clear marking of equipment is therefore essential so that the user can assess the suitability of the equipment for its intended location. There have been numerous changes to the marking requirements of the various types of explosion protected apparatus over the years but in Europe the markings have now been standardised and can be found in BS 5501 (EN 50 014 to EN 50 020).

All equipment should be fitted with a plate which carries the following information;

a) manufacturer's name or trademark.

b) equipment type.

c) identification of the type of protection, e.g. *e, d, p* etc.

d) name or symbol of the test house and the certificate reference. (This will include the EEx mark. Additionally, when either a *Certificate of Conformity* to one of the harmonised standards listed in the EEC Directive 79/196/EEC, or an *Inspection Certificate* has been issued by an approved body recognised by the EEC, the Distinctive Community Mark may also be affixed.)

e) apparatus group.

f) temperature T class or maximum surface temperature.

g) any other relevant information.

4.6.2 CERTIFICATION

All hazardous area equipment must be accompanied by a manufacturers certificate which references the type approval tests of the component concerned and the prototype test certificate number which will have been issued by an approved test house following type approval tests.

There must be a certificate reference number and manufacturers name on the equipment to tie it back to the approval certificate.

When conformity with EEC Directives is required, certification is to be carried out by an *approved body* as listed by the Commission.

5. COMMISSIONING, INSPECTION, MAINTENANCE, TESTING AND REPAIR

5.1 PRE-COMMISSIONING

All electrical apparatus, systems and installations should be inspected prior to commissioning to confirm that the selection and installation of the equipment is appropriate for the location and in accordance with the specifications.

5.2 TESTING

All electrical installations should be tested prior to commissioning and these tests should include the following:

a) insulation resistance measurement;

b) earth electrode resistance measurement;

c) earth loop impedance measurement;

d) setting, and where appropriate, operation of protective devices.

The results of the tests should be recorded and included in the record of inspections.

5.3 MAINTENANCE AND INSPECTION

To maintain the hazardous area protection provided by a piece of equipment it must be effectively inspected, maintained and tested. Correct functional operation does not provide an indication that the equipment is still in good protective order. Inspections should also be carried out following any repairs or modifications.

The scheduling of routine inspections will depend on the type of equipment, the factors governing its deterioration and the consideration of the findings from previous inspections.

All these activities should be carried out by suitably trained personnel.

GUIDANCE NOTES AND REFERENCE STANDARDS

1. Department of Energy Offshore Installations: Guidance on Design Construction and Certification.

2. Institute of Petroleum: Code of Safe Practice Part 1, 1965.

3. BS 476 — Fire tests on buildings and structures

4. BS 5345 — Code of practise for the selection, installation and maintenance of electrical apparatus for use in potentially explosive atmospheres

5. BS 5501 — Electrical apparatus for potentially explosive atmospheres

6. BS 5490 — Specification for degrees of protection provided by enclosures.

7. CP 1003 — Electrical apparatus and associated equipment for use in explosive atmospheres of gas or vapour other than mining applications

8. CP 1013 — Earthing

9. CP 1021 — Cathodic Protection

10. API RP 500 — Recommended Practice for classification of areas for electrical installations at drilling rigs on land and on Marine, Fixed and Mobile platforms.

11. The IEC Rules for Offshore Structures provide practical guidance and recommendations on electrical equipment and installation practises for offshore work.

12. BS IS EN 60079 — 17 Code of Practise for repair of Ex equipment.

ALTERNATIVE EQUIPMENT TESTING AUTHORITIES

France	CERCHAR, LCIE
Italy	CESI
Belgium	INIEX
Germany	PTB, BVS
Denmark	DEMKO
U.S.A.	Underwriters Laboratory (UM), Factory Mutual (FM)
Canada	CSA

SAFETY SYSTEMS

Part 6. EMERGENCY SYSTEMS

1.0 EMERGENCY SUPPORT SYSTEMS (ESS)

Emergency Support Systems (ESS) are designed to minimise the effects of escaped hydrocarbons. It is a term which originates from the document *API RP 14C, — Recommended Practise for Analysis, Design, Installation and Testing of Basic Surface Safety Systems for Offshore Production Platforms.* The document provides guidance on a number of emergency support systems, including:

i) **GAS DETECTION**

The gas detection system provides a means of sensing the presence of escaped hydrocarbons and of initiating alarms and platform shutdown systems before gas concentrations reach the lower explosive limits (LEL).

ii) **CONTAINMENT SYSTEMS**

The containment systems will collect escaped liquid hydrocarbons and initiate platform shutdown systems.

iii) **FIRE LOOP**

The fire loop provides a very basic and full proof heat activated means of initiating the emergency shutdown system (ESD).

iv) **FIRE DETECTION**

Fire detection equipment is used to enhance the fire detection capability on an offshore installation and the equipment includes flame, smoke and thermally activated devices.

v) **EMERGENCY SHUTDOWN SYSTEM (ESD)**

In its most basic form the ESD system provides a manual means of initiating a platform shutdown by personnel observing abnormal conditions or undesirable events.

vi) **SUB-SURFACE SAFETY VALVE (SSSV)**

The sub-surface safety valves which may also be referred to as the *surface controlled sub-surface safety valves* (SCSSV) are located in the wells, the closure of which may be effected by the *fire loop* or the ESD system.

A more complete description of fire and gas detection systems can be found in chapter three and the fire loop and SCSSVs are discussed in chapter eight. Containment tends to relate primarily to the provision of bunds around storage tanks and machinery from which an escape of hydrocarbons could result.

Emergency shutdown systems (ESD) will now be dealt with in more detail as will emergency shutdown valves (ESDV), a safety system introduced in the UK by legislation in 1989.

2.0 EMERGENCY SHUTDOWN SYSTEM (ESD)

The *emergency shutdown system* is a name given to control equipment which is installed for the purpose of shutting down the production process and venting it to atmosphere in order to reduce the onboard hazard.

Level 1. – Unit shut down – Close B2 and B3
Level 2. – SPS – Surface process shutdown – Close B1, B2, B3 and ESDV
Level 3. – TPS – Total process shutdown – Close B1, B2, B3 and ESDV. Open all vent valves.

V = Vent valves.   B = Block valves

SCHEMATIC – EMERGENCY SHUT DOWN SYSTEM (shows only one train)

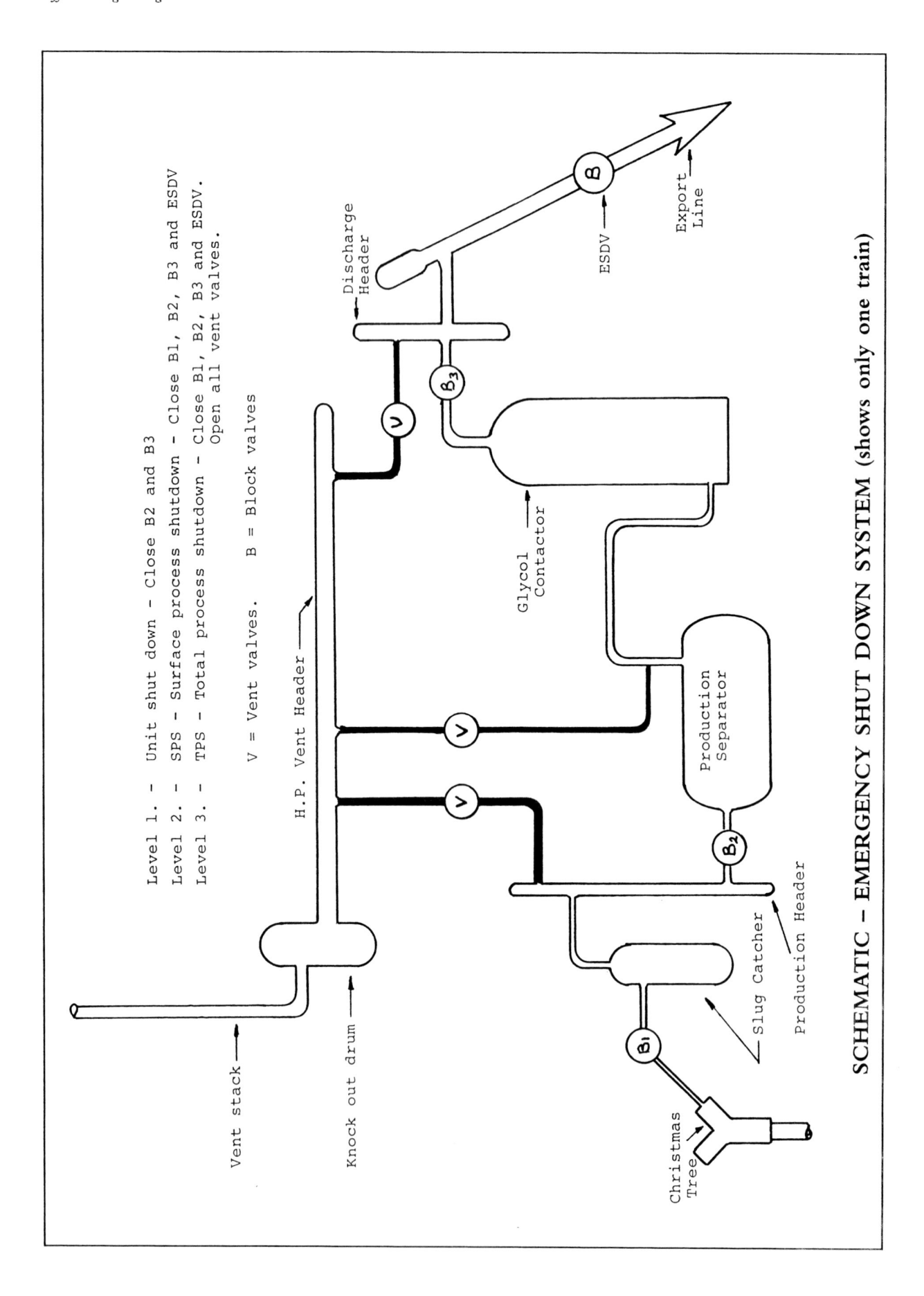

SAFETY SYSTEMS

The ESD system should operate automatically and independently of all other control and monitoring systems and be designed to *fail-safe*. It should be possible to initiate an ESD from manual push buttons located at strategic positions on the installation, such as the control room, wellheads, helideck and lifeboat stations. Most modern installations also have the facility to activate the ESD system from a remote location such as the onshore reception facility via the *telemetry* system.

The principal aim of the ESD system is to protect personnel, plant and equipment and minimise the risk of environmental pollution from the production process.

The design of the ESD system will vary from platform to platform depending on equipment complexity and production requirements. Hence an installation with a compression facility will have a more complex system than a basic oil or gas producing satellite platform.

The ESD system operates selectively in stages but these final events must occur:

(1) Shutdown of all production and associated test facilities.

(2) Closure of all wellhead valves.

(3) Opening of all blowdown (vent) valves.

(4) Closure of mudline (sub-surface) safety valves.

The number of shutdown stages will vary from platform to platform as will the cause and effect philosophy. There are general guidance notes available to the industry but the ESD system philosophy is normally specific to a particular company and the systems vary enormously.

A typical ESD system would operate on three levels of shutdown.

1. **UNIT SHUTDOWN**

    This shuts down individual process and utility systems and may be initiated manually and/or automatically. It gives the operator a chance to take corrective action prior to losing all production.

    A unit shutdown can occur due to a process or an internal trip caused by an event such as a flame failure on a glycol re-boiler or by activation of a fire and gas signal (25% LEL gas — lower explosive limit).

2. **SURFACE PROCESS SHUTDOWN (SPS)**

    All process and chemical injection systems will shutdown on an SPS but utilities such as main power generation remain available. It can be initiated automatically, manually and via telemetry but does not vent the process equipment.

    An SPS can be initiated by disturbances such as a high level gas alarm (50% — LEL), main power generation failure or a high or low export line pressure.

3. **TOTAL PLATFORM SHUTDOWN (TPS)**

    All platform process and utility systems are shutdown on a TPS. Wellheads and sub-surface safety valves close and main power generation shuts down although emergency power generator will normally still run. The blowdown valves open and the platform is depressed.

    A TPS is normally initiated from a fire detection signal such as a heat detector or combination of ultra-violet (U/V) detectors.

FIRE LOOP

As a further safety precaution a pneumatic fire loop is often included as part of the ESD system. This can either be a stainless steel instrument piping loop fitted with fusible plugs or it may simply consist of a small bore plastic pipe loop.

The loop extends throughout the wellhead and process areas and is pressured to 3.5–7.0 bar (50–100 psi). Should a fire occur, the *fusible plugs* or plastic pipe will melt and release the air pressure. This will then activate the ESD system via an electrical, pneumatic or hydraulic solenoid.

3.0 EMERGENCY SHUTDOWN VALVES – PIPELINES

As a result of the explosion and eventual destruction of the *Piper Alpha* installation on 6th July 1988 which claimed 167 lives, the Government passed the *Offshore Installations (Emergency Pipe-Line Valve) Regulations 1989* (SI 1989 No. 1029). The regulations stipulated that all fixed pipelines bringing toxic or flammable substances to, or from a platform be fitted with an *emergency shutdown valve* (ESDV).

The Emergency Pipe-Line Valve Regulations were introduced due to the fact that many offshore installations share common sub-sea pipelines for the conveyance of produced oil and gas to a reception facility on the mainland. The ESDV is designed to isolate an installation should an emergency situation such as a fire occur, and hopefully prevent a re-occurrence of the Piper Alpha disaster where the initial fire was fuelled by oil and gas pumped in by adjoining platforms.

The Emergency Pipe-Line Regulations were revoked by the *Pipelines Safety Regulations* (SI 1996 No. 825) in 1996 but the philosophy behind the installation of the valves remains the same under the new legislation and the requirements are essentially unchanged and outlined in Schedule 3 of the *Guidance On Regulations*.

The emergency shutdown valve should be located below the lowest deck of the installation and as far down the pipeline riser as is practicable whilst still being above the highest anticipated wave crest and available for maintenance. The ESDV must be of a rapid acting, fail-safe design which will operate automatically on signals generated by the ESD system, and which can be operated at a control point adjacent to the valve.

The valve and its actuating mechanism must be protected from damage arising from fire, explosion or impact from dropped objects, the aim being to ensure that the valve will close under all foreseeable conditions. In order to define the type and extent of fire protection required the operator will need to consider the type, severity and duration of anticipated fires as well as the minimum duration for which the integrity and operability of equipment to be protected must be maintained.

In addition to the requirement for an ESDV to be fitted to pipeline risers, the Pipelines Safety Regulations also cover the sub-sea pipeline outboard of the ESDV. The regulations require that a *Major Accident Prevention Document* (MAPD) be prepared, essentially a management tool designed to ensure that the operator has assessed the risks from major accidents and has introduced an appropriate safety management system to control those risks.

The original SI 1029 Regulations were accompanied by *Guidance Notes In Support Of The Offshore Installations Emergency Pipe-Line Valve* which laid down strict rules to ensure that regular maintenance and tests of the ESDVs were carried out and officially recorded and most operators have elected to follow a similar format for the maintenance of the ESDVs under the new Regulations. The guidance notes are still available and provide practical advice on maintenance, function and leak testing of the ESDV.

SAFETY SYSTEMS

GUIDANCE NOTES AND REFERENCE STANDARDS

1. API RP 14C, Recommended Practise for Analysis, Design, Installation and Testing of Basic Surface Safety Systems for Offshore Production Platforms.

2. Guidance on Regulations, Schedule 3 — The Pipelines Safety Regulations (SI 1996 No. 825).

3. Department of Energy Offshore Installations: Guidance on Design Construction and Certification.

4. Technical requirements for ESD systems for Offshore Installations by I. Turner of Lloyd's Register of Shipping (paper commissioned by UK Department of Energy in 1982 and prepared by Det Norske Veritas and Lloyd's Register of Shipping).

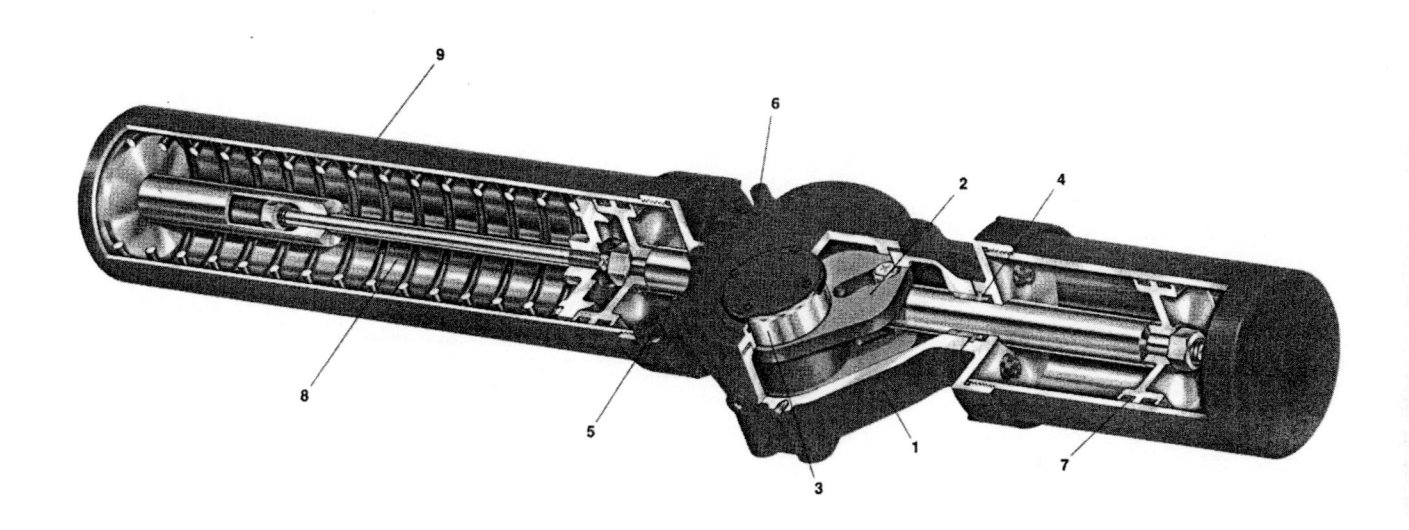

1. Housing
2. Scotch-Yoke Mechanism
3. Journal Bearings
4. Dual Rod Bushings
5. Bi-directional Travel Stops
6. Vented Fill Plug
7. Piston Seal(s)
8. Xylan Cylinder Coating
9. Spring Cartridge

FAIL SAFE VALVE ACTUATOR
(Reproduced with kind permission of Bettis Actuators and Controls)

Chapter Four

PIPING SYSTEMS AND PROCESS PRESSURE VESSELS

PART 1.   PIPING SYSTEMS

    1.   Hydrocarbon Process

    2.   Utility

PART 2.   PROCESS PRESSURE VESSELS

    1.   Gas/oil/water Separators

    2.   Gas/liquid Separators

    3.   Oil/water Separators

    4.   Holding tanks

PART 3.   PIPING AND PRESSURE VESSEL DESIGN

    1.   Piping Systems – Installation and Layout

    2.   Piping Systems – Design

    3.   Pressure vessels – Design

PART 4.   PIPING SYSTEMS — CONSTRUCTION

    1.   Pipe

    2.   Fittings

    3.   Valves

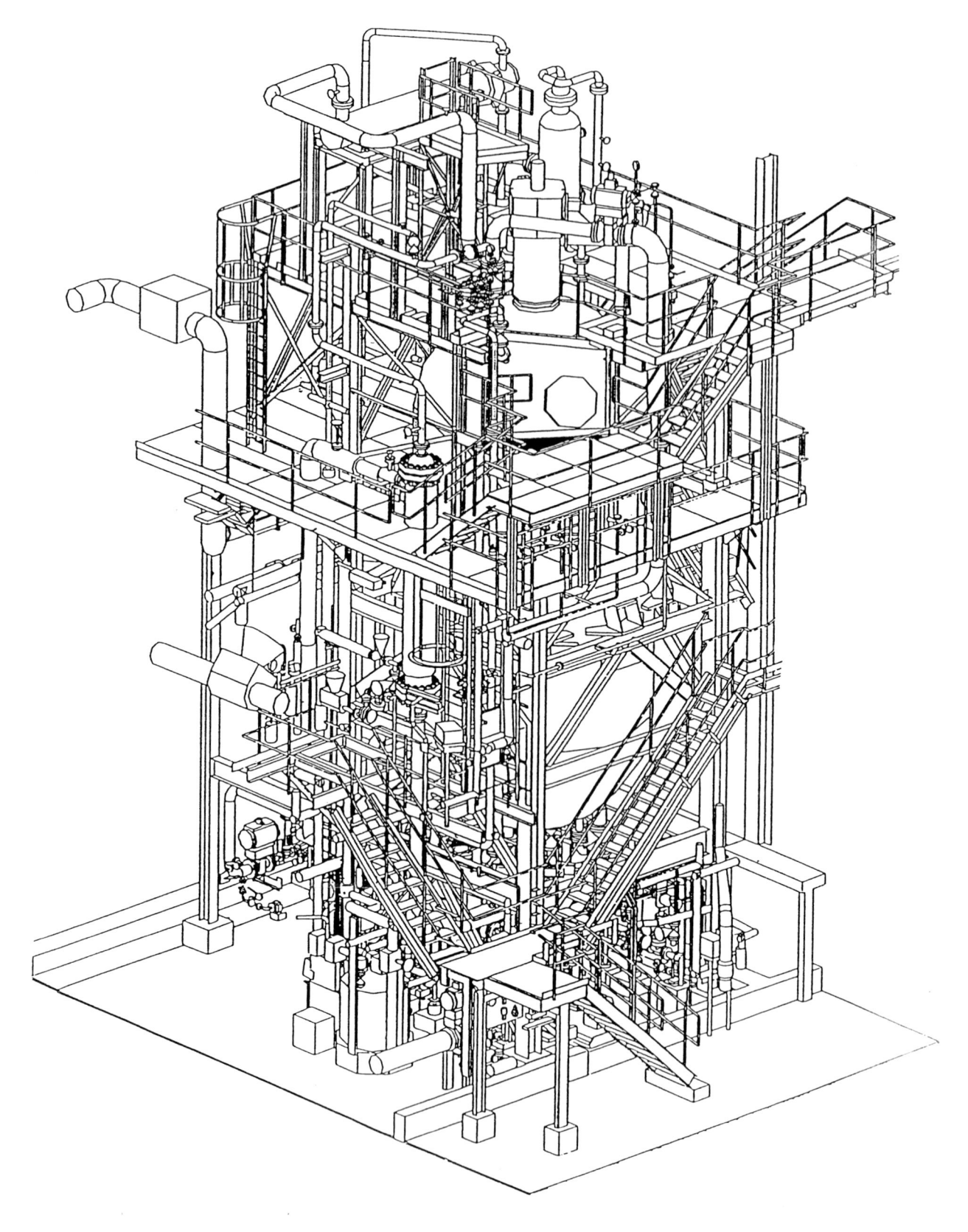

COMPUTER GENERATED DRAWING OF PROCESS PRESSURE VESSEL AND PIPEWORK
(Reproduced with permission of Kraerner Cape Surveys)

Part 1. PIPING SYSTEMS

The piping systems on an offshore installation may be divided into two basic categories:

1. Hydrocarbon process
2. Utility

All pipework should be colour coded to assist in system identification and the coding system particulars and P&IDs (piping and instrumentation diagrams) should be available for review as supporting documents to the platform Safety Case or the Statutory Operations Manual.

A brief explanation as to the function of each system will now be given.

1. HYDROCARBON PROCESS

i) GAS/OIL PROCESS

The main objective of an offshore installation is to separate hydrocarbon products into liquid or gaseous forms and to remove any impurities that will inhibit transportation. A more detailed account of the gas and oil production processes can be found in chapter five.

ii) CONDENSATE

Condensate is a clear highly volatile liquid produced as a by-product of the gas production process. The condensate system is primarily concerned with the removal of water prior to the injection of the condensate into the gas sub-sea pipeline for transportation to the onshore reception facility.

iii) FUEL GAS

Electrical power on a large offshore installation is generated by either gas turbine driven, or gas engine driven alternators. Both systems run on process gas as do the glycol reboilers. The fuel gas is taken either from the main outlet header, or produced as a by-product of the various refinement processes and is subject to further separation, drying and filtration before entering the fuel gas main.

iv) POWER GAS

Older installations frequently utilise process gas to provide the motive power for the actuation of valves. New installations tend to favour hydraulics or compressed air systems which create fewer restrictions on hazardous area boundaries.

v) VENT SYSTEM

Vent systems are generally associated with gas production installations and both high (HP) and low (LP) pressure systems are employed.

   a) The HP vent system is used primarily to depressure process pressure vessels during an emergency shutdown (ESD) situation, or to relieve excess vessel pressure via pressure safety valves (PSV).

The system consists of a large diameter ring main, a knock out (KO) drum or separator vessel designed to remove entrained liquids, and a vertical pipe or vent stack that will release the gas to the atmosphere at a safe location above the installation. The vent system is continuously purged with gas to ensure a rich, and thus non-explosive atmosphere prevails, and it is protected against fire by a dry powder, carbon dioxide, or halon equipped fixed fire fighting system.

    b)   The LP vent system is virtually a duplicate of the HP vent system designed to operate at lower pressures. The system provides a route for the release of gases from PSVs and atmospheric vents.

vi) FLARE SYSTEM

The flare system fulfils a similar function to the vent system, the obvious difference being that the gas is ignited as it leaves the end of the flare boom.

The flare system will only be encountered on oil production installations where the quantities of gas requiring disposal are often considerable and would constitute a hazard if simply released into the atmosphere through an open vent.

vii) CORROSION INHIBITOR

The use of a corrosion inhibitor provides some protection to the internal surfaces of carbon steel pipework. The inhibitor, frequently Cronox 638S is mixed with methanol or MEG (20%/80%) and may be injected into wells, process plant and sub-sea pipelines.

viii) HYDRATE INHIBITOR

Both *methanol* and *monoethylene glycol* (MEG) are used extensively by the offshore industry to combat *hydrate* formation. They may be injected into the process plant on an occasional basis during periods of peak gas production, or continuously metered into the sub-sea pipeline, a practise frequently encountered on small satellite platforms devoid of water separation equipment. The inhibitors are eventually recovered from the *wet gas* at the onshore reception facility or the mother platform.

When employed on an occasional basis stocks can be replenished (bunkered) by supply boat but the quantity consumed by continuous pipeline metering necessitates the installation of a small diameter (3 inch/75 mm) sub-sea pipeline which is attached to (piggy-backed) the main gas export line so that inhibitor can be supplied direct from the onshore reception facility or the mother platform. Prior to use the inhibitors are stored in atmospheric tanks, the methanol tanks frequently employing a blanket of fuel gas to reduce the air space and thus the flammability.

    a)   **Methanol.** Flash point $16°C$ ($60°F$). Methanol is without doubt the most effective of the hydrate inhibitors and it is the only inhibitor capable of melting a hydrate, should one occur. For this reason reserves of methanol may be encountered on installations which rely primarily on MEG for routine hydrate inhibition. The effectiveness of methanol as an inhibitor is due primarily to its willingness to vaporise, for to be effective an inhibitor must be delivered to the location where condensation of water is likely to occur. Unfortunately this characteristic hinders the separation of the methanol at the onshore reception facility and up to 50% of the methanol is unrecoverable. Heavy methanol consumption can prove both expensive and upsetting to the effectiveness of subsequent items of process equipment.

PIPING SYSTEMS AND PROCESS PRESSURE VESSELS

b) **Monoethylene glycol (MEG).** Flash point 116°C (266°F), boiling point 197°C (388°F), freezing point −13°C (9°F). Monoethylene glycol is hygroscopic, that is it has a strong affinity for water. Up to 90% of the MEG injected into the gas stream can be recovered at the onshore reception facility and as such it makes an efficient and economic hydrate inhibitor.

c) **Triethylene glycol (TEG).** Flash point 160°C (320°F), boiling point 288°C (550°F), freezing point −7°C (20°F). Triethylene glycol is not generally employed as a hydrate inhibitor being more suited as a drying agent in a dehydration process where it is chosen in preference to other glycols because of its ability to withstand the elevated temperatures associated with the re-generation process.

The subjects of hydrates and hydrate formation are discussed more fully in chapter five.

2. UTILITY SYSTEMS

i) FIREWATER

The firewater system can be extremely complex, particularly on large installations and the subject is discussed in greater detail in chapter three.

ii) SEAWATER SERVICE

In its simplest form the seawater service system consists of a small diameter (4 inch/100 mm) ring main which is maintained at a constant pressure by an electrically powered deepwell submergible pump. The water is used primarily for wash down purposes and for the flushing of domestic toilets but it may also be used to maintain the firewater ring main at a constant pressure of approximately 100 psi (7 bar) to insure an instantaneous supply of water prior to the activation of a fire pump.

The general service seawater system associated with a large oil production installation is considerably more complex than that previously described. It is often the largest single piping system on the platform, supplying water to the utility and production process coolers and also to the enhanced oil recovery plant.

iii) SODIUM HYPOCHLORITE

Sodium hypochlorite solution is frequently injected into seawater pump suctions to combat the growth of marine organisms and the algae that contribute towards the fouling of filters and pipelines.

The sodium hypochlorite is produced by the electrolysis of seawater within a hypochlorite generation cell, the seawater being forced to flow between two concentric titanium tubes which are connected to a DC (direct current) power supply. The resulting sodium hypochlorite solution is extremely corrosive and must be stored in stainless steel or GRP (glass reinforced plastic) tanks prior to use. The electrolysis of seawater also produces hydrogen gas but the quantities generated are insufficient to constitute a hazard provided that the storage tanks are located in a well ventilated open deck area.

iv) POTABLE WATER

Potable water for domestic consumption is stored in fresh water tanks and supplied to the accommodation via a pressurised surge vessel. Fresh water may be bunkered from supply boats, or produced from seawater by desalination, or a *reverse osmosis* plant.

v) COMPRESSED AIR

Compressed air is normally supplied from large air receivers which are maintained at a constant pressure of approximately 100 psi (7 bar) by multi-stage reciprocating air compressors. The discharge manifold connecting the air receivers divides into two separate systems.

- **a) Instrument air.** It is most important that the vast array of delicate instruments employed on a modern offshore installation are provided with a supply of clean, dry air if they are to function reliably. Consequently, air drawn from the main air receivers passes through an electrically heated desiccant drying medium and fine filtration package prior to entry into the instrument air main.

- **b) Utility air.** Utility air is provided for various operations such as the starting of diesel engines and pressurising of water distribution vessels. Hand tools, pumps and other items of portable equipment can be operated from hose stations located throughout the installations.

vi) DIESEL FUEL

Cranes, lifeboats, standby and emergency generators are all powered by diesel engines and diesel consumption can be quite considerable, particularly during process shut-down periods when fuel gas is unavailable. The diesel oil is bunkered from supply boats and may be stored in purpose built tanks or in structural void spaces such as crane pedestals and jacket legs.

vii) HELIFUEL

A number of offshore installations provide helicopter refuelling facilities. The aviation spirit is delivered and stored in purpose built fuel tanks by supply vessels.

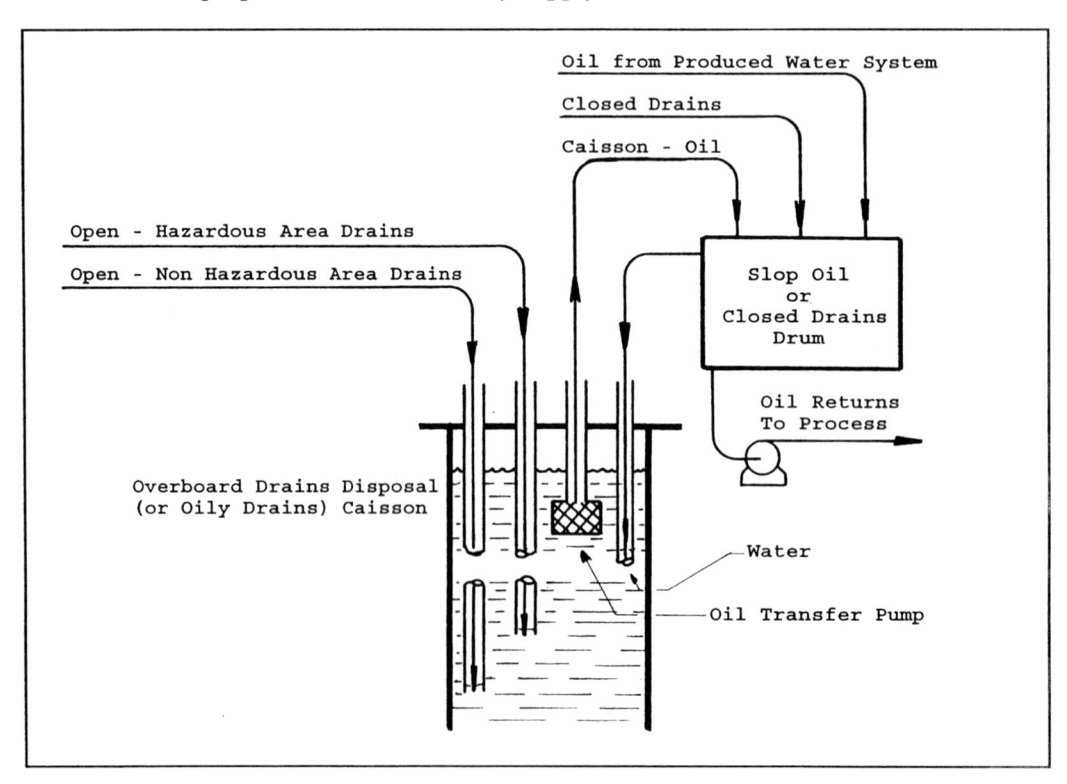

DRAINS SYSTEMS — Older Installations

PIPING SYSTEMS AND PROCESS PRESSURE VESSELS

viii) DRAINS

The design of a drainage system on an offshore installation is yet another area which tends to reflect individual company philosophy and the lack of standardisation often leads to confusion. However, the basic function of the system is to ensure that pollutants are removed from waste waters prior to their discharge into the sea.

The drainage system may be divided into three categories. Open drains, closed drains and the sewage disposal system.

a) **Open drains – Non-hazardous.** A less confusing title would be *safe area open drains* because the system simply provides for the removal of surface deck water from locations designated as safe areas. The waters are relatively free of contaminants and are piped directly into the overboard drains disposal caisson, a discharge pipe extending into the sea a distance of approximately 25 metres (90 feet).

Open drains – Hazardous. In practice, these drains rarely constitute a hazard but they are named as such because they originate from within locations designated as hazardous areas. They are used primarily for the disposal of water wash, rain and deluge waters and one or other of two systems may be used, depending on the age of the installation.

Generally speaking, the hazardous areas open drains on installations constructed prior to 1985 are discharged directly into the *overboard drains disposal caisson* alongside the non hazardous open drains. In order to reduce the possibility of contaminating safe areas with potentially hazardous substances the hazardous area drains discharge pipe is terminated above the non-hazardous area drain discharge pipe, both being submerged in sea water. On newer installations the hazardous open deck drains are collected in a dedicated *hazardous area open deck drains sump tank*, thus entirely eliminating the possibility of contaminating safe and hazardous area drains. The hazardous area open drains sump tank acts purely as a holding vessel, the contents being transferred to the *oily water separator* (OWS) where the differences in the specific gravity of the oil and water are used to effect separation. The water gravitates into the open drains caisson whilst the oil is pumped into the gas export line or returned to the process plant.

The open drains disposal caisson provides a final opportunity for any remaining condensate/oil to separate from the water. The oil floats to the top of the caisson where it can be transferred into the *slop oil tank* or the *oily water separator* by a pump which operates automatically, activated by an oil/water interface sensor. An oil content of less than 40 ppm is required before the water can be discharged into the sea.

b) **Closed drains –** The *closed drains system* provides the means by which a process vessel may be manually emptied and should not be confused with the *produced water* system which provides for the automatic removal of liquids separated from the gas/oil production process. The vessel drains are used on an occasional basis whilst the produced water system is in constant use and may be used to dispose of quantities of water varying from several tons to several thousand tons a day.

The closed drains system consists of a number of separate collection headers (pipelines) which are graded in accordance with the pressure ratings of the pressure vessels that they serve, the headers being protected against over pressure by restrictive orifice plates located in the pressure vessel outlets. The headers terminate within the slop oil tank, or a dedicated closed drain drum, depending on the design of the system.

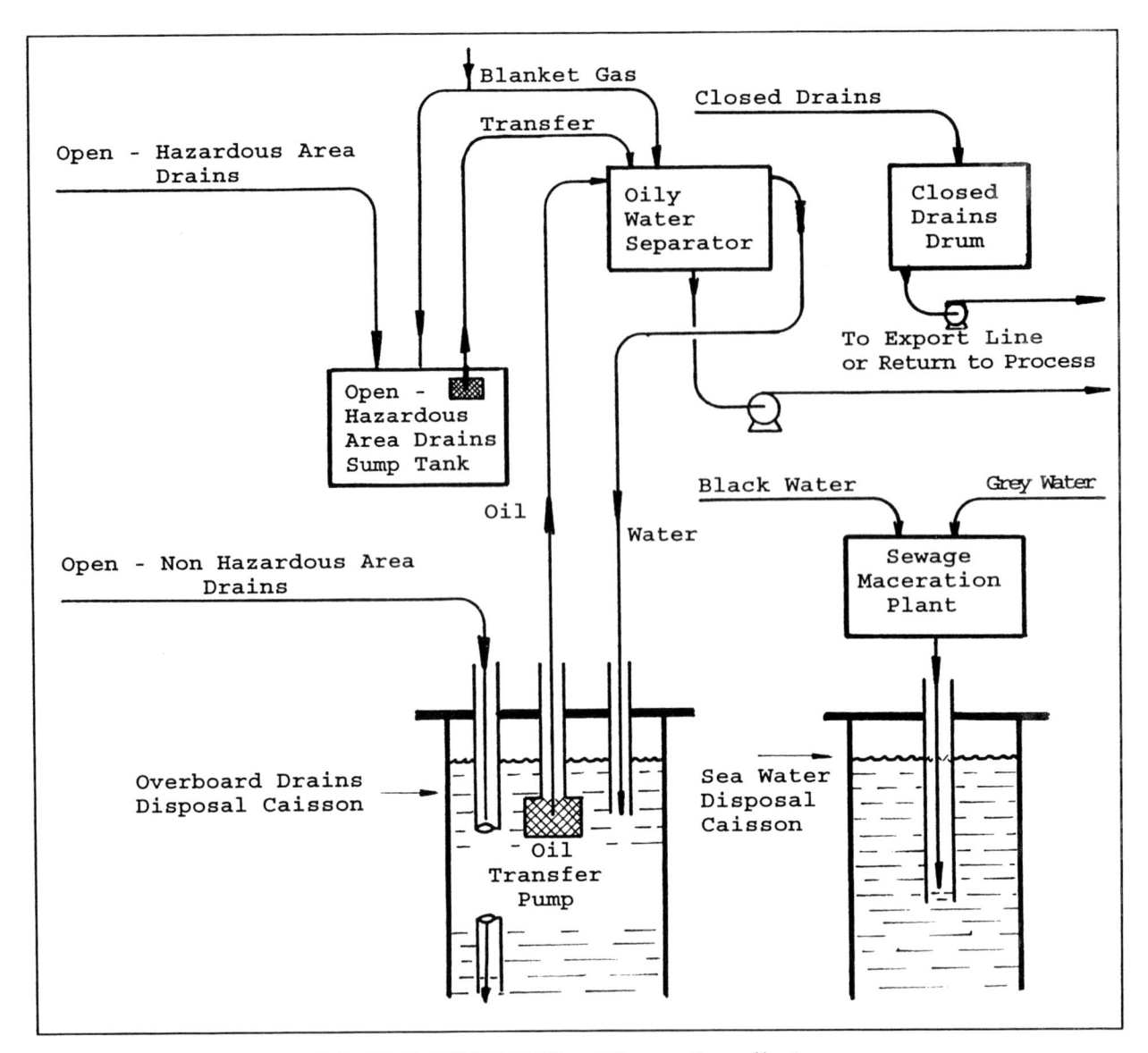

DRAINS SYSTEMS — Newer Installations

The slop oil tank featured on older installations frequently permits a degree of water/oil separation with the water gravitating to the overboard drain disposal caisson whilst oil products are disposed of in the same manner as the contents of the closed drains drum, that is pumped into the export line or back into the process system for re-circulation.

c) **Sewage system** — The sewage system can be divided into *grey water* which consists of drainage water from showers and sinks, and *black water* or raw sewage. Both systems combine prior to their entry into the sewage maceration plant and subsequent discharge into the seawater disposal caisson.

Part 2. PROCESS PRESSURE VESSELS

The vast majority of process vessels employed on an offshore installation are remarkably simple in design and rely on very basic principles of fluid dynamics to effect the separation of oil (or condensate), gas, water and sand.

Generally speaking the vessels are common to both oil and gas production installations and with the exception of the specialist dehydration, NGL removal, stabilisation and sweetening system vessels, they fall into four main categories:-

1. Gas/oil(condensate)/water separators

    i) Production separator
    ii) Test separator
    iii) Condensate separator
    iv) Flash/degassing/vacuum separations

2. Gas/liquid separators

    i) Slug catcher
    ii) Scrubber/suction drum
    iii) Knock out drum
    iv) Suction boot

3. Oil/water separators

    i) Oily water separator
    ii) Flotation unit
    iii) Skimmer
    iv) Coalescer
    v) Hydrocyclones

4. Holding tanks.

1. GAS/OIL(CONDENSATE)/WATER SEPARATORS

The production separators lie at the very heart of the oil and gas production processes and are virtually identical to each other in design and to the other separators included in this category.

i) a) **Production separator-gas process**

The production separator (gas) is primarily concerned with the removal of liquids (water and condensate) from the gas stream and their subsequent separation prior to discharge from the vessel. The process gas entering the separator is forced to undergo a sharp change in direction and the momentum possessed by the entrained liquids ensures that they are thrown out of the gas stream and into the base of the vessel. Given time, the separation of the two liquids will proceed naturally due to the variation in their specific gravities. The liquid levels are monitored by an oil/water interface sensor to ensure that the water is maintained at a level which permits the flotation of oil over a weir located in the end of the vessel.

Offshore Engineering

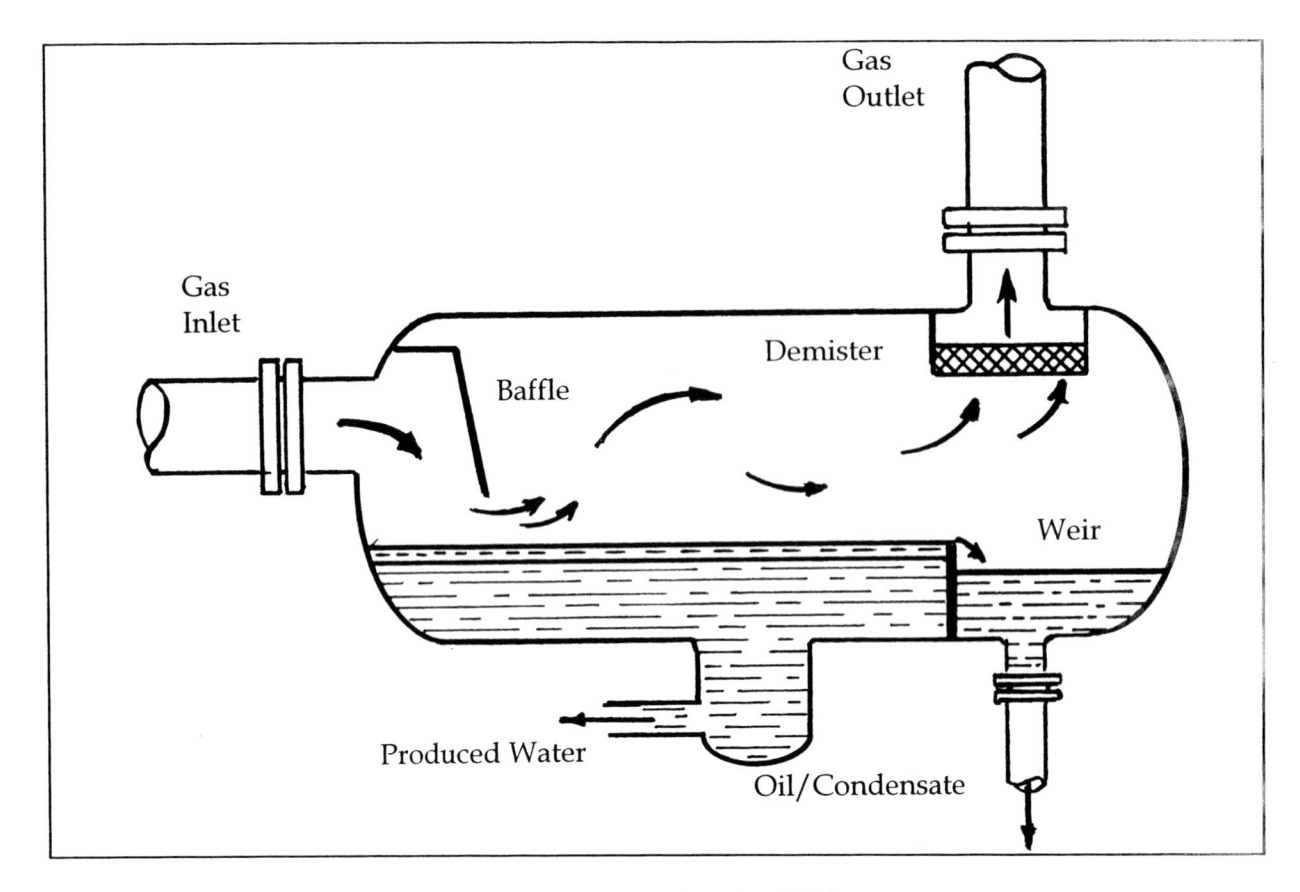

PRODUCTION SEPARATOR

b) Production separator-oil process

The production separator (oil) is primarily concerned with the separation of gas and water from the crude oil. As the crude oil enters the separator the reduction in pressure facilitates the release of dissolved gases which are removed from the top of the vessel for further processing. The separation of oil from water relies on the variation in specific gravities of the two liquids in a process identical to that employed by the gas process production separator.

ii) TEST SEPARATOR

The test separators are identical to the production separators in all but size being designed to process the contents of only one well at a time. A manifold arrangement provides the means by which each individual well can be routed into the separator whilst metering devices located in the vessel discharge nozzles permit the measurement of oil/condensate, gas and water. The results obtained provide an indication as to the performance of a particular well, and to the condition of the reservoir.

iii) CONDENSATE SEPARATOR

The condensate separator is identical in operation to the gas process production separator and facilitates the removal of natural gas liquids (NGL) from the process gas stream.

iv) FLASH/DEGASSING/VACUUM SEPARATORS

These vessels are all smaller versions of the oil process production separator which operate at, near or below atmospheric pressure.

PIPING SYSTEMS AND PROCESS PRESSURE VESSELS

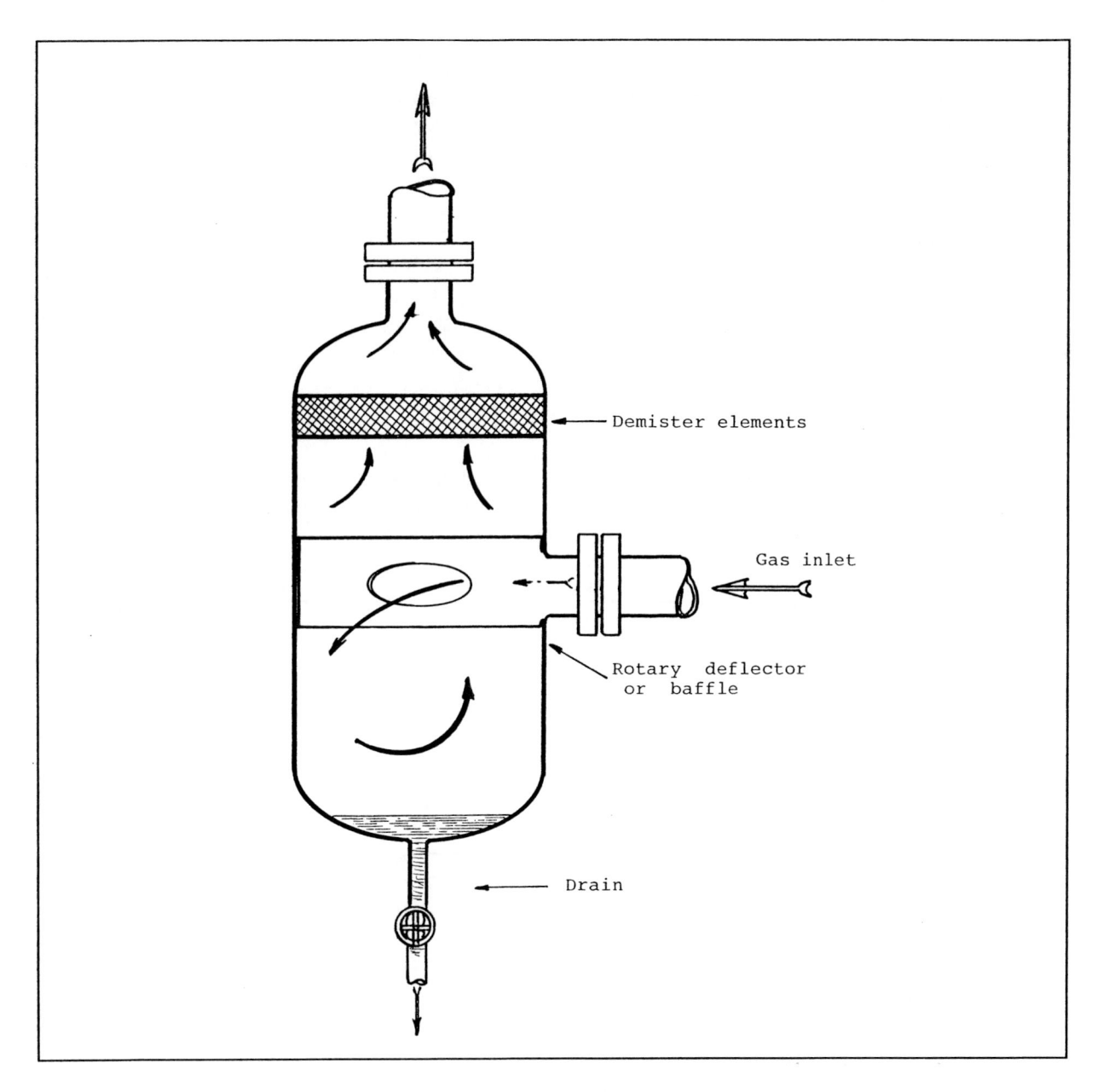

GAS SCRUBBER OR SUCTION DRUM

2. GAS/LIQUID SEPARATORS

The vessels in this category are concerned primarily with the removal of liquids from a gas stream and employ a primitive baffle or diffuser to divert the gas and eject the heavy liquid phases. No attempt is made to separate the liquids thus removed, they are simply discharged into the produced water system or routed back for further processing.

i) SLUG CATCHER

The slug catcher will be found on the end of sub-sea pipelines and on gas production installations where the reservoir generates large quantities of water and/or sand. It prevents slugs (surges) of water or sand flooding the gas process plant and upsetting the more delicate items of equipment.

Offshore Engineering

ii) SCRUBBERS/SUCTION DRUMS

Essentially identical to one another, the scrubber is located after a heat exchanger whilst a suction drum precedes a compressor. They are a more refined version of the slug catcher operating on identical principles but further employing demister pads (metal mesh screens) more suited to the removal of condensed liquid droplets rather than large liquid slugs.

iii) KNOCK OUT DRUMS (OR POT)

The K.O. drum is identical to the slug catcher but is employed at the end of vent headers (manifolds) to remove hydrocarbon liquids that could otherwise damage the vent stack or constitute a hazard if released into the atmosphere.

iv) SUCTION BOOT

The suction boots are normally associated with gas compressor suction pipelines and can best be explained with the aid of a sketch. The vertical suction pipe is extended downwards so that moisture deposited during the ascent of the gas into the compressor can be safely removed from the system.

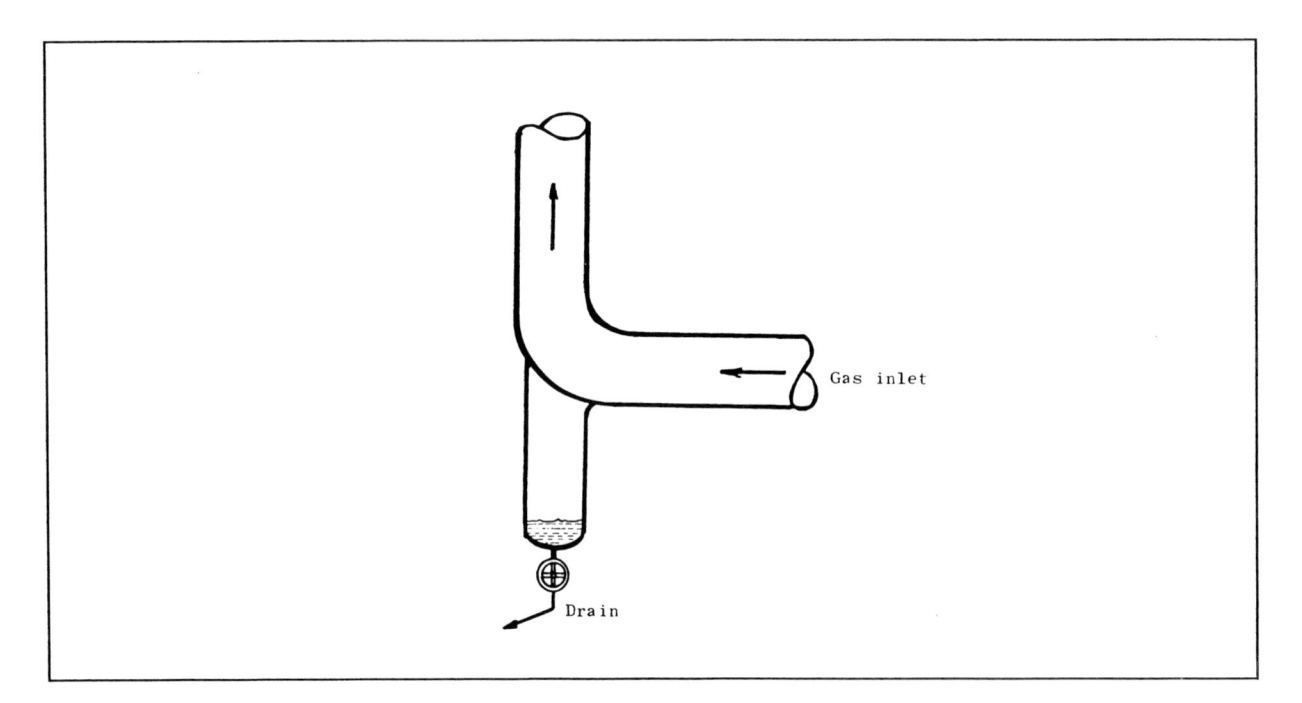

SUCTION BOOT

3. OIL/WATER SEPARATORS

The vessels which fall into this category are used primarily for the separation of oil or condensate from water and with the exception of the coalescer, are associated with the produced water and drains systems. The vessels rely on the variation in the specific gravities of the two liquids to effect separation and again with the exception of the coalescer (and the hydrocyclone), they operate at atmospheric or near atmospheric pressure.

PIPING SYSTEMS AND PROCESS PRESSURE VESSELS

i) OILY WATER SEPARATORS

The sketch is self explanatory, the various baffles channel the separated water to the outlet connection whilst oil is *skimmed* from the surface and drained into a holding tank.

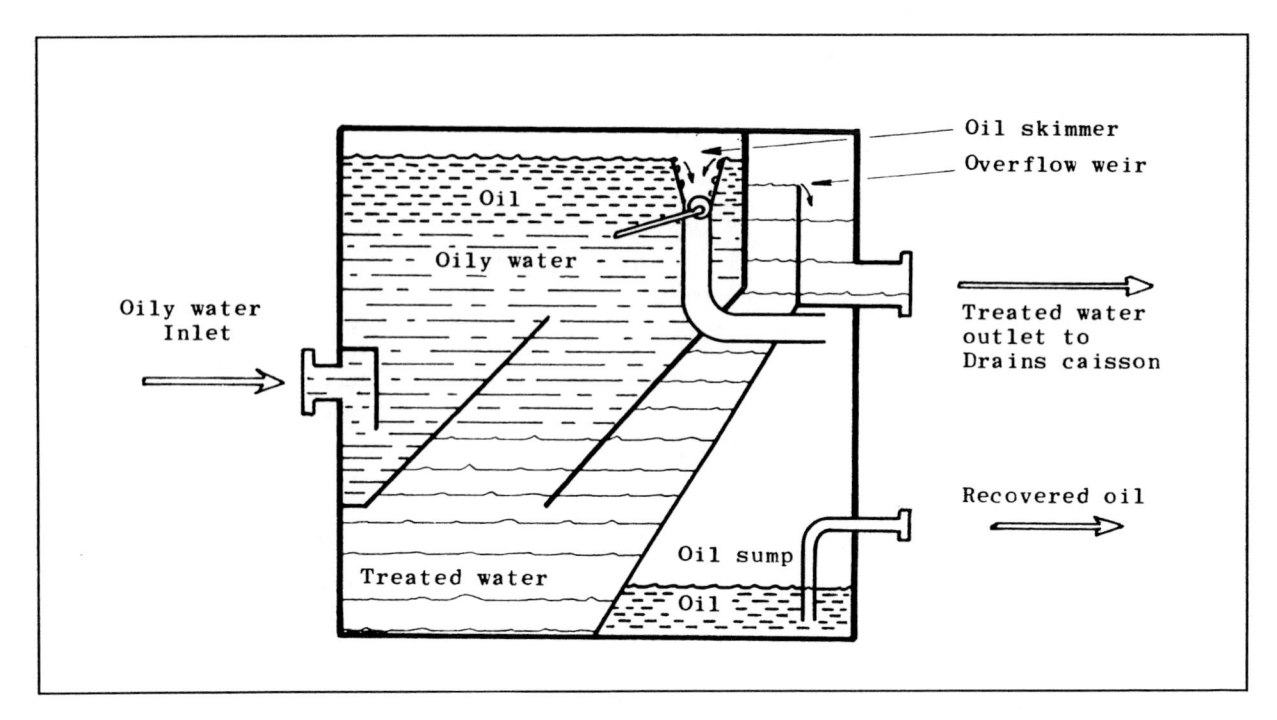

OILY WATER SEPARATOR (OWS)

ii) FLOTATION UNIT

The operation of the flotation unit is basically identical to that of the OWS.

iii) SKIMMER

The flotation unit and skimmer are essentially variations on the same theme. Some skimmers employ a separation weir whilst others are devoid of all internal equipment and rely on a level switch and an oil/water interface meter to effect separation of the two liquids via the control of drain valves.

iv) COALESCER

Condensate received by the coalescer should be relatively water free. The coalescer performs a final *polishing* operation on the condensate prior to it's injection into the sub-sea pipeline. The vessel operates at process pressure and contains metal screens, corrugated steel rolls or anthracite which are all designed to increase the surface area of the vessel and encourage entrained moisture to coalesce and separate from the condensate.

v) HYDROCYCLONES

The operating principles of a hydrocyclone differ from those of a conventional separator although they still rely on the variation in the specific gravities of the two liquids to effect separation. They can be configured to effect the separation of two liquids, or for the separation of solids from a liquid.

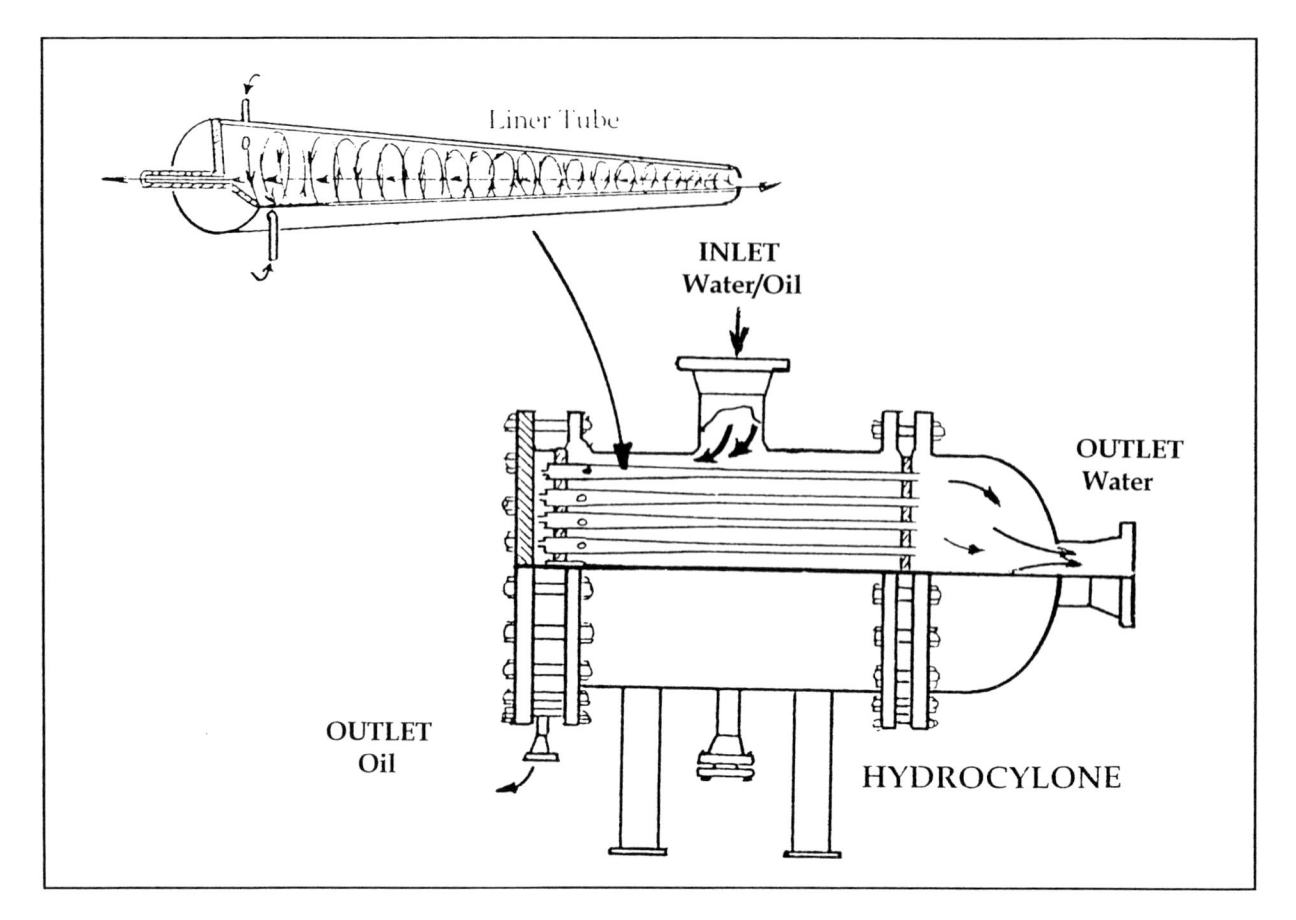

Hydrocylone

Hydrocyclones operate under system pressure, and utilize pressure drop as the primary source of energy for effecting the separation of oil from water. Within the vessel, a spinning motion is imparted to the water as it enters the tapered tubular liners through tangential ports and the velocity of the water is further accelerated by the tapered shape of the liner. The water, being heavier than oil, is thrown to the walls of the liner by centrifugal force and exits the vessel through the discharge nozzle whilst the oil forms a core at the axis of the column of water and is discharged from the inlet end of the vessel.

4. HOLDING TANKS

A number of holding tanks are incorporated in the design of the various piping systems and the *bad oil, drains sump* and *drains drums* are typical examples of the type. Generally speaking the tanks consist of atmospheric pressure vessels devoid of all internal equipment which are used as temporary storage for oil, water and condensate products prior to subsequent processing. The vessel contents are maintained between pre-determined limits by a transfer pump activated by a level sensing device.

Part 3. PIPING AND PRESSURE VESSEL DESIGN

The roots of the modern oil industry lie in the USA where the birth of the mass produced automobile gave rise to a phenomenal demand for gasoline. The ensuing black gold rush necessitated the development of equipment and working practices that reflected the specific requirements of the oil and gas industry. It is therefore not surprising that the standards, codes and recommended working practises produced by the Americans have been adopted by, or used as the basis of new standards by all countries now involved in the oil and gas industry.

There are a number of national standards organisations (NSO) and institutions that have developed specifications and codes of practise relating to the design, materials, fabrication, installation and testing of offshore pipeline systems and pressure vessels and these may be listed as:

1. BS British Standards Institute
2. IP Institute of Petroleum (UK)
3. API American Petroleum Institute
4. ANSI American National Standards Institute
5. ASME American Society of Mechanical Engineers
6. NACE National Association of Corrosion Engineers (USA)
7. ASTM American Society for Testing and Materials
8. MSS Manufacturers Standardisation Society (USA)
9. AISI American Iron and Steel Institute

Specifications may be loosely divided into three categories.

1. PIPING SYSTEMS — INSTALLATION AND LAYOUT

The American Petroleum Institute (API) produce a number of *Recommended Practises* (RP) which provide guidance in the design and layout of offshore piping systems. They are informative, easy to read and are used as the basis for industry standards world-wide. The more frequently used RPs are:

(i) API RP 2G, Recommended Practice for Production Facilities on Offshore Structures. (Now discontinued)

(ii) API RP 14E, Recommended Practice for Design and Installation of Offshore Production Platform Piping Systems.

(iii) API RP 1111, Recommended Practice for Design, Construction, Operation and Maintenance of Offshore Hydrocarbon Pipelines.

(iv) API RP 14C, Recommended Practice for Analysis, Design, Installation and Testing of Basic Surface Safety Systems for Offshore Production Platforms.

(v) API RP 520, Recommended Practice for the Design and Installation of Pressure Relieving Systems in Refineries — Parts I and II.

(vi) API RP 521, Guide for Pressure and Depressuring Systems.

Offshore Engineering

2. PIPING SYSTEMS — DESIGN

In practise, we find that American standards predominate, regardless of the geographic location of the installations.

(i)	ANSI/ASME B31.3.	Chemical Plant and Petroleum Refinery Piping.
(ii)	NACE MR-01-75	Sulphide Stress Cracking Resistant Material for Oilfield Equipment.
(iii)	BS 4515	Welding of Steel Pipelines on Land and Offshore.
	BS 8010	Code of Practise for Pipelines.
(iv)	BS 3351	Specification for Piping Systems for Petroleum Refineries and Petrochemical Plants. (Withdrawn).

It would be true to say that virtually all hydrocarbon process pipework is manufactured in accordance with the requirements of ANSI/ASME B31.3. Reference to BS 3351 may occur on some older installations and it provides some very valuable guidance but this standard has since been withdrawn.

When fabricating pipework for the conveyance of *sour* oil or gas then the requirements of the NACE MR–01–75 must be complied with in addition to ANSI/ASME B31.3. Sour reservoir products contain sulphur compounds, notably *hydrogen sulphide* (H_2S) which is a highly toxic, corrosive gas with the odour of rotten eggs. The UK fields produce predominantly *sweet gas* but where hydrogen sulphide does occur, the requirements of NACE must be strictly adhered to, particularly in the selection of materials and in the approval of welding procedures.

Whilst it is permissible to use American Codes such as ANSI/ASME for process pipework on UK registered installations, the sub-sea pipelines come under the jurisdiction of the Pipeline Department of the *Health and Safety Executive* (HSE), and must be manufactured to the requirements specified in BS 4515 and BS 8010.

3. PRESSURE VESSELS — DESIGN

Today, almost all new vessels destined for use on UK registered installations are manufactured in accordance with the requirements specified in BS 5500 although small air receivers may have been constructed in accordance with BS 5169. However, BS 5500 was not introduced until 1976 and consequently up to about 1980, the majority of pressure vessels were manufactured in accordance with the ASME Boiler and Pressure Vessel Code.

3.1 ASME BOILER AND PRESSURE VESSEL CODE

The ASME Code exists as a self-contained reference library which covers not only the design and manufacture of pressure vessels (Section VIII), but also the specifications for non-destructive examination (Section V), and approval of welding procedures and welders (Section IX).

PIPING SYSTEMS AND PROCESS PRESSURE VESSELS

The ASME Code consists of 11 volumes or sections which cover all aspects of fired and unfired pressure vessel construction. They are:

(i) Power Boilers

(ii) Material Specifications
Part A — Ferrous Materials
Part B — Non-ferrous Materials
Part C — Welding Rods, Electrodes, and Filler Metals
Part D — Properties

(iii) Subsection NCA — General Requirements for Division 1 and Division 2

(iii) Division 1
Subsection NB — Class 1 Components
Subsection NC — Class 2 Components
Subsection ND — Class 3 Components
Subsection NE — Class MC Components
Subsection NF — Component Supports
Subsection NG — Core Support Structures
Appendices

(iii) Division 2 — Code for Concrete Reactor Vessels and Containments

(iv) Heating Boilers

(v) Non-destructive Examination

(vi) Recommended Rules for Care and Operation of Heating Boilers

(vii) Recommended Guidelines for the Care of Power Boilers

(viii) Pressure Vessels
Division 1
Division 2 — Alternative Rules

(ix) Welding and Brazing Qualifications

(x) Fibre-Reinforced Plastic Pressure Vessels

(xi) Rules for In-service Inspection of Nuclear Power Plant Components

The Sections covering NDE and welding qualification tests are referenced by numerous piping specifications, including ASME/ANSI B31.3.

3.2 BS 5500 UNFIRED FUSION WELDED PRESSURE VESSELS

The BS 5500 is not as self contained as the ASME Code and must be used in conjunction with a number of British Standards to obtain information pertaining to NDE and welding requirements. The main supporting documents are:

(i)	BS EN 287	Approval testing of welders for fusion welding. Replaces BS 4871[1]
(ii)	BS EN 288	Specification and approval of welding procedures for metallic materials. Replaces BS 4870[1]
(iii)	BS 2910[2]	Radiographic examination of fusion welded circumferential butt joints in steel pipes.
(iv)	BS 2600[2]	Radiographic examination of fusion welded butt joints in steel.
(v)	BS 3923[2]	Ultrasonic examination of welds.
(vi)	BS 4416[2]	Penetrant testing of welded or brazed joints in metals.
(vii)	BS 6072[2]	Methods for magnetic particle flaw detection.

Note 1
Whilst BS 4870 and 4871 were superseded in 1992, they were in use for over 20 years and will be encountered repeatedly, particularly as the new standards do not invalidate previous approvals.

Note 2
Refer to Appendix III for changes to these specifications.

Part 4. PIPING SYSTEMS – CONSTRUCTION

Offshore piping systems must be designed and constructed in accordance with a recognised fabrication specification and ANSI/ASME B31.3 *Chemical Plant and Petroleum Refinery Piping* tends to be used as the basis for the vast majority of process and utility piping systems. The specification stipulates parameters which cover the design, selection of materials, construction techniques, non-destructive testing (NDT) and hydrostatic pressure test requirements.

A piping system must be designed to comply with a specific pressure rating or class in accordance with the following table:

CLASS	WORKING PRESSURE		TEST PRESSURE	
	psig	bar	psig	bar
150	285	20	450	30
300	740	51	1125	78
400	990	68	1500	104
600	1480	102	2225	154
900	2220	152	3350	230
1500	3705	255	5575	383
2500	6170	423	9275	639

The figures quoted are for maximum non shock working pressures at temperatures between $-20°F$ ($-29°C$) and $250°F$ ($121°C$).

A piping system contains a selection of components categorised as either pipe, fittings or valves and these components are further categorised by the method of assembly. The various methods of assembly will be dealt with first.

i) THREADED

Threaded fittings are used extensively for low pressure utility systems and may be used for hydrocarbon service up to 2 inch (50 mm) nominal bore (NB).

ii) SOCKET WELDED

Socket welded components employ a fillet weld as the means of attachment and they may be used for utility pipework up to 3 inch (75 mm) NB and for hydrocarbon service up to 2 inch (50 mm) NB.

iii) BUTT WELDED

Butt weld fittings are preferred for critical service applications such as high pressure hydrocarbon process systems. A butt weld offers maximum security and permits the application of more searching NDT techniques.

The various component parts of a piping system will now be discussed.

Offshore Engineering

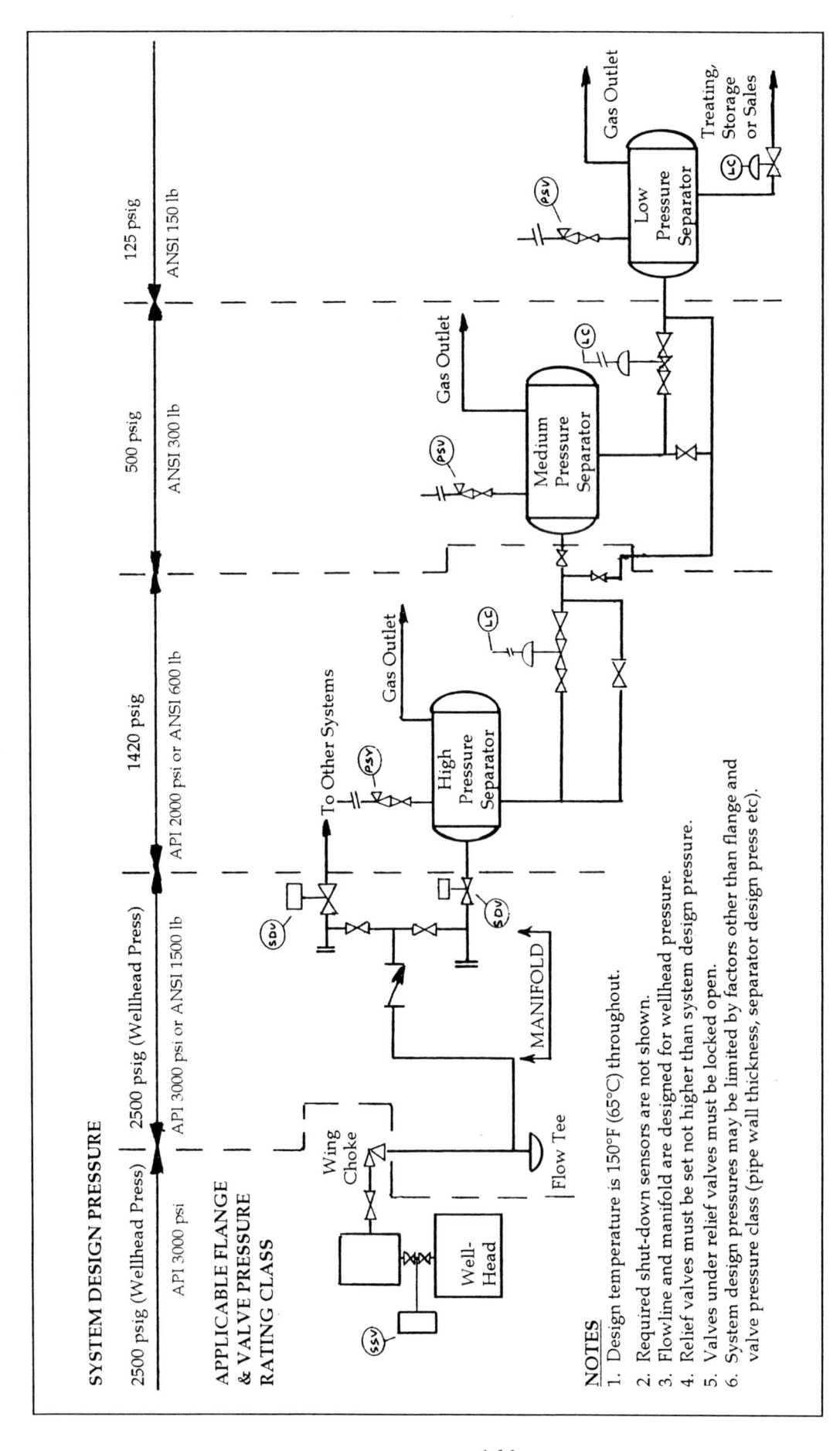

SYSTEM DESIGN PRESSURE

2500 psig (Wellhead Press)	2500 psig (Wellhead Press)	1420 psig	500 psig	125 psig

APPLICABLE FLANGE & VALVE PRESSURE RATING CLASS

API 3000 psi	API 3000 psi or ANSI 1500 lb	API 2000 psi or ANSI 600 lb	ANSI 300 lb	ANSI 150 lb

NOTES
1. Design temperature is 150°F (65°C) throughout.
2. Required shut-down sensors are not shown.
3. Flowline and manifold are designed for wellhead pressure.
4. Relief valves must be set not higher than system design pressure.
5. Valves under relief valves must be locked open.
6. System design pressures may be limited by factors other than flange and valve pressure class (pipe wall thickness, separator design press etc).

FROM API 14E – RECOMMENDED PRACTISE FOR THE DESIGN AND INSTALLATION OF OFFSHORE PRODUCTION PLATFORM PIPING SYSTEMS

1. PIPE

Prior to the selection of a pipe material specification the service conditions of the system must first be established. The service conditions primarily relate to operating pressure, temperature and the ability to resist corrosion. Approximately 95% of offshore piping systems are manufactured from Grade B pipe which complies with ASTM A106, or API 5L specifications. These and other relevant specifications will now be discussed.

i) ASTM A106 – Specification for Seamless Carbon Steel Pipe for High Temperature Service

The vast majority of utility and process piping systems are fabricated from pipe complying with this specification. Suitable for applications ranging in temperature from −29°C to 343°C (−20°F to 650°F) it is also used for systems operating within the normal operating temperature range of −29°C to 204°C (−20°F to 400 °F) and is available in sizes ranging from 2 inch (50 mm) NB to 24 inch (609 mm) NB.

ii) API Spec 5L – Specification for Line Pipe

Line pipe to specification 5L Grade B is generally used for the larger diameter pipeline systems operating within the ambient temperature range and is available in sizes up to 80 inch (2,032 mm) outside diameter. The Specification includes a range of high strength steel pipes such as API Spec 5L Grade X 52, a material that will be frequently encountered in sub-sea pipeline service. Due to the higher carbon content of these steels they may require specialised welding procedures and close supervision during production welding if sound welds are to be achieved.

Note: The numerical designation indicates the yield strength of the pipe

e.g. X 52 = 52,000 psi.
X 65 = 65,000 psi

iii) ASTM A333 – Specification for Seamless and Welded Steel Pipe for Low Temperature Service

There are a limited number of applications which require a pipe suitable for low temperature service and A333 is used extensively for high pressure vent systems because of the low temperatures generated by the gas as it expands during venting. The temperature range of the basic grade pipe is −46°C to 343°C (−50°F to 650°F) but grades are available for temperatures as low as −196°C (−320°F).

iv) ASTM A312 – Specification for Seamless and Welded Austenitic Stainless Steel Pipe

Stainless steels are used where resistance to corrosion or clinical cleanliness are required and typical service applications are chemical injection and heli-fuel distribution systems where grade 316L is used extensively.

v) ASTM A790 – Specification for Seamless and Welded Ferritic/Austenitic Stainless Steel Tube for General Service

The A790 range of stainless steels are normally referred to as *duplex* having a significantly higher chromium content then the A322 stainless steels. However, the addition of Nitrogen and Molybdenum greatly enhance the mechanical properties of the steel virtually doubling the yield strength and increasing the resistance to both corrosion and stress corrosion cracking. The price of the material increases in a similarly dramatic manner.

Offshore Engineering

The combination of high strength, erosion and corrosion resistance has made the duplex steels a popular choice for corrosive hydrocarbon service and they are being used increasingly for sub-sea pipelines and wellhead manifolds, particularly where sour reservoir products are encountered.

vi) CUNIFER

Cunifer is a 90/10 copper nickel alloy which is used extensively for the construction of fire mains and general service sea water systems due to its ability to resist the corrosive effect of sea water.

GENERAL

Having established the most suitable material for a particular piping system and the diameter required to produce a satisfactory flow rate, the wall thickness or pipe schedule required to withstand the design pressure and temperature can be determined from tables reproduced within the relevant specification.

It is recommended that a minimum schedule 80 pipe thickness be employed for hydrocarbon process systems of 3 inch (75 mm) NB or less in order to combat the effects of vibration and corrosion. The wall thickness quoted in the piping specification includes a corrosion allowance of 0.05 inch (0.2 mm).

All pipe should be clearly marked with the material specification and grade and supported by suitable documentation.

2. FITTINGS

The term *fittings* includes items such as flanges, elbows, bends, reducers and branches and they should be manufactured in accordance with one of the following specifications.

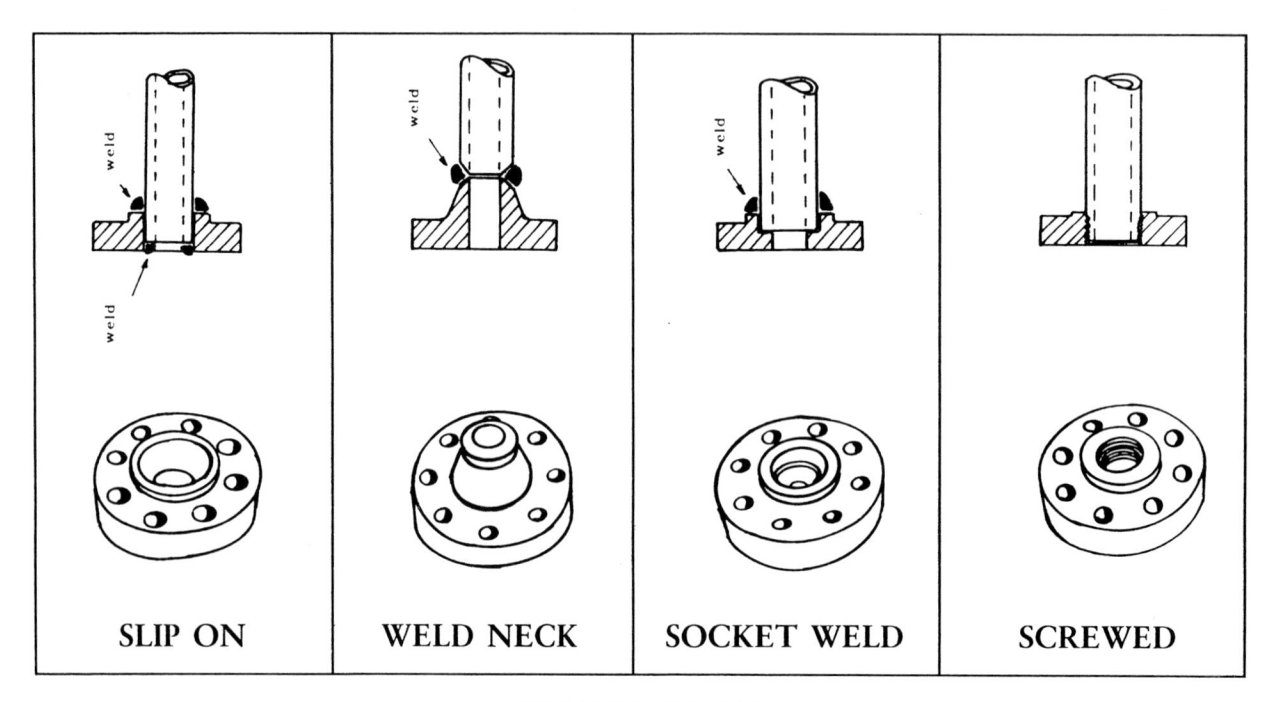

| SLIP ON | WELD NECK | SOCKET WELD | SCREWED |

FLANGE TYPES

PIPING SYSTEMS AND PROCESS PRESSURE VESSELS

i) **FLANGES**

a) **ANSI B16.5 – Steel Pipe Flanges and Flanged Fittings**

There are 7 flange ratings in the AISI/ASA series and they are used for virtually all utility and process services. These flanges are supplied with a raised face or profile (RP) sealing area up to and including AISI 400 rating whilst the higher pressure flanges are machined to accommodate ring joint (RJ) seals.

The 7 available ratings are:

AISI 150, 300, 400, 600, 900, 1500, 2500.

The numerical part of the designation indicates the maximum non-shock loading working pressure rating in psi at a temperature of 454°C (850°F).

The temperature range that most offshore piping systems are designed for is −29°C to 37°C (−20°F to 100°F) and within these temperatures the working pressures can be increased to the corresponding values shown in the Table at the beginning of the Chapter.

A rule of thumb method of calculating the ambient working pressure of an AISI flange or valve is to multiply the designated class rating by 2.4.
e.g. AISI 300 × 2.4 = 720 psi

Similarly the maximum test pressure can be approximated by multiplying by 3.5. e.g. AISI 300 × 3.5 = 1050 psi

Note: This rule does not apply to Class 150 components.

b) **API Spec 6A – Wellhead Equipment**

The API range of flanges are used for wellhead service and are supplied in five pressure ratings, the numerical designation indicating the maximum working pressure within the operating temperature range of −29°C to 121°C (−20°F to 250°F). The flanges are all of the ring groove type and are available in the following pressure ratings:

API 2000, 3000, 5000, 10000, 15000

c) **MSS SP-44 – Steel Pipe Line Flanges**

The *Manufacturers Standardisation Society* (MSS) developed the *Standard Practise* (SP) 44 due to the demand for flanges in sizes larger than the 24 inch diameter covered by ANSI B16.5. The Standard contains a selection of flanges varying in size from 12 inches to 38 inches and in pressure ratings corresponding to the customary ANSI classes, namely 150, 300, 400, 600 and 900.

d) **Compact Flanges**

During the last five years a number of specialised flange manufactures have developed what are referred to as *compact flanges* and these do not comply with any of the recognised flange specifications such as ANSI B16.5 or API 6A. They rely on a metal sleeve or bush seal rather than a ring joint or gasket and on spherical faced nuts to ensure accurate alignment.

Compact flanges are physically much smaller than the equivalent ANSI/API flanges and weigh up to 30% less and whilst more expensive to manufacture, the saving in material costs, particularly when duplex and titanium materials are used tends to redress the balance. The reduction in the size of the flanges has been achieved by taking advantage of computer aided design techniques and improved manufacturing methods to optimize the design.

The main advantage of compact flanges is their light weight and they have found favour in sub-sea applications where the reduced weight simplifies installation. They are also used extensively as riser flanges attached to *floating production storage and off-loading* vessels (FPSO) where the reduction in wave induced inertia forces on the flanges and considerably reduces loads on the risers.

e) Gaskets

Spiral wound asbestos gaskets are used as a sealing medium on AISI raised face (RP) flanges up to and including Class 400. The higher pressure rated flanges and all the API flanges employ ring joint seals. Soft iron rings are used for 600 and 900 class flanges and low carbon steel rings for all other ratings.

ii) FITTINGS – OTHER THAN FLANGES

The remaining fittings are covered by the following specifications which are self explanatory:

a) ANSI B16.9 — Wrought Steel Butt Welded Fittings.

b) ANSI B16.11 — Forged Steel Fittings, Socket Welded and Threaded.

Fittings intended for use with Grade B pipe (A106 or 5L) should be manufactured from materials which comply with:

i) ASTM A234 — Specification for Piping Fittings of Wrought Carbon Steel and Alloy Steel for Moderate and Elevated Temperatures.

or

ii) ASTM A105 — Specification for Forgings, Carbon Steel, for Piping Components

3. VALVES

The choice of valve for a particular piping application will be dictated by the service conditions of the piping system and the mode of operation required. A number of valve types are available and may be briefly described as follows:

i) BALL VALVE

The ball valve is normally associated with hydrocarbon service as it provides an efficient on/off operation. The valve is ideally suited to automation because of the short distance that the actuator has to travel to effect full operation.

PIPING SYSTEMS AND PROCESS PRESSURE VESSELS

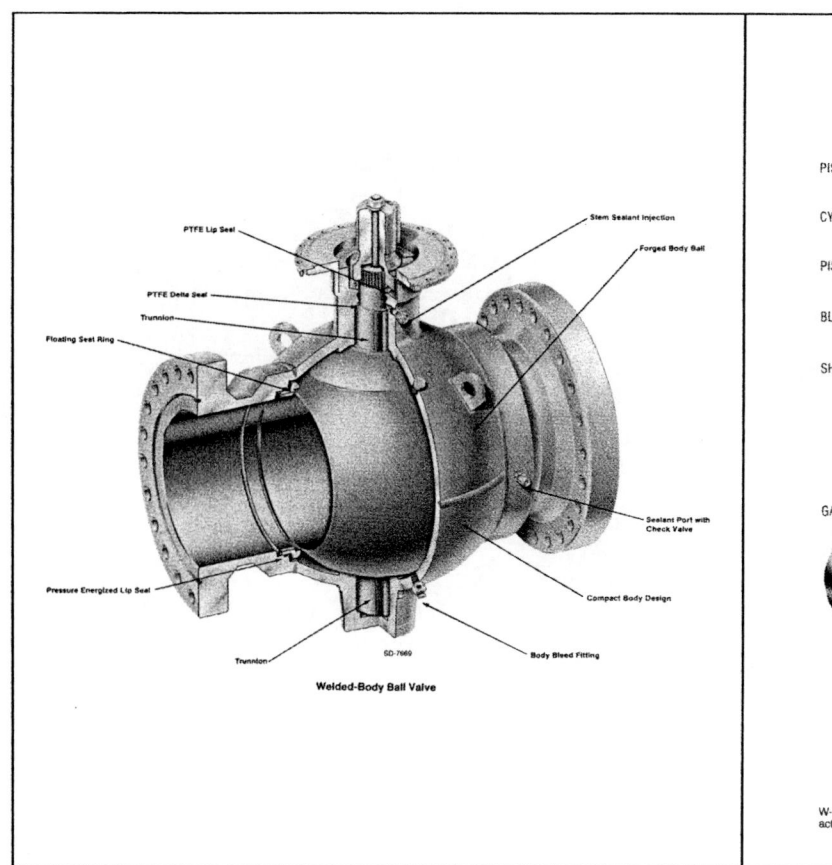

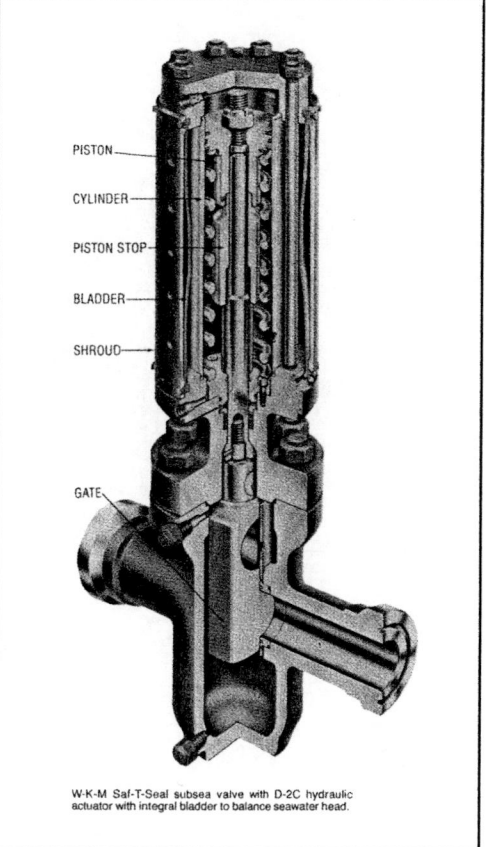

WELDED-BODY BALL VALVE   SUB-SEA GATE VALVE
AND ACTUATOR

(Reproduced with permission of Cooper Oil Tools, Houston Texas)

ii) GATE VALVE

Like the ball valve, the gate valve is used extensively in hydrocarbon service applications and in particular for Christmas tree valves. Whilst the valve seats are prone to damage they are relatively cheap and easy to replace.

iii) PLUG VALVE

The plug valves have a similar action to a ball valve and are widely used for hydrocarbon systems of less than 3 inches (75 mm) NB. They are particularly suitable for drain lines because of their quick and efficient sealing action.

iv) CHOKE VALVE

Manual, and automatic chokes are used extensively to regulate the flow of hydrocarbons in both process and drilling applications. On production installations automatic chokes are frequently fitted to the Christmas tree wing valves in order to reduce well pressures to that required by the process plant.

Offshore Engineering

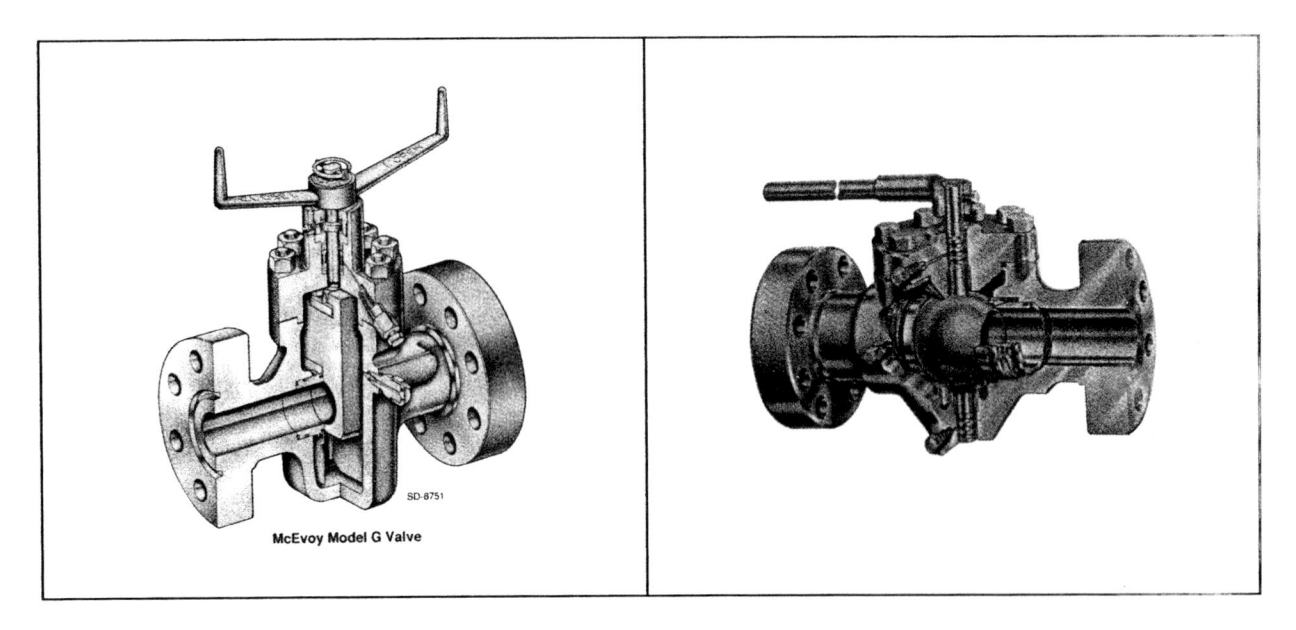

GATE VALVE

McEvoy Model G Valve

BALL VALVE

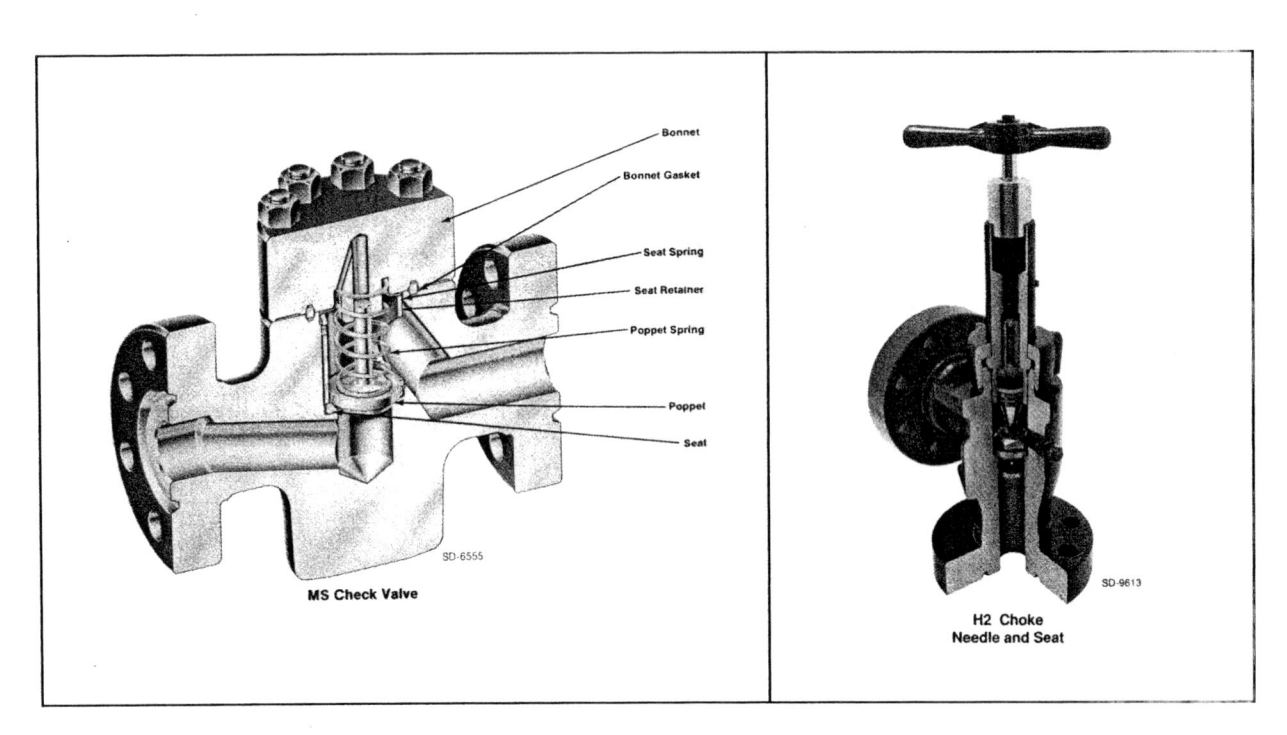

CHECK VALVE

MS Check Valve

- Bonnet
- Bonnet Gasket
- Seat Spring
- Seat Retainer
- Poppet Spring
- Poppet
- Seat

CHOKE

H2 Choke Needle and Seat

v) BUTTERFLY VALVE

Butterfly valves can be difficult to make leak tight but the inclusion of a neoprene sealing face greatly enhances the sealing effectiveness. They are used predominately as isolation valves on fire mains.

vi) DIAPHRAGM VALVE

The diaphragm valves tend to be restricted to low pressure sea water service where they provide an enduring efficient seal.

vii) CHECK VALVE

Check valves may also be described as reflux, non-return, back pressure, retaining and clack valves. The purpose of a check valve is to permit flow in one direction and to prevent it in the reverse direction and they are used extensively in both hydrocarbon and utility piping systems.

Valves intended for hydrocarbon process or firemain service must be type approved in accordance with a fire test such as API RP 6AF or BS 6755. Other relevant specifications pertaining to valves are:

a) STUD BOLTS

ASTM A193 – Specification for Alloy-Steel and Stainless Steel Bolting Materials for High-Temperature service.

ASTM A320 – Specification for Alloy Steel Bolting Materials for Low-Temperature Service.

b) NUTS

ASTM A194 — Specification for Carbon and Alloy Steel Nuts for Bolts for High-Pressure and High-Temperature Service.

Bolts and nuts should be protected against corrosion by cadmium plating, hot dip galvanizing or resin coatings.

The acid test of the finally completed piping system is the *hydrostatic pressure test*. This should be carried out at a pressure of 1.5 × design pressure and is intended to highlight any weaknesses in the structural integrity of the constituent components of the piping system. It should not be confused with a *leak test* which may be carried out at 1.1 × design pressure the object of which is to simply ensure there are no leaks from valves, gaskets, plugs and other temporary fittings.

Chapter Five

PRODUCTION

PART 1. GAS PRODUCTION

    1. The Gas Process

    2. The Liquid Process

PART 2. OIL PRODUCTION

    1. The Oil Process

    2. Associated Gas

PART 3. ENHANCED OIL RECOVERY AND THE OIL DRIVE MECHANISM

    1. The Oil Drive Mechanism

    2. Enhanced Oil Recovery

    3. Water Injection

    4. Gas Injection

    5. Gas Lift

    6. Deepwell/Submergible Pumps

OIL AND GAS PRODUCTION

The main function of the process plant on an offshore installation is to separate hydrocarbon products into liquid and gaseous phases and to remove any impurities that could inhibit transportation, such as water. The refinement process is loosely described as production.

Before embarking on the description of the oil and gas production processes, the definition of what constitutes an oil field and a gas field must be clearly understood.

Basically speaking the underground rock formations which create conditions conducive towards the formation of an oil field are somewhat different to those which constitute a gas field and for this reason offshore installations are generally categorised as either oil or gas producing. In the UK sector of the North Sea, all the gas fields are located in the southern sector where there is absolutely no oil. In the northern sector the fields are predominately oil producing but the oil is also associated with gas and water. Consequently, an oil production installation usually features a gas process plant in addition to the oil processing facility.

The production processes rely on very basic principles of fluid flow and thermo-dynamics and the reader would be justified in expecting to find a virtually standard design of offshore installation. In practice, this is certainly not the case for whilst to a certain extent the process equipment must be tailored to suit individual reservoir conditions, it is often Company policy which exerts the greatest influence on platform design, an aspect of offshore engineering which can be particularly frustrating. However, whilst equipment layout may vary from installation to installation, the general principles of operation remain the same.

In order to simplify the explanation of the oil and gas production processes they have been dealt with separately.

PRODUCTION

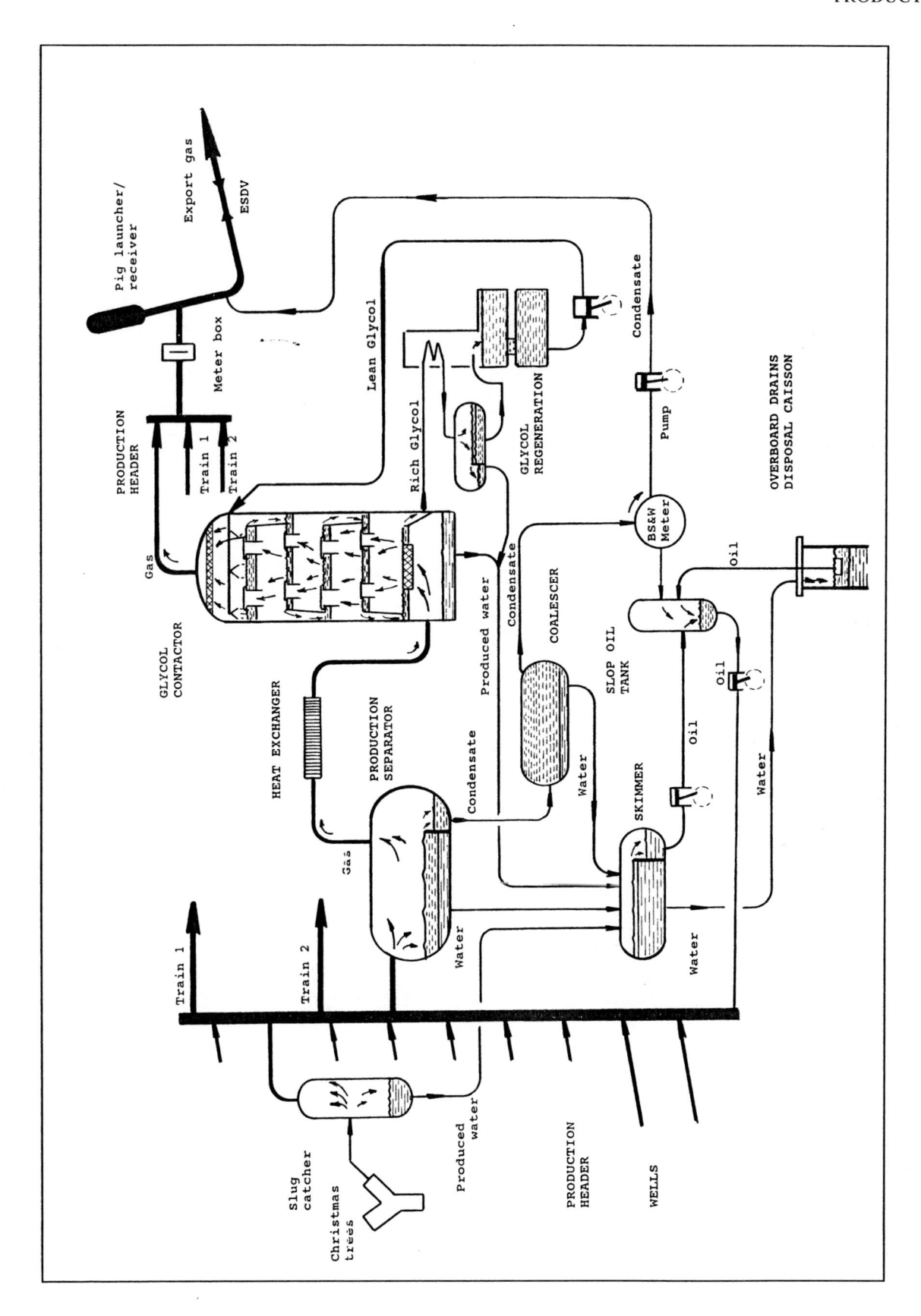

GAS PRODUCTION PROCESS

Part 1. GAS PRODUCTION

The basic function of the process equipment on a gas producing offshore installation is to remove water prior to the transportation of the gas to the onshore reception facility. The gas and liquid process arrangements will now be discussed.

1. THE GAS PROCESS

A sketch has been prepared to assist in the explanation of the gas production process. A more detailed description of the operating principles of the process vessels can be found in chapter four.

i) WELLHEADS

The gas process commences at the *Christmas tree*. This is the main isolation valve assembly through which wet gas at full reservoir pressure 2,000–3,000 psi (140–200 bar) must pass before entering the production flowline. From the flowline the gas enters a slug catcher where the first phase of liquid removal takes place.

ii) SLUG CATCHER

The slug catcher design facilitates the removal of large slugs of water and sand prior to the gas entering the production header.

In common with most gas process equipment, the slug catcher relies on very basic principles of fluid dynamics for its operation. The incoming gas is forced to make a sharp change of direction within the vessel and in so doing, the heavy liquid phases are thrown out into the base of the vessel. The liquids are disposed of by the *produced water system*.

iii) METER RUN AND TEST SEPARATOR

Having removed any large slugs of liquid or sand that could prove injurious to subsequent items of process equipment the gas may flow through a metering orifice box prior to entry into the production header. Individual metering of each well permits output to be regulated and provides an indication as to the performance of the well. When more specific information is required the well can be routed through a test separator which can measure the exact quantities of *gas, water* and *condensate* produced. Regular condition monitoring of the reservoir provides an early indication of any problems likely to effect gas production and permits corrective action to be taken.

iv) PRODUCTION HEADER

The production header is essentially a manifold which receives the gas from all the slug catchers (maybe 12) and redistributes it into a number of process trains (maybe 3). For reasons of clarity, only one process train is shown in the diagram.

Some installations do not require slug catchers so the production header receives gas direct from the Christmas tree flowlines.

Offshore Engineering

The next step in the liquid removal process takes place in the production separator. However, before the gas enters the production header it must be reduced in pressure to less than 1,440 psi (98 bar), the maximum operating pressure of the process train. The gas is reduced to this pressure due to the impracticality of constructing plant and equipment capable of operating at the full reservoir pressure.

Pressure reduction and the regulation of gas flow is effected by a control valve or choke which is normally fitted between the slug catcher and production header. In the absence of slug catchers the choke will be installed on the Christmas tree outlet or wing valve.

v) PRODUCTION SEPARATOR

The production separator facilitates gas and liquid separation in a similar fashion to the slug catcher. However, the liquids thrown out as the gas changes direction settle and separate in the base of the vessel. The water enters the produced water system whilst any condensate (condensed gas liquids) present is discharged in to the coalescer.

The gas continues its journey by entering a heat exchanger.

vi) HEAT EXCHANGER

The purpose of the heat exchanger is to produce gas at a temperature at which the water absorption process within the glycol contactor will be at its most efficient. The type of heat exchanger used will depend on reservoir conditions.

Some reservoirs produce gas at temperatures as high as 60°C (140°F) and under these conditions the gas will require cooling before it enters the contactor. This is normally carried out within a thermostatically controlled fan assisted *fin fan* tube type cooler. Cooling the gas also assists in the removal of liquids which condense out as the temperature falls.

Once the gas is at the correct temperature it can be passed into the glycol contactor for final drying.

vii) GLYCOL CONTACTOR (OR ABSORPTION TOWER)

This vessel may function as a combined scrubber and contactor, or the scrubber may exist as a separate vessel through which the gas passes prior to entering the contactor.

At this stage in the process the gas should be relatively dry. The scrubber will entrain any remaining water droplets within a series of fine metal mesh screens before the gas enters the contactor for the final mopping up operation.

The contactor consists of a vertical pressure vessel fitted with a number of horizontal trays over which *triethylene glycol* (TEG) cascades. As the gas flows upwards through the vessel it *contacts* the TEG which acts like liquid blotting paper absorbing any remaining moisture. The gas then passes through a mist extractor designed to remove any entrained glycol before entering the discharge header. The TEG leaves the base of the vessel for circulation through the glycol regeneration plant. (The dehydration process is described in greater detail under *oil production*.)

PRODUCTION

viii) DISCHARGE HEADER

The discharge header is essentially a manifold which receives gas from the various process trains and redirects it into the export line.

ix) EXPORT LINE

The export line usually contains a sphere launcher/receiver, a metering box, a condensate re-injection point and an emergency shut down valve (ESDV).

- (a) **Sphere (or pig) launcher/receiver** — Over a period of time, liquids, scale and debris will accumulate in the horizontal section of the sub-sea pipeline and eventually gas flow will be affected. To counter this problem large polypropylene spheres or pigs are periodically launched under gas pressure to sweep the line. The operation is referred to as *pigging the line*, the debris being pushed ahead of the pig into a slug catcher at the receiving installation or the reception facility on the beach.

    Modern technology has spawned a new addition to the pig family, the *intelligent pig*. These sophisticated animals contain a magnetic eddy current measuring device energised by an intrinsically safe power source and are employed to carry out an ultrasonic inspection of the sub-sea pipeline. The pig is propelled by gas (or oil) pressure and can quickly and efficiently detect any abnormalities in the condition of the pipeline.

- (b) **Condensate re-injection point** — Condensate is pumped into the export gas for delivery to the beach where it will again be separated for sale as a separate commodity to the gas.

- (c) **Metering box** — Gas leaving the installation enters a sub-sea pipeline which is invariably shared with other installations in the field. A metering box enables gas production to be calculated to ensure that an installation is credited with its due rewards.

- (d) **ESDV** — The emergency shut down valve consists of a fail-safe, fire-proof valve designed to isolate the platform from the sub-sea distribution system. This ensures that should an accident occur, the installation will not be fed with gas from other platforms. These valves were installed because this very situation arose on the *Piper Alpha* installation, a serious problem turning into a disaster as gas from other platforms fuelled Piper's fires.

To complete the section on gas processing a brief explanation of gas compression is required.

x) GAS COMPRESSION

Reservoir depletion eventually necessitates the installation of compressors to assist in the extraction of the gas (down to a pressure of less than 50 psi/3 bar) and to ensure that the gas is delivered to the national grid at a pressure of at least 1,000 psi (70 bar).

Frequently one platform in the field will be provided with a compression facility and it will receive the gas from various satellite installations to compress for onward transmission.

Offshore Engineering

Where space permits, a self contained compressor package may be located within the confines of the existing installation. However, the larger developments normally require an additional *compression jacket* to accommodate the compressor trains and associated process equipment. Compressor size and type will vary depending on field requirements but the larger installations tend to use rotating compressors powered by marinised aero engines manufactured by companies such as Rolls Royce. The gas engines run on *fuel gas* produced on the installation from the process gas.

2. THE LIQUID PROCESS

As previously stated, the basic function of the process equipment on an offshore installation is simply to remove water prior to the transportation of the gas to the onshore reception facility.

The bulk of the water is deposited in the *production separators* in a relatively straightforward operation but the process is complicated by the fact that the water is often accompanied by *condensate*, a valuable hydrocarbon by-product. In the production separators the condensate and oil separate by gravitation prior to their discharge through separate outlets. The condensate is directed to the coalescer where any remaining water is removed whilst the water enters the produced water system where any remaining oil is removed.

The systems employed for the recovery of the condensate and the purification of the water will now be discussed, after first explaining why it is imperative that the water is removed.

i) PREVENTION OF HYDRATES

Under certain conditions of temperature and pressure water particles separate from the gas stream and freeze, trapping hydrocarbon molecules to form a solid ice like substance known as a *hydrate*. Restrictions such as valve chests, orifice plates, pipeline reducers and bends exacerbate the problem as they create a throttling effect which further cools the gas and accelerates hydrate formation. If left unchecked the hydrates will eventually form a blockage at these restrictions and in extreme cases complete sections of the sub-sea pipeline have been known to freeze.

Methanol and *monoethylene glycol* (MEG) are used extensively by the offshore industry to depress the dew point of the gas stream and thus reduce the possibility of hydrate formation. They may be injected into the process plant, sub-sea pipeline and the wells (methanol only) during periods of peak production when conditions of temperature and pressure approach values known to produce hydrates. The choice of inhibitor is largely dictated by individual Company preference.

Should a hydrate blockage occur it can be left to thaw naturally or melted by injecting methanol. Prior to thaw the gas pressures either side of the blockage must be equalised if an explosive collapse of the hydrate is to be avoided. Propelled by gas pressure a hydrate can cause a tremendous amount of damage smashing valve chests, pipe bends and puncturing pressure vessels.

In time the natural reservoir pressure will decline and remove the conditions associated with hydrate formation. This will permit the removal of certain items of process plant such as the glycol regeneration system.

Further information on hydrate inhibitors can be found in chapter four.

PRODUCTION

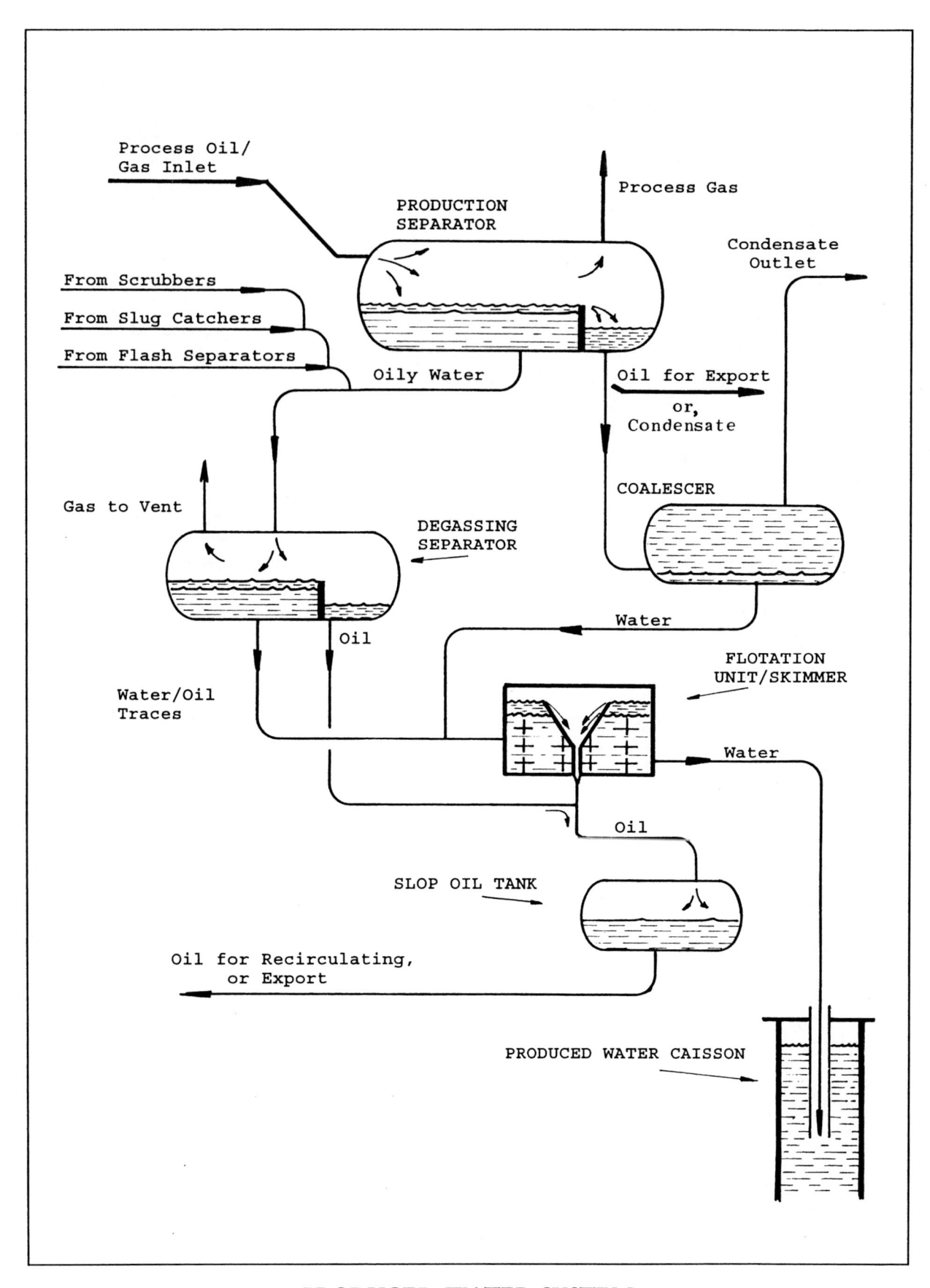

PRODUCED WATER SYSTEM

ii) PREVENTION OF CORROSION

Due to the colossal costs associated with the installation and replacement of sub-sea pipelines, every effort must be made to eliminate corrosion. The composition of natural gas produced from the North Sea varies considerably from field to field but consists primarily of methane (60% to 92%) and ethane (3% to 15%) with smaller percentages of *propane, butane, pentanes* and *hexanes*. Two of the least desirable constituents which may also be found in the gas are *carbon dioxide* (CO_2) and *hydrogen sulphide* (H_2S), the latter creating a condition referred to as *sour gas*.

Carbon dioxide can account for up to 25% of the composition of the gas and the *carbonic acid* produced in the presence of water is the primary cause of sweet corrosion in sub-sea pipelines.

Hydrogen sulphide is extremely toxic, highly corrosive and whilst prevalent elsewhere in the world it is fortunately relatively rare in the North Sea. If present, it must be removed if considerable damage to process equipment is to be avoided and the welfare of platform personnel safeguarded. The removal process is referred to as *sweetening* and is described under oil production.

2.1 CONDENSATE SYSTEM

Condensate is a clear, highly volatile liquid consisting of the heavier hydrocarbon factions that condense out of the wet gas as it leaves the well. Liquid processing involves the removal of water from the condensate received from the production separators. The dry condensate can then be re-injected into the gas for delivery to the onshore reception facility (*the beach*) whilst the water is discharged into the sea, via the produced water system.

i) COALESCER

The coalescer operates at process pressure and its function is to refine the condensate received from the production separator. The condensate passes through fine wafer pads or demisters onto which the water droplets coalesce before migrating into the base of the vessel.

The condensate passes through a *BS&W* (basic sediment and water) detection meter as it leaves the coalescer. If free of water, the condensate is pumped into the gas export line for transportation to the beach. If an unacceptable water content is recorded by the meter, the condensate will be diverted into the slop oil tank for further processing.

ii) SLOP OIL TANK

The slop oil tank is simply a holding vessel for water contaminated condensate, known as *slop oil* or *bad oil*. It is fed from the *coalescer, skimmer* and *caisson drains* and the contents will eventually be pumped back into the production header for re-circulation through the process equipment.

2.2 PRODUCED WATER SYSTEM

Produced water consists essentially of oil or condensate contaminated water removed from the production separators and other process vessels and the function of the produced water system is to clean the water prior to discharge into the sea.

Generally speaking the production of water increases as the reservoir ages. A basic gas production installation could produce less than 100 tonnes of water a day whilst a large oil production installation may have to cope with the disposal of in excess of 20,000 tonnes of water a day. Whilst the quantities of water may vary considerably the basic equipment and operating principles remain the same for both oil and gas producing installations.

i) PRESSURE REDUCTION

The first operation entails a reduction in the water pressure from that of the oil or gas process to atmospheric, or slightly above atmospheric pressure. This may be achieved by the insertion of orifice plates into the process vessel produced water outlet connections and a degassing drum may also be included in the system.

ii) DEGASSING DRUM

The degassing drum facilitates the separation of oil, water and gas. The reduction in pressure permits the release of dissolved gases which are disposed of at the HP vent or flare stack whilst the liquids are separated using an oil/water interface. The oil or condensate is transferred to the slop oil tank whilst the water is directed to a polishing unit for further refinement.

iii) POLISHING UNIT

The polishing unit which may also be referred to as a *produced water separator, flotation unit* or *skimmer* once again utilises the different specific gravitates of oil and water to achieve separation by gravitation. any remaining oil is transferred to the slop oil tank whilst the water is discharged into the sea via a dedicated *produced water caisson*, or the *overboard drains disposal water caisson*.

iv) OVERBOARD DRAINS DISPOSAL WATER CAISSON

The overboard disposal drains caisson provides the final opportunity for any remaining condensate or oil to separate from the water. The oil floats to the top of the caisson where it is returned to the slop oil tank by a transfer pump which is activated automatically by an oil/water interface sensor. An oil content of less than 40 ppm (parts per million) is required before the water can be discharged into the sea.

v) HYDROCYCLONES

The produced water system described above is typical of what will be found on the vast majority of offshore installations. However, in recent years, hydrocyclones have been installed by some operators who are faced with the problem of disposing of very large volumes of produced water and they have proven to be extremely effective and reliable in service, there being no moving parts. They are installed immediately after the production separator. Water exiting the vessel should be routed through a degasser/skimmer prior to discharge in order to remove any entrained gasses.

2.3 WET GAS

Whilst the preceding text explains at some length the reasons why it is necessary to remove the water from the gas, there are situations where *wet gas* systems will be encountered.

Wet gas production is particularly common on small unmanned, minimal facilities installations which produce back to a central processing facility (CPF). The platform typically accommodates from two to four wells and following pressure reduction at the Christmas tree, the gas enters the sub-sea pipeline with no further processing taking place. The pipeline is dosed with corrosion inhibitors and MEG to prevent hydrate formation, the liquids being removed on the CPF.

The operating life of these platforms varies from 5 to 10 years and adequate corrosion allowances are included in the design to support this mode of operation.

Part 2. OIL PRODUCTION

It would appear that Mother Nature has a far greater affinity for Her oil than Her gas for whilst the exploitation of a gas field proceeds with the ease of deflating a large balloon, the recovery of oil is fraught with problems and resisted at every turn.

Crude oil exists in an underground reservoir accompanied by large quantities of hydrocarbon gas and formation water and it is the gas, and to a lesser extent the water that provide the driving force required to bring the oil to the surface. Whilst the oil may be difficult to entice from the ground, the subsequent processing required prior to export is minimal. However, the same cannot be said for the refinement of the gas associated with the oil and it becomes apparent why in the formative years of the oil industry the gas was simply burnt off at the flare stack. Today, energy conservation policies and the realisation of the value of the gas in terms of both operational and financial benefits have tended to reduce the routine flaring of gas, at least in Europe.

If an equipment plan of an offshore installation is studied a complex picture emerges involving a number of process systems intermingled with one another. However it should be remembered that the basic object of the exercise is simply to separate oil, gas and water and to refine them to an acceptable level of purity prior to their discharge from the installation. These separate processes will now be discussed together with the enhanced oil recovery equipment, a most important feature of oil production designed to loosen Mother Natures grip on Her oily treasures.

1. THE OIL PROCESS

The refinement of crude oil to a quality suitable for transportation by sub-sea pipeline or oil tanker is a remarkably simple operation consisting primarily of the removal of associated gas and formation water within a series of production separators.

A sketch has been prepared to assist in the explanation of the hydrocarbon process equipment employed on an oil production installation. More detailed descriptions of the process vessels and principles of operation can be found in chapter four.

i) WELLHEADS

The oil process commences at the Christmas tree where oil at full reservoir pressure is admitted to the installation and reduced in pressure to approximately 450 psig (30 bar) prior to entry into the production header or the well test separator.

ii) PRODUCTION HEADER

The production header is essentially a manifold which receives oil from all the wells on the installation (up to 60) and redistributes it to a number of process trains (up to 4). The process trains are identical and only one has been shown on the sketch in the interests of clarity.

From the production header the oil enters the first of a series of production separators.

Offshore Engineering

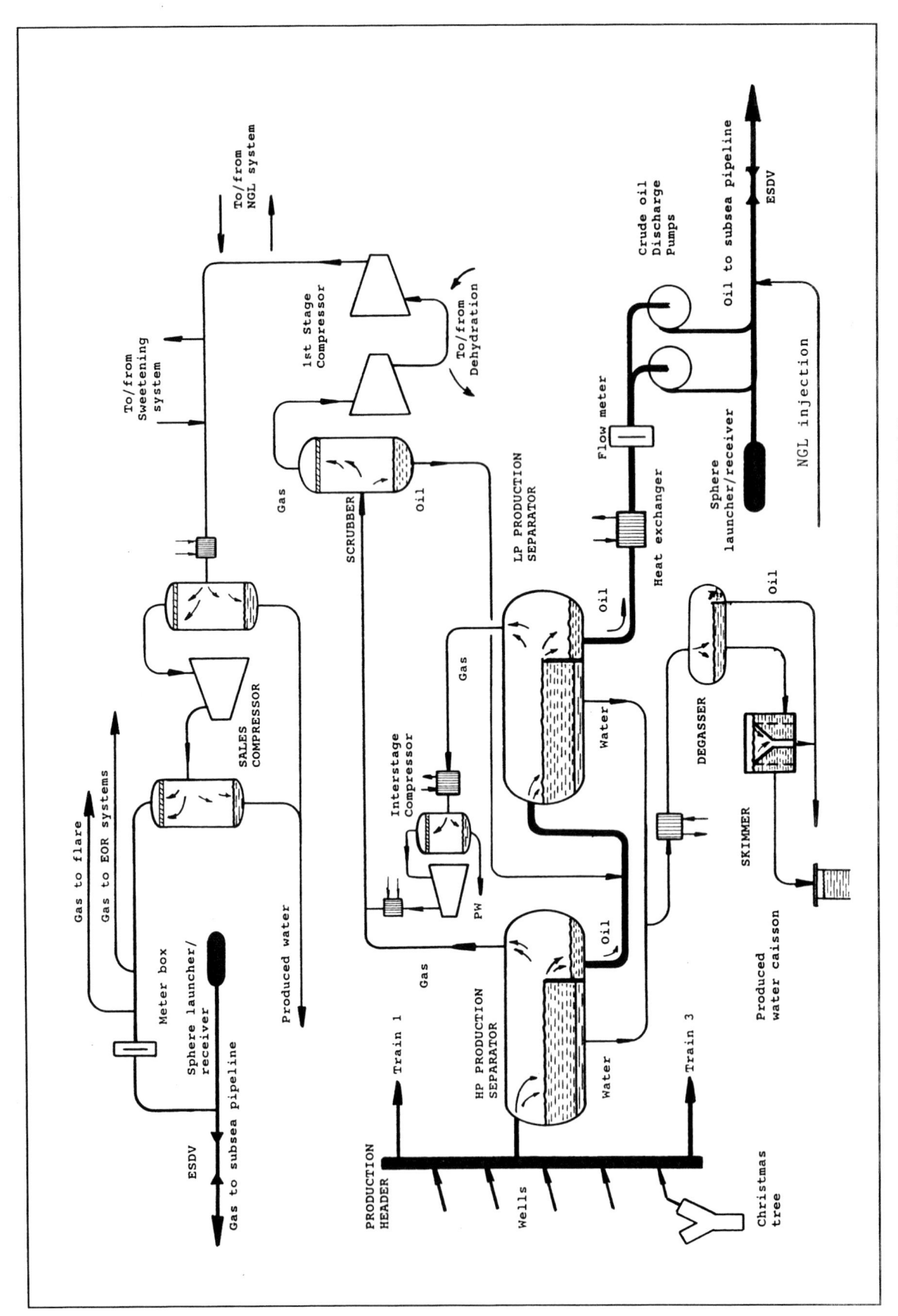

PRODUCTION OF OIL AND ASSOCIATED GAS

iii) **PRODUCTION SEPARATORS**

The number of production separators employed will be dependant on the volume and pressure of the liquids entering the system but most installations require at least two and occasionally three per train which operate at successively lower pressures.

The production separators facilitate the separation of oil, associated gas and water. The removal of gas is assisted by the reduction in pressure to atmospheric or near atmospheric conditions (45 psi–3 bar), whilst the separation of oil from the water relies on an oil/water interface. The oil may be heated prior to entry into the separator to lower its viscosity and enhance the separation process.

The water is disposed of via the produced water system whilst the oil is virtually ready for export. The export oil is referred to as *dead crude* if processing was completed at atmospheric pressure whilst *live crude* describes an oil processed at a pressure slightly above atmospheric.

iv) **TEST SEPARATOR**

The test separator is identical to the production separator in all but size and fulfils the same function as the test separator described in the Gas Process.

v) **CRUDE OIL EXPORT**

Prior to export the crude oil will normally be subjected to cooling, metering, injection of natural gas liquids (NGL) and a significant increase in pressure by the discharge pumps.

a) **Cooling**

The temperature of oil produced from a deep reservoir can be considerable, occasionally in excess of 180°C (380°F). Consequently, a number of heat exchangers may be deployed prior to, between and after the various stages of separation to effect changes in the temperature of the oil to assist in the separation of the oil from the water, and in the stabilisation of the crude prior to export.

b) **Metering**

Due to the fact that oil is normally transported by a sub-sea pipeline that is frequently shared with other installations, it is most important that the quantity of oil exported is measured in order that each installation may be accredited with it's just deserves. The oil is metered prior to the injection of (NGL) which is measured separately.

c) **NGL injection**

The disposal of NGL by injection into the crude oil prior to export is analogous to the re-injection of condensate encountered on gas production installations. The process is known as spiking and transforms a *dead crude* into a *live one*.

The practice of spiking crude can only be performed on pipelines which terminate at a refinery suitably equipped to handle live oils.

Offshore Engineering

d) Pumping

Crude oil must be subjected to a considerable increase in pressure (750 psi/50 bar) to facilitate transportation by sub-sea pipeline. The discharge pumps are generally of a two stage centrifugal type, electrically powered through a synchro-torque variable speed coupling which can be adjusted to suit fluctuating oil production. Alternatively, gas turbine driven pumps may be installed.

In common with gas producing installations the export pipeline is fitted with a sphere launcher/receiver and an emergency shutdown valve (ESDV).

2. ASSOCIATED GAS

The quantities of gas associated with crude oil production can be considerable with some of the larger northern sector oil installations producing more gas than some of the unmanned satellite installations in the southern sector of the North Sea. The quantity of gas produced dictates both the processing arrangements, and the method of disposal.

2.1 THE GAS PROCESS

The main objective of gas processing is to produce a gas which complies with the *dew point* (hydrocarbon and moisture) criteria specified for the sub-sea pipeline to ensure the avoidance of problems associated with *corrosion, hydrate* formation and the build up of condensed liquids. Basically this involves the removal of water, natural gas liquids (NGL) and gaseous impurities such as *Hydrogen Sulphide* (H_2S) and *Carbon Dioxide* (CO_2), should they occur.

A brief description of the arrangements for processing associated gas will now be given.

i) PRODUCTION SEPARATOR

The gas process commences at the production separator where the reduction in pressure facilitates the vaporisation of gas from the crude oil. Two stages of separation are shown in the sketch but three successive stages of pressure reduction are frequently employed.

Gas from the HP separator passes directly to the compressor inlet scrubber whilst gas liberated from the LP separator must undergo compression, cooling and scrubbing (liquid removal) before recombining with the HP gas.

ii) INLET SCRUBBER

The inlet scrubber is concerned primarily with liquid removal prior to the gas entering the first full stage of compression. Liquids removed may be returned to the inlet of the LP production separator, or they may be disposed of via the produced water system, depending on the design of the plant.

iii) COMPRESSION

Gas processing involves a considerable number of compression stages, due to the fact that associated gas is released from the crude oil at relatively low pressures and requires elevation to in excess of 1,000 psi (70 bar) for export and 6,000 psi (420 bar) if it is to be re-injected in to the reservoir for enhanced oil recovery.

The inclusion of gas compressors greatly complicates the process plant as each unit must be provided with suction and discharge scrubbers, and heat exchangers in order to avoid the dangers associated with condensed liquid carry over into the machinery. Both electrically powered reciprocating, and gas turbine driven centrifugal compressors are used extensively offshore, the centrifugal types favoured for high volume, low to medium pressure service whilst the reciprocating variety are more suited to low volume, high pressure duties such as gas re-injection in to the reservoir.

iv) INTERMEDIATE PROCESSING

At this stage in the process the gas is relatively moisture free and the subsequent processing required will be dependant on the condition of the gas as it left the reservoir. Whilst water removal and compression equipment are common to all installations, dehydration, NGL removal, stabilisation and sweetening facilities are only incorporated as and when required. These intermediate process arrangements will be discussed on completion of this section and their take off points (if required) can be seen on the *Production of Oil and Associated Gas* sketch.

v) EXPORT

Depending on the quantities produced the process gas may be sold, used for enhanced oil recovery (EOR), fuel gas, or flared.

a. Sale

When substantial quantities of associated gas are produced, and it can account for up to 20% of the installations total hydrocarbon output, the financial returns may justify the construction of a sub-sea pipeline to facilitate the transportation and sale of the gas.

b. Enhanced Oil Recovery (EOR)

Enhanced oil recovery is an essential feature of the oil production process and associated gas can be utilised for re-injection into the reservoir or for gas lift operations. Such is the demand for gas for EOR that installations frequently import gas through a sub-sea pipeline when their own reserves are inadequate.

c. Fuel Gas

Virtually all installations are self-sufficient in fuel gas which is used to fire regeneration boilers, gas turbine driven alternators, pumps and compressors. The fuel gas is normally produced as a by-product of processes such as NGL removal and stabilisation.

d. Flaring

The flare on an oil producing installation burns continuously providing the means of disposing of unsaleable and toxic gases, and any gas surplus to requirements produced during periods of plant start-up or maintenance.

Where the gas is destined for sale it will undergo re-compression prior to being transported to an onshore reception facility by sub-sea pipeline. The pipeline will include a metering box, sphere/launcher receiver and an emergency shutdown valve (ESDV) identical to the arrangements employed on a gas production installation.

Offshore Engineering

2.2 THE DEHYDRATION PROCESS

The function of the *glycol contactor*, dehydration or absorption tower is to dehydrate process gas to ensure that the *dew point* temperature is below that point at which moisture condensation and thus *hydrate* formation may occur in the sub-sea pipeline.

The dehydration process is based on the strong affinity that glycols have for water, they are said to be hygroscopic and *Triethylene glycol* (TEG) is chosen in preference to other glycols because of its greater ability to resist the high temperatures associated with the regeneration process. At the commencement of the dehydration process the *dry glycol* is referred to as being in the Lean condition whilst on completion of water absorption it is described as being Rich.

The sketch shows a typical dehydration/regeneration plant.

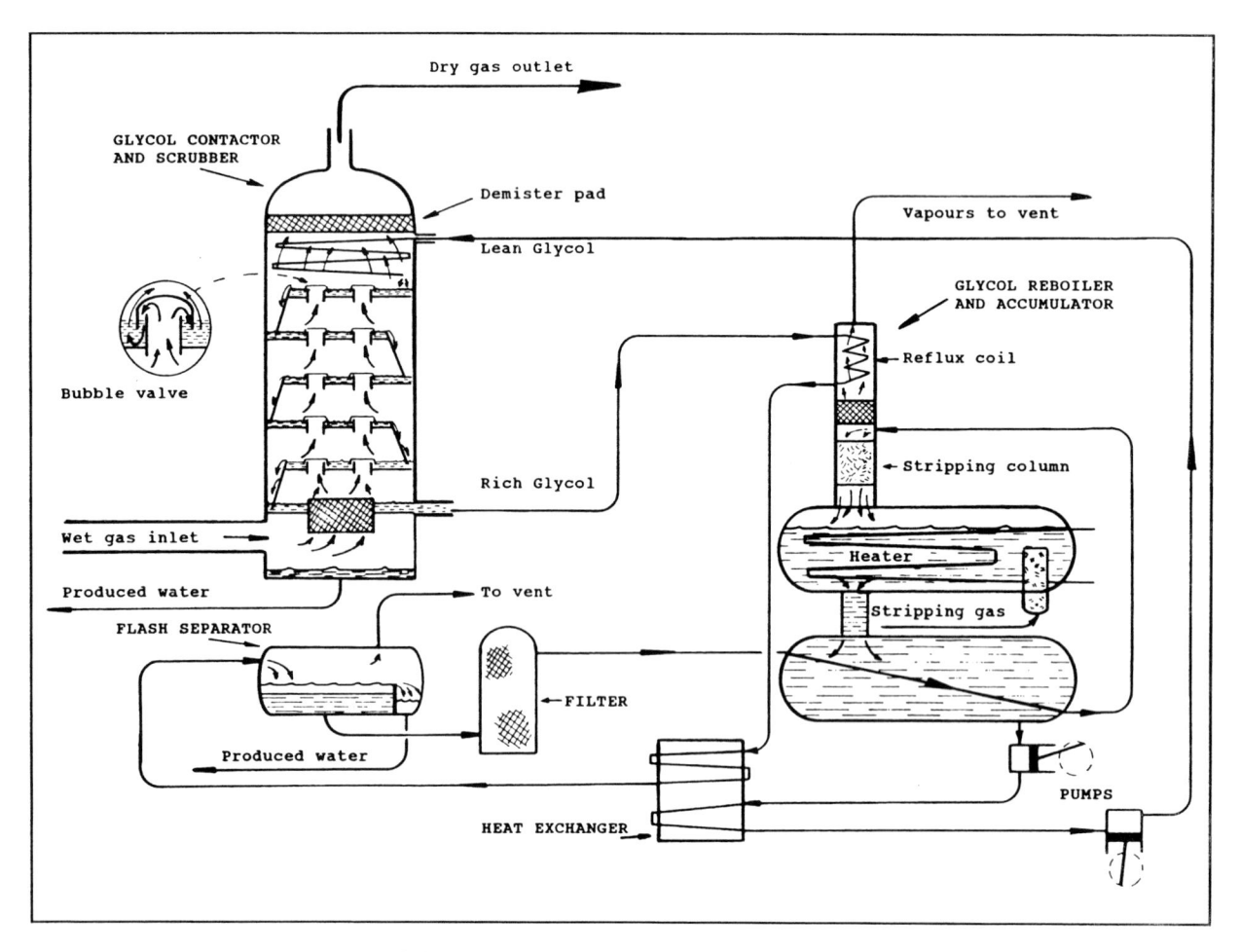

GAS DEHYDRATION PROCESS

i) DEHYDRATION

The main function of the contactor is to provide conditions conducive to the water absorption process, that is the thorough mixing and agitation of the glycol and process gas. It consists of a tall, vertical pressure vessel that is divided into numerous sections by horizontal trays fitted with *non-return* or *bubble valves*. Lean glycol entering the top of the vessel cascades downwards flooding each tray in turn whilst the process gas flows upwards passing from section to section through the non-return valves.

The non-return valves actually force the gas to bubble through the glycol and the intimate contact that this affords greatly enhances the water absorption process. Eventually dry process gas emerges from the top of the contactor after having first passed through a demister pad designed to remove any entrained glycol. The rich glycol accumulates in the base of the vessel prior to entry into the regeneration plant.

Throughout the dehydration process the temperature of the glycol must be maintained at approximately 2°C to 3°C above that of the process gas to prevent the condensation of liquid hydrocarbons into the glycol.

ii) REGENERATION

The dehydration process benefits from high process pressure and low process temperature, conditions which increase the water absorption capacity of the glycol. In the regeneration process the conditions are reversed in order to encourage the release of water vapour and a return of the glycol to the lean condition suitable for re-circulation.

The main components of the regeneration plant are the flash separator which reduces the glycol pressure to atmospheric, and the reboiler where the actual drying out or purification process takes place.

With reference to the sketch.

a. The Reflux Coil

The reflux coil, or to give it its correct title the *Reboiler Vapour Condenser* is located in the top of the *stripping column* and is essentially a heat exchanger whose function is to pre-heat the rich glycol entering the regeneration system and to reduce glycol losses by encouraging the condensation of glycol from the hot gases and water vapours leaving the reboiler drum.

b. Flash Separator

The flash, or *condensate separator* operates at atmospheric pressure and provides 3 phase separation. Process gas is released to the *LP vent stack*, water and condensate to the produced water system, and the glycol proceeds to the filtration package.

c. Filtration Package

Glycol being a viscous liquid has a tendency to foam and impurities, be they in liquid or solid form accentuate this tendency. The flash separator provides for the removal of liquid impurities whilst the filtration package removes solid particles.

d. Heat Exchangers

The glycol regeneration process involves a considerable transfer of heat both to and from the glycol and a number of heat exchangers are employed to increase the overall efficiency of the operation.

e. Glycol Reboiler

The purification of rich glycol is effected within a glycol reboiler or regenerator. The boiler, which operates at atmospheric pressure employs a combination of heat and *stripping gas* to remove entrained water.

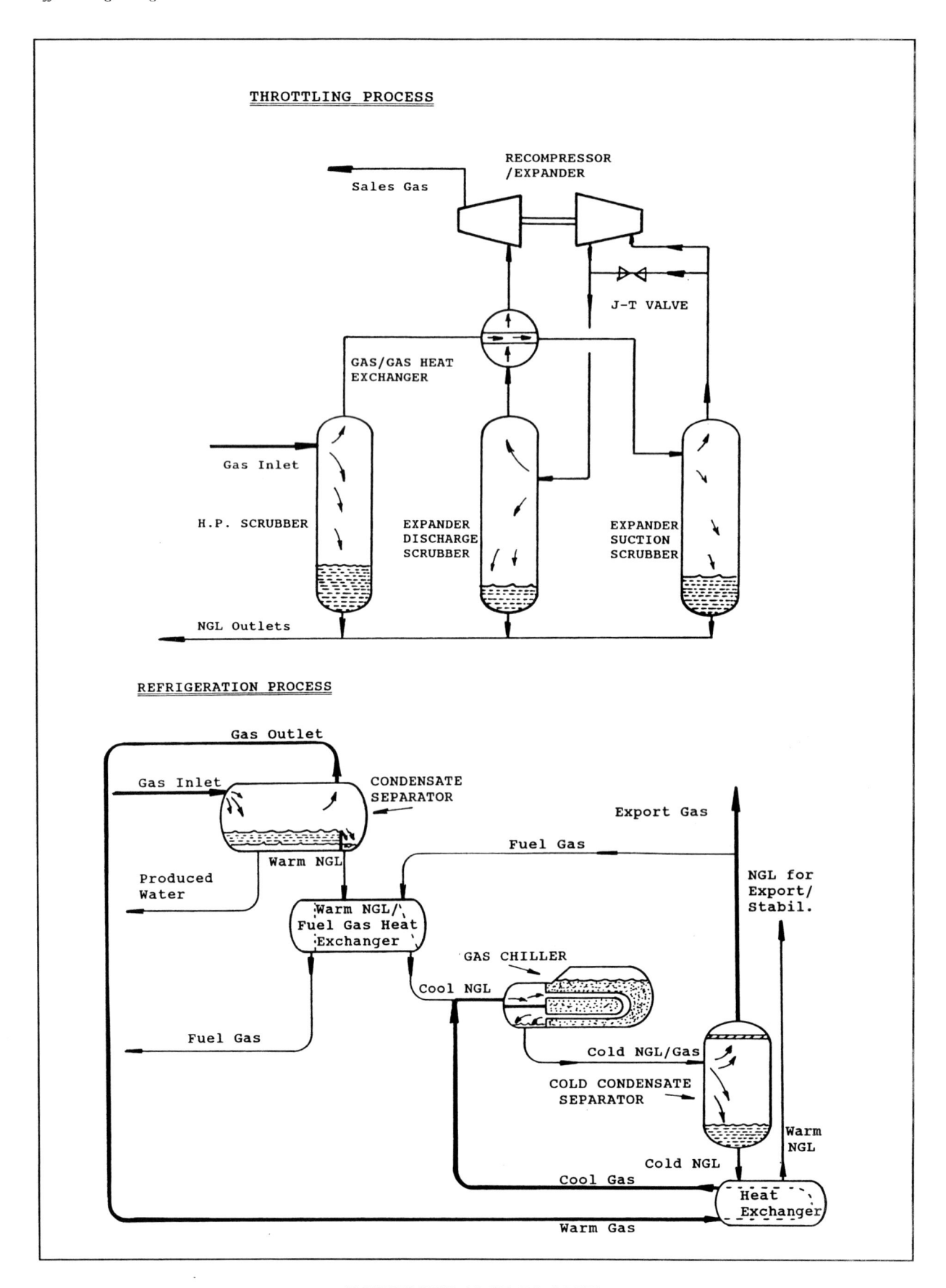

REFRIGERANT PLANT

Rich glycol enters the reboiler drum at the base of the stripping column where it is heated to a temperature of 200°C (418°F) by gas fired burners (or electric heaters). Ideally a slightly higher temperature would be preferred because a temperature of 200°C (418°F) will only guarantee a glycol purity of 98.5%. Unfortunately chemical breakdown of the glycol commences at 204°C (425°F) and the oxalic and glycolic acids produced are highly corrosive. To compensate for the temperature limitations of the glycol a stream of stripping gas (fuel gas) is admitted into the base of the reboiler drum. The gas enhances the separation effect by reducing the partial pressure of the water vapour which permits an improvement in glycol purity to 99.9%, the value required by the dehydration process.

The lean glycol leaving the base of the reboiler drum is temporarily stored in the glycol accumulator, prior to cooling and subsequent transfer to the glycol contactor.

2.3 NATURAL GAS LIQUID (NGL) REMOVAL

Natural gas is a complex mixture of hydrocarbons, primarily *methane* and *ethane* but including heavier gaseous hydrocarbon fractions such as butane and propane which possess boiling points within the ambient temperature range. These heavier fractions readily condense, are referred to as *natural gas liquids* (NGL) and can account for as much as 10% of the total oil output of an oil production installation. The more volatile of these liquid fractions are referred to as condensates and tend to be associated more with gas producing installations.

The function of the NGL system is to create conditions conducive towards the condensation and subsequent separation of the heavier gaseous fractions (NGL) from the export gas to ensure that it complies with the hydrocarbon *dew point* criteria specified by the sub-sea pipeline.

In common with the majority of hydrocarbon process operations the removal of NGL is effected using very basic principles of fluid and thermodynamics, namely a reduction in temperature to encourage condensation, followed by a sharp change in direction designed to throw out the liquid phases thus formed. Sea water cooled heat exchangers provide the means by which the initial reduction in temperature is achieved with subsequent stages employing a throttling process, or a dedicated refrigeration plant. Both methods will now be described.

i) THROTTLING

The NGL removal system employs two preliminary stages of cooling and NGL separation prior to the gas being subjected to a more drastic temperature reduction within either an expander/re-compressor, or a *Joule-Thomson* valve.

The Joule-Thomson (J-T) valve is quite simply a throttling valve named after the pioneering scientists James Joule (whose name was given to the SI unit of energy) and William Thomson (later Lord Kelvin after whom the absolute temperature scale was named). It is used in preference to the expander/re-compressor during periods of off-peak gas production.

The expander/re-compressor is an extremely efficient piece of machinery which utilises the expansion of process gas to impart rotational movement to a turbine wheel which in turn drives the re-compressor attached to the opposite end of the shaft.

Cold gas leaving the expander (−20°C/−5°F) or J-T valve (4°C/40°F) enters the discharge scrubber where any remaining NGL is deposited prior to re-compression and export.

Offshore Engineering

ii) REFRIGERATION

The sketch shows a simplified layout of a refrigerated NGL removal process. The majority of vessels employed are heat exchangers designed to improve the efficiency of the operation for whilst NGL removal benefits from a reduction in temperature the subsequent stabilisation process (if installed) relies on heat for its effectiveness.

The initial separation of NGL and process gas takes place within the condensate separator. Gas leaving the separator is cooled prior to re-combination with NGL at entry to the refrigeration vessel where the subsequent reduction in temperature ($-30°C/-30°F$) accelerates the condensation of any remaining NGL. The contents of the refrigeration vessel are discharged into the cold condensate separator where NGL is once again separated from the gas, the gas destined for export leaving the top of the vessel.

The disposal of NGL will be dependant on the quantity produced. The liquids may be re-injected into the crude oil export pipeline, into a dedicated NGL pipeline or re-injected into the reservoir for recovery at a later date. When transportation by sub-sea pipeline is intended the NGL may be subjected to a stabilisation process in order to prevent problems created by the vaporisation of dissolved gases.

2.4 STABILISATION

The function of the stabilisation process is to further refine the NGL by removing the more volatile of the dissolved gases. Removal of gases likely to vaporise in the sub-sea pipeline will ensure trouble free transportation and operation of equipment at the onshore reception facility.

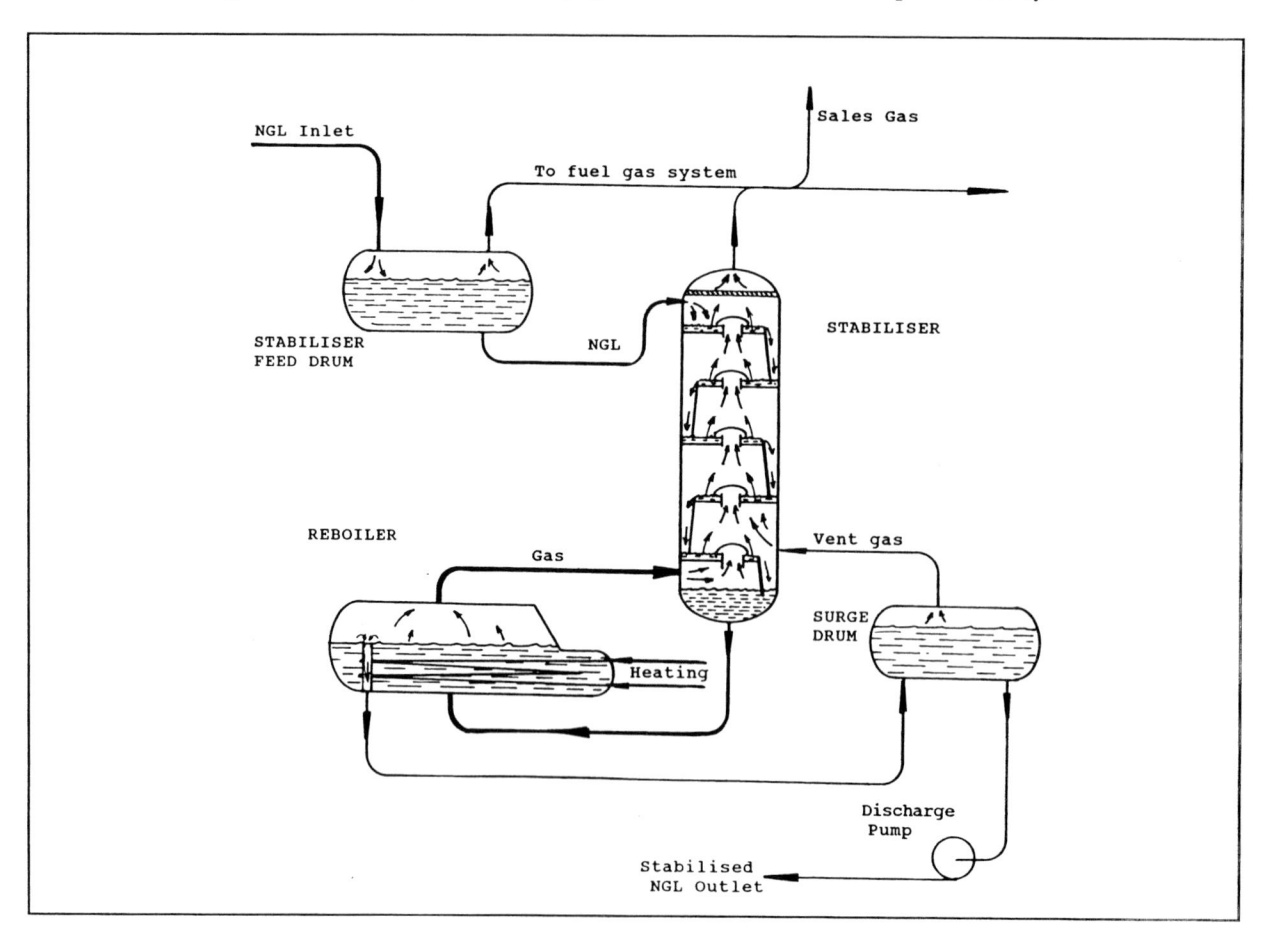

STABILISATION PROCESS

PRODUCTION

Stabilisation is a distillation process which permits the composition of the NGL to be closely controlled by the addition of heat in a reboiler. The plant shares many similarities with the gas dehydration equipment.

A sketch has been prepared to assist in the explanation of the process.

The NGL entering the top of the stabiliser gravitates downwards from tray to tray whilst hydrocarbon gases released from within the reboiler travel upwards effecting a stripping action designed to assist in the release of dissolved gases. The NGL eventually accumulates in the base of the vessel prior to transfer to the reboiler.

The temperature of the NGL in the reboiler is set at a value dependant on which of the dissolved gases are to be removed from the NGL by vapourisation, normally methane and ethane. The gases are returned to the stabiliser where they transfer heat into the NGL prior to commencing the journey up through the vessel. Gas expelled from the top of the stabiliser may be routed back to the sales compressor, or used as *fuel gas*.

Stabilised NGL is removed from the surface of the reboiler via a stand pipe and transferred to a surge drum for cooling prior to metering and re-injection into the crude oil export line.

2.5 THE SWEETENING PROCESS

Two of the least desirable gases which may occur in a reservoir are *carbon dioxide* (CO_2) and *hydrogen sulphide* (H_2S), the latter creating a condition referred to as a *sour* gas or oil. The function of the Selexol system is to *sweeten* the gas, that is to remove the hydrogen sulphide, and to a lesser extent the carbon dioxide contaminants prior to export, or re-injection of the gas in to the reservoir.

Selexol is a proprietary solvent, a complex *glycol* (dimethyl ether of polyethylene glycol) which is used in an absorption/regeneration plant which operates on similar principles to the glycol dehydration/regeneration process.

A sketch has been prepared to assist in the explanation of the process.

i) ABSORPTION PROCESS

The Selexol Tower contains two packing elements designed to promote intimate contact between the process gas and Selexol liquid. The upper packing element is flooded with lean (pure) Selexol and removes the H_2S content of the process gas whilst CO_2 is absorbed by the semi-lean Selexol within the lower packing element.

The Selexol accumulates in the base of the vessel in what is referred to as a rich condition, that is containing dissolved H_2S and CO_2 gases prior to regeneration whilst the gas proceeds to the next stage of refinement.

ii) REGENERATION PROCESS

The regeneration process relies on a combination of pressure reduction and stripping gas to effect the purification of rich Selexol.

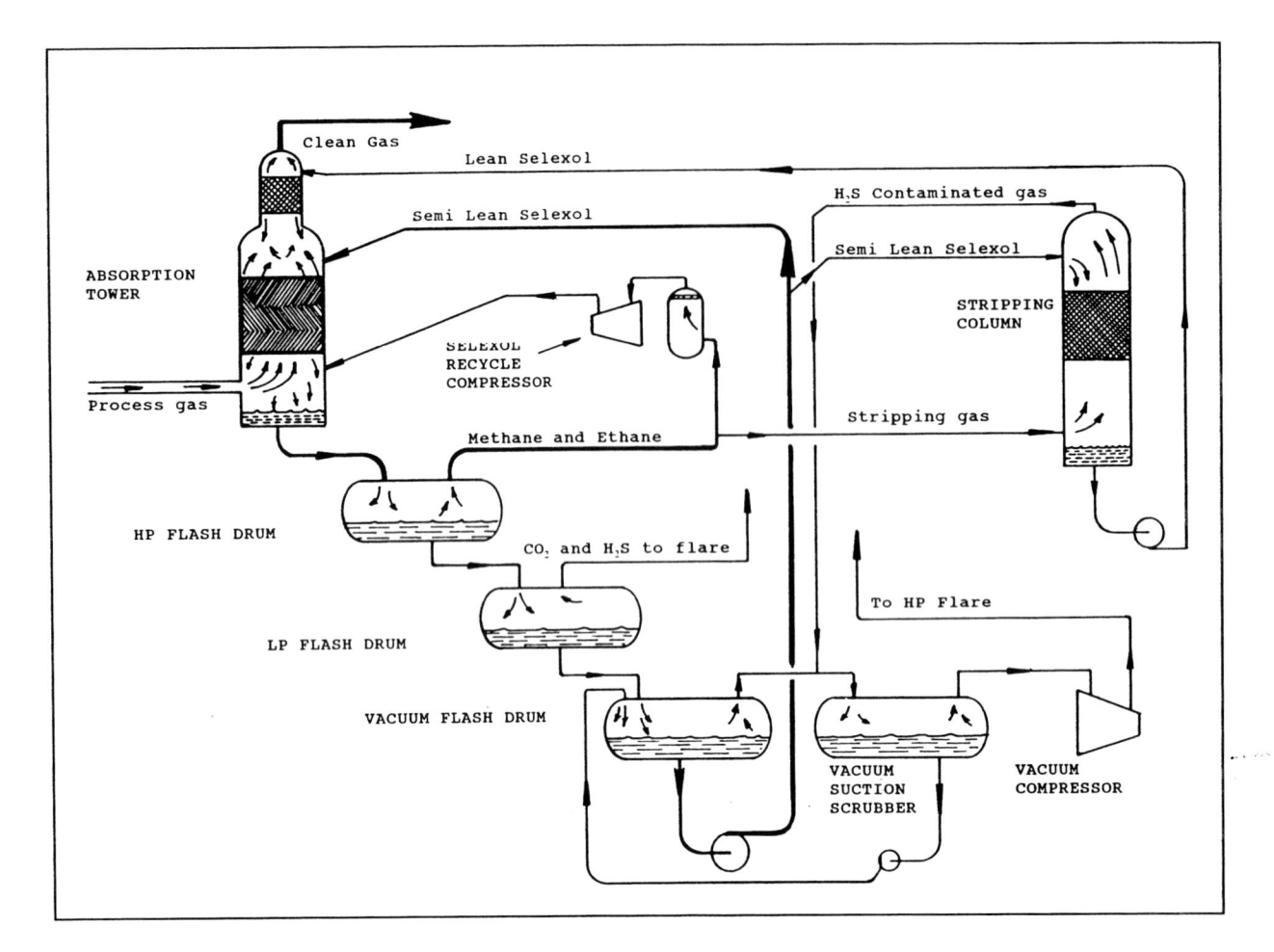

SELEXOL SYSTEM

Selexol entering the regeneration system is reduced in pressure in three successive stages by a series of flash separation drums. Gases released from the HP flash drum consist primarily of the lower boiling point fractions such as methane and ethane which are routed back to the absorption tower for recycling, or admitted to the Selexol stripper as stripping gas.

Gases released from the L.P. flash drum and Vacuum flash drums consist primarily of CO_2 and H_2S and being of no commercial value they are disposed of via the LP flare header. The Selexol emerging from the vacuum flash drum is in a semi-lean condition and suitable for re-entry into the lower packing element of the absorption tower.

To obtain the lean Selexol required by the upper packing element approximately 5% of the semi lean Selexol is subjected to further refinement within a stripping column. The intermingling of the Selexol with stripping gas completes the purification process. The H_2S contaminated stripping gas leaves the top of the vessel for disposal at the LP flare header whilst lean Selexol accumulates in the base of the vessel prior to re-circulation in the absorption tower.

Part 3. ENHANCED OIL RECOVERY AND THE OIL DRIVE MECHANISM

1. THE OIL DRIVE MECHANISM

The pressure contained within a new oil field can be considerable anything from 3,000 to 15,000 psi (204 to 1,000 bar) with 5,000 to 8,000 psi (340 to 545 bar) being representative of conditions prevailing in the North Sea. Unfortunately, being a liquid, oil is virtually incompressible and once production has commenced the natural reservoir pressure deteriorates rapidly and artificial means must be employed to enhance oil recovery if output is to be maintained at economic levels.

Basically, enhanced oil recovery (EOR) involves either the injection of water or gas into the formation in order to increase the reservoir pressure, or the location of equipment within the *production tubing* to assist in the extraction of oil. Frequently a combination of methods are employed.

To fully understand the principles on which EOR rely we must first familiarise ourselves with the natural mechanisms which drive the crude oil from the reservoir. As stated previously it is the reservoir pressure which provides the motive force for oil production and the reservoir pressure is created by associated gas and/or formation water.

The driving force for the production of oil may be any one of four mechanisms depending on the rock formation surrounding the reservoir and the quantity of water and associated gas accompanying the oil.

i) **DEPLETION/SOLUTION DRIVE**

Depletion drive describes the situation encountered when reservoir pressure exceeds the *bubble point* pressure of the hydrocarbon gas. This means that there is no *free gas* available as it is all dissolved within the crude oil. The gas is produced with the oil and it is the expansion of the gas which provides the driving force.

Solution drive is extremely inefficient and can only be relied on to recover 5% to 25% of available oil reserves.

ii) **GAS DRIVE**

When the reservoir pressure is less than the gas bubble point pressure, a layer or pocket of free gas exists above the oil. The gas acts like an accumulator, pushing the oil into the well bore and up to the surface, and assists in the drainage of the upper reaches of the reservoir.

Gas drive permits the recovery of 20% to 40% of the oil contained within the reservoir.

iii) **WATER DRIVE**

Water drive is the most efficient of the natural recovery mechanisms relying on the expansion of the underlying water zone or aquifer to provide the driving force for oil production. Water is virtually incompressible so to be effective the volume of the water must be considerable in relationship to the volume of oil. Theoretically water drive can be used to recover 40% to 80% of the reservoir, unfortunately the large volumes of water produced with the oil from an ageing field limit the commercial viability of the process.

Offshore Engineering

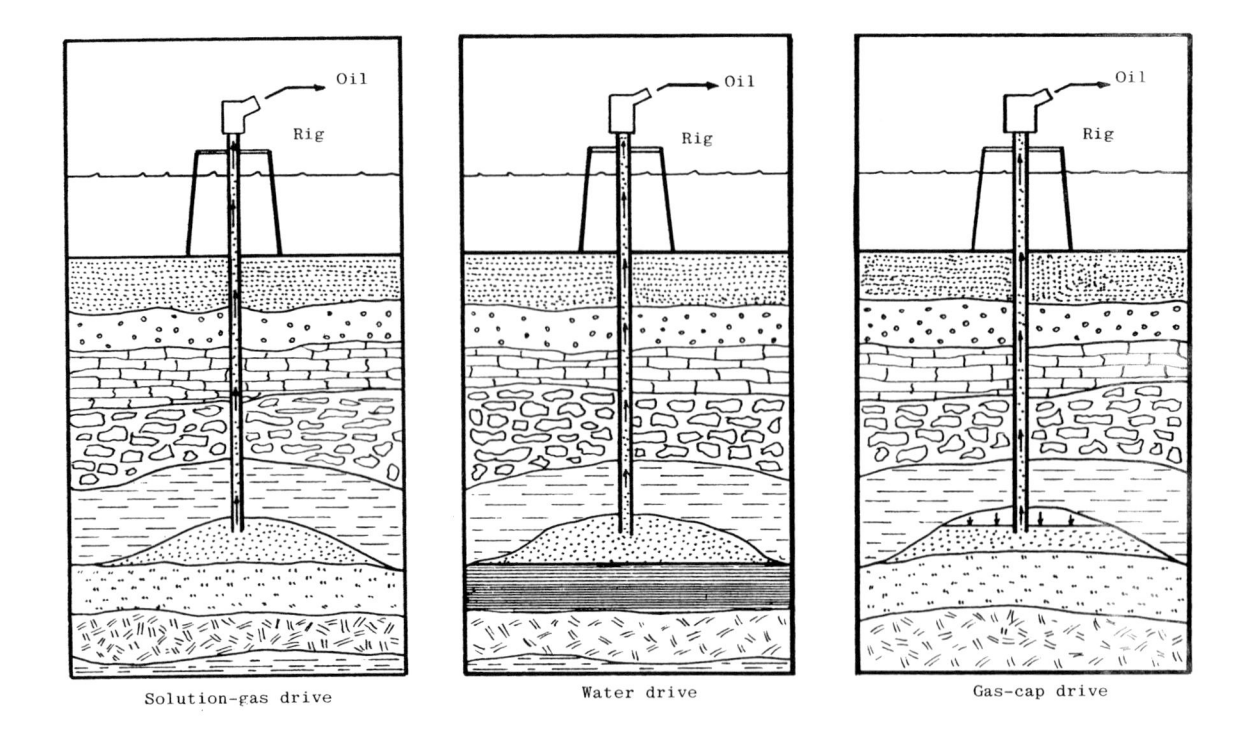

| Solution-gas drive | Water drive | Gas-cap drive |

RESERVOIR DRIVE MECHANISMS

iv) GRAVITY DRAINAGE

Gravity drainage is associated with steeply *dipping* reservoirs and this type of rock formation is conducive towards efficient oil output.

Initially oil production commences under the influence of depletion drive but as pressures fall the development of an overlying layer of free gas creates conditions more associated with gas drive.

2. ENHANCED OIL RECOVERY SYSTEMS

As previously stated, enhanced oil recovery may be effected by artificially increasing the reservoir pressure in order to supplement the natural oil drive mechanism, or equipment may be located within the production tubing of a well to help overcome the static head created by the column of oil.

The equipment associated with EOR considerably complicates life on board an oil production installation, the drilling derrick being in almost constant use involved in frequent modifications to the production tubings and in the drilling of new injection wells. For all that effort, the most efficient EOR systems can only economically recover 35% to 45% of the total oil contained within the reservoir, hence the description of oil field capacity in terms of *recoverable reserves*. This contrasts sharply with the exploitation of a gas field which can be almost literally sucked dry.

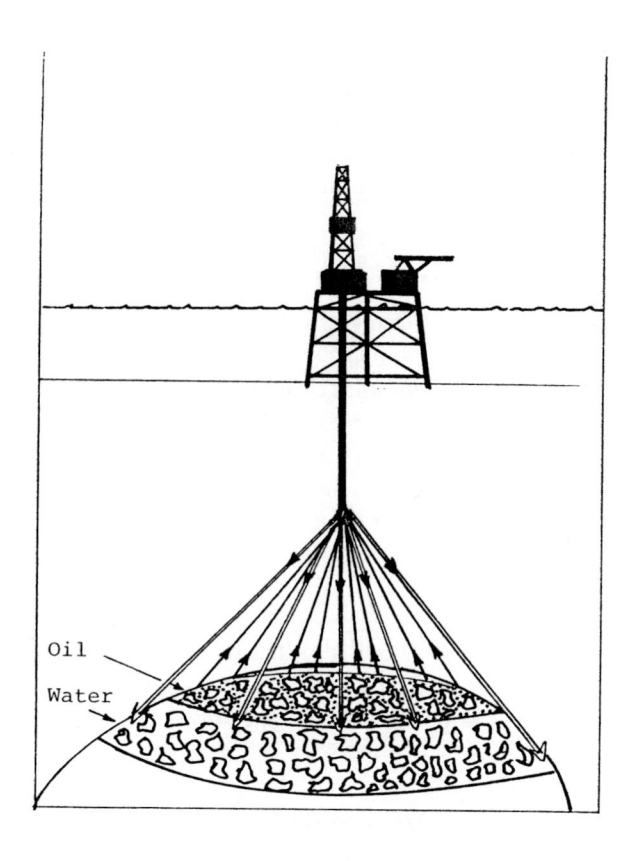

WATER INJECTION

The four most widely used EOR techniques are:-

1. Water Injection
2. Gas Injection
3. Gas Lift
4. Submergible deepwell pump

3. WATER INJECTION

Water injection provides a relatively cheap and efficient means of improving oil production from a depleted field and it is used extensively in the North Sea. The water is injected under pressure into the flanks of the oil bearing strata through purpose drilled wells (or redundant oil wells). Water entering the oil bearing rock displaces any remaining particles of oil and the reduction in free space increases the reservoir pressure.

Water for the injection system may be drawn from the sea or from the *produced water* system. Produced water consists of water removed from the crude oil during processing and is particularly suitable for re-injection due to the absence of dissolved oxygen. A mature oil field produces a considerable quantity of formation water, typically four times as much water as oil and the disposal of up to 150,000 barrels (21,000 tonnes) a day can present problems. If the water is to be pumped in to the sea, it must first be cleaned to ensure that all traces of oil and condensate are removed to avoid the risk of pollution and so re-injection greatly simplifies the problem of water disposal, and provides water which is ideal for enhanced oil recovery.

A sketch has been prepared to assist in the explanation of a typical *general service sea water system* that provides water for reservoir injection, process and utility equipment. Relatively complex filtration and purification arrangements are employed to ensure that the water entering the reservoir is scrupulously clean. The fine pores contained within the hydrocarbon bearing rocks must be protected against impurities that could restrict the drainage of oil and reduce the output of the reservoir.

i) SEA WATER LIFT PUMPS

Multi-stage, centrifugal submergible pumps installed within separate caissons provide the sea water to the system, the pump suctions being dosed with *sodium hypochlorite* solution to combat marine growth and discourage microbiological activity.

ii) FILTRATION PACKAGES

The sea water system contains two filtration packages, the first for general filtration of all sea water prior to entry into the deaerator and the second purely for the filtration of water destined for injection in to the reservoir.

The first filtration vessel contains a graded filter bed consisting of typically anthracite, garnet and pea gravel which will remove solid impurities greater than 5 microns (0.0002 inch) in size. The addition of a filtration aid (a positively charged polyelectrolyte) further enhances the efficiency of the operation.

The second filtration package employs a series of perforated stainless steel tubes covered with a woven fabric. Both filtration packages are periodically back-washed in order to clean the filtration medium of deposited solids.

iii) DEAERATOR

Removal of dissolved oxygen greatly reduces the corrosive effect of sea water and will increase the operating life of the plant considerably. The deaerator vessel may utilise *stripping gas* (fuel gas) or *evacuation* to encourage the release of dissolved oxygen and the injection of an oxygen scavenger such as *ammonium bisulphate* ($NH_4 HSO_3$) into the base of the vessel further improves the process.

Efficient deaeration can reduce the oxygen content of sea water from 9.25 parts per million to less than 30 parts per billion.

Water leaving the deaerator is subject to further conditioning by the addition of a scale inhibitor designed to prevent the formation of insoluble sulphates and carbonates that can occur when sea water eventually mixes with the natural waters in the reservoir.

iv) DEGASSING

The degassing vessel basically provides agitation to the sea water to assist in the removal of excess hydrocarbon gas. Water leaving the vessel supplies both platform utilities and the sea water injection system.

v) INJECTION PUMP

The sea water is injected into the reservoir by immensely powerful (2,000 kW) electric motor driven centrifugal multi-stage pumps which can generate water pressures ranging from 3,000 to 5,000 psi (205 to 340 bar).

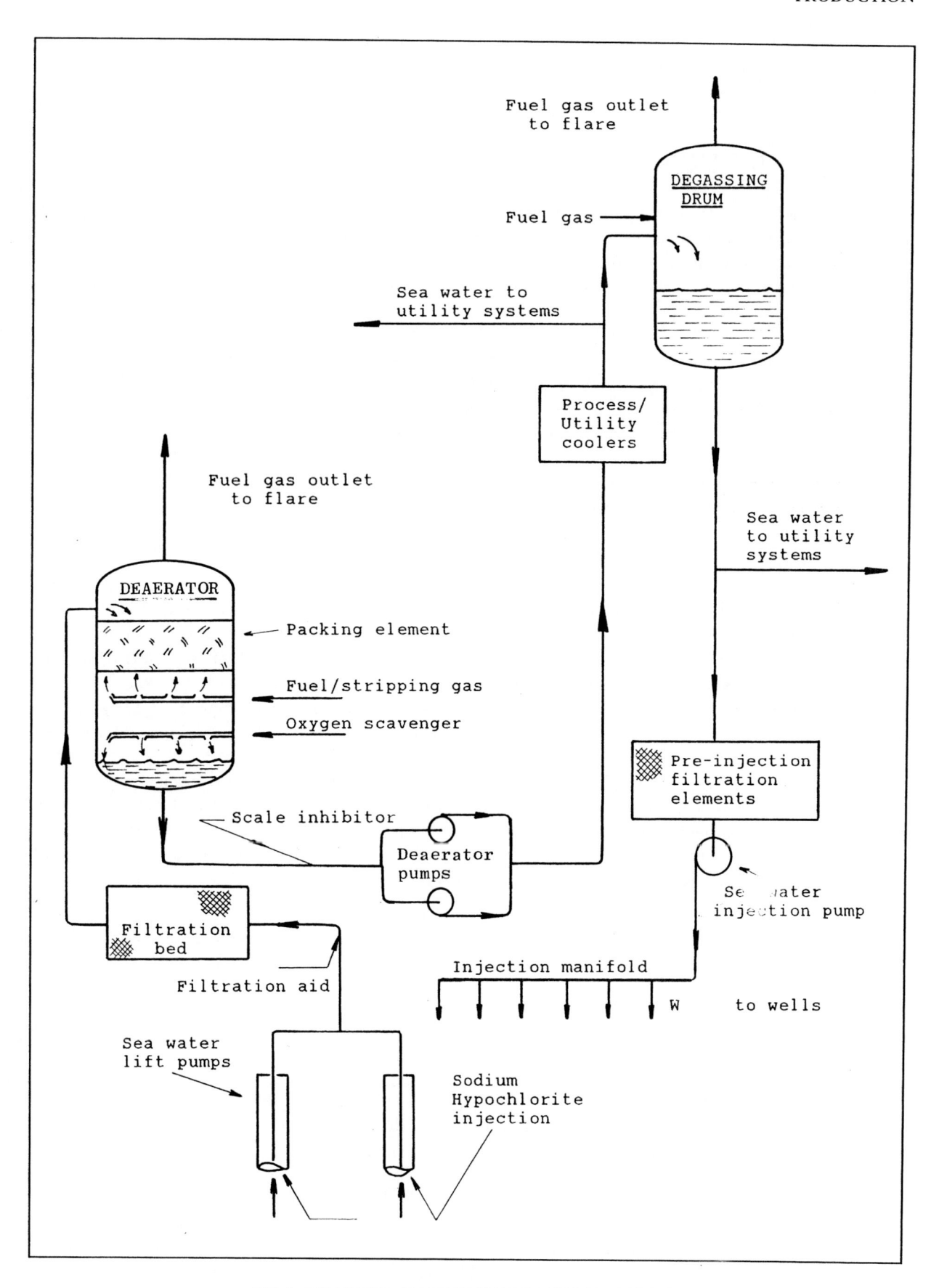

SEA WATER INJECTION SCHEMATIC

The water injection wells are to all intents and purposes identical to production wells and include a *Christmas tree* and *mudline safety valve*. They are also connected to the *emergency shutdown* (ESD) system.

4. GAS INJECTION

Gas injection operates on similar principles to water injection, that is to artificially increase the reservoir pressure. The process also provides the ideal means of disposing of the associated gas removed from crude oil when the quantities produced do not justify the installation of a sub-sea pipeline. The gas should not be thought of as lost because it can be recovered at a later date on completion of oil production.

The gas injection system consists of a high pressure compressor and distribution manifold which deliver gas to the reservoir at pressures of up to 6,500 psi (450 bar) through purpose drilled wells. Gas destined for re-injection is subjected to the full purification process that sales gas would receive in order to avoid contamination of the reservoir and injection equipment.

In the future, nitrogen injection could provide an alternative to process gas, the nitrogen being produced from a cryogenic air separation plant.

EOR Equipment

In order for oil to reach the surface, the reservoir pressure must exceed the static head pressure created by the column of oil within the well bore. The pressure required can be considerable, approximately 4,000 psi (310 bar) for a well of depth 12,000 feet (3,650 metres). Frequently the reservoir pressure produced by an ageing field is considerably less than this and the installation of a submergible pump or introduction of gas lift valves into the production tubing can greatly enhance oil output.

Gas lift valves and submergible pumps achieve the same end result by differing means and it should be remembered that both methods assist the reservoir oil drive mechanism and are not designed to produce oil from a field devoid of all natural reservoir pressure. The gas lift valves effect a reduction in static head pressure by reducing the specific gravity of the oil, whilst the submergible pump simply provides the oil with a helping hand in the form of a pressure boost.

5. GAS LIFT

Gas lift involves the re-injection of process gas into the production tubing in order to produce a foaming gas/oil mixture with a specific gravity considerably less than the naturally occurring crude oil.

Process gas is introduced into the annulus space surrounding the production tubing through a connection on the wellhead. It then passes into the production tubing through a series of gas lift non-return valves, the flow being regulated by the pressure in the annulus.

The sketch show a general arrangement of the gas lift equipment. The non-return valves are located within eccentric tubular housings referred to as *side pocket mandrels* (SPM) which are run (installed) as part of the production tubing. The SPMs ensure that an unrestricted channel is maintained within the production tubing which permits the maximum flow of oil, and the passage of the wireline tools employed to install and replace the gas lift valves.

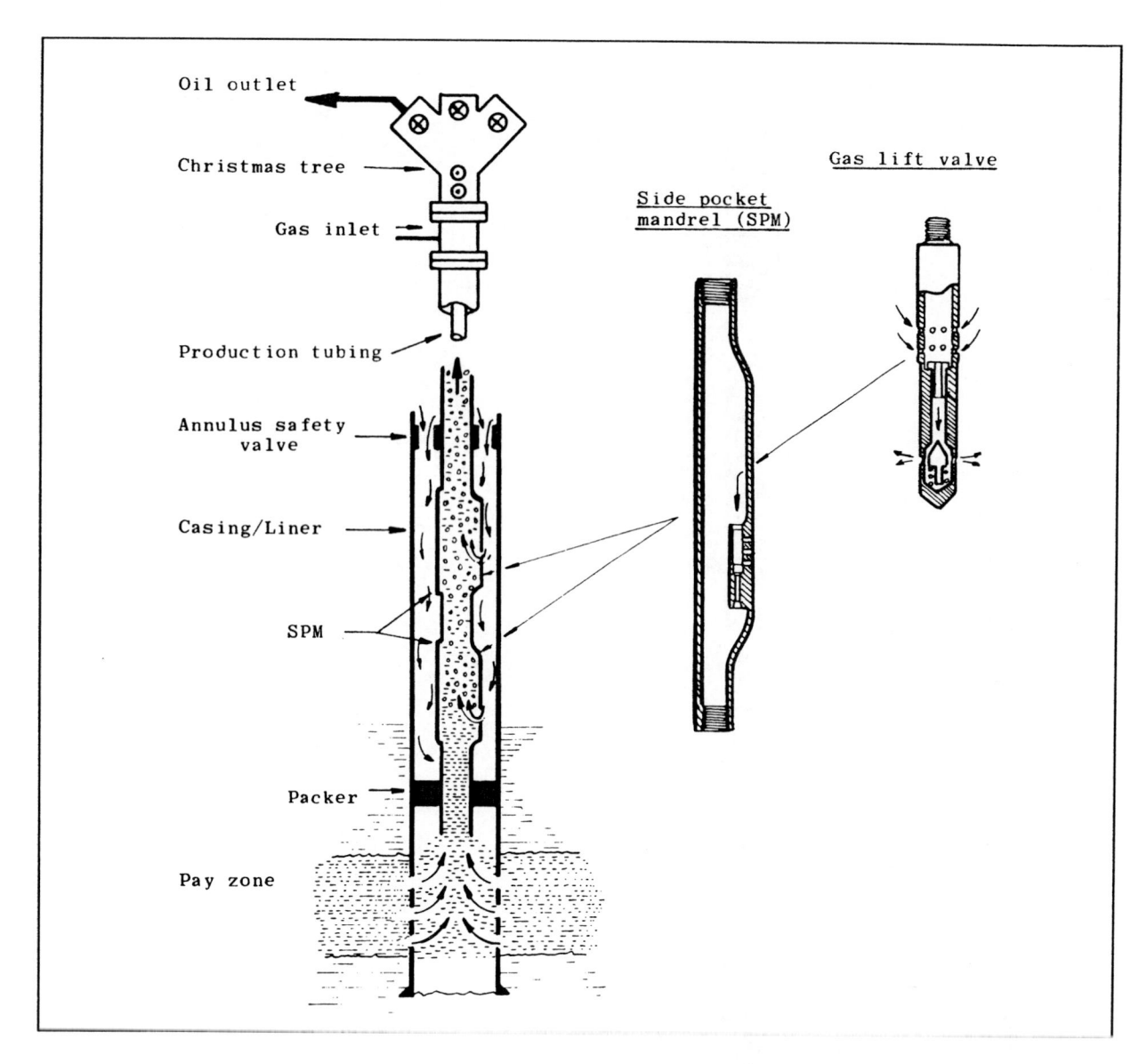

GAS LIFT ARRANGEMENTS

6. THE DEEPWELL/SUBMERGIBLE PUMP

6.1 ELECTRICAL SUBMERGIBLE PUMPS (ESP)

On fixed installations, the oil industry has for many years used electrically powered submersible pump sets as a means of boosting production rates beyond that achievable by natural reservoir pressure. These pumps are referred to as *downhole, submergible* or *submersible pumps* and they are installed in the well as part of the *production tubing*.

The pump units are individually tailored to suit the conditions pertaining to a particular well and up to 200 impellers may be used. A typical pump designed for service in a North Sea well would comprise 100 pump stages powered by a 150 to 300 kW electric motor, output being 13,000 barrels of oil a day at a surface pressure of between 7 to 15 bar (100 and 300 psi).

Offshore Engineering

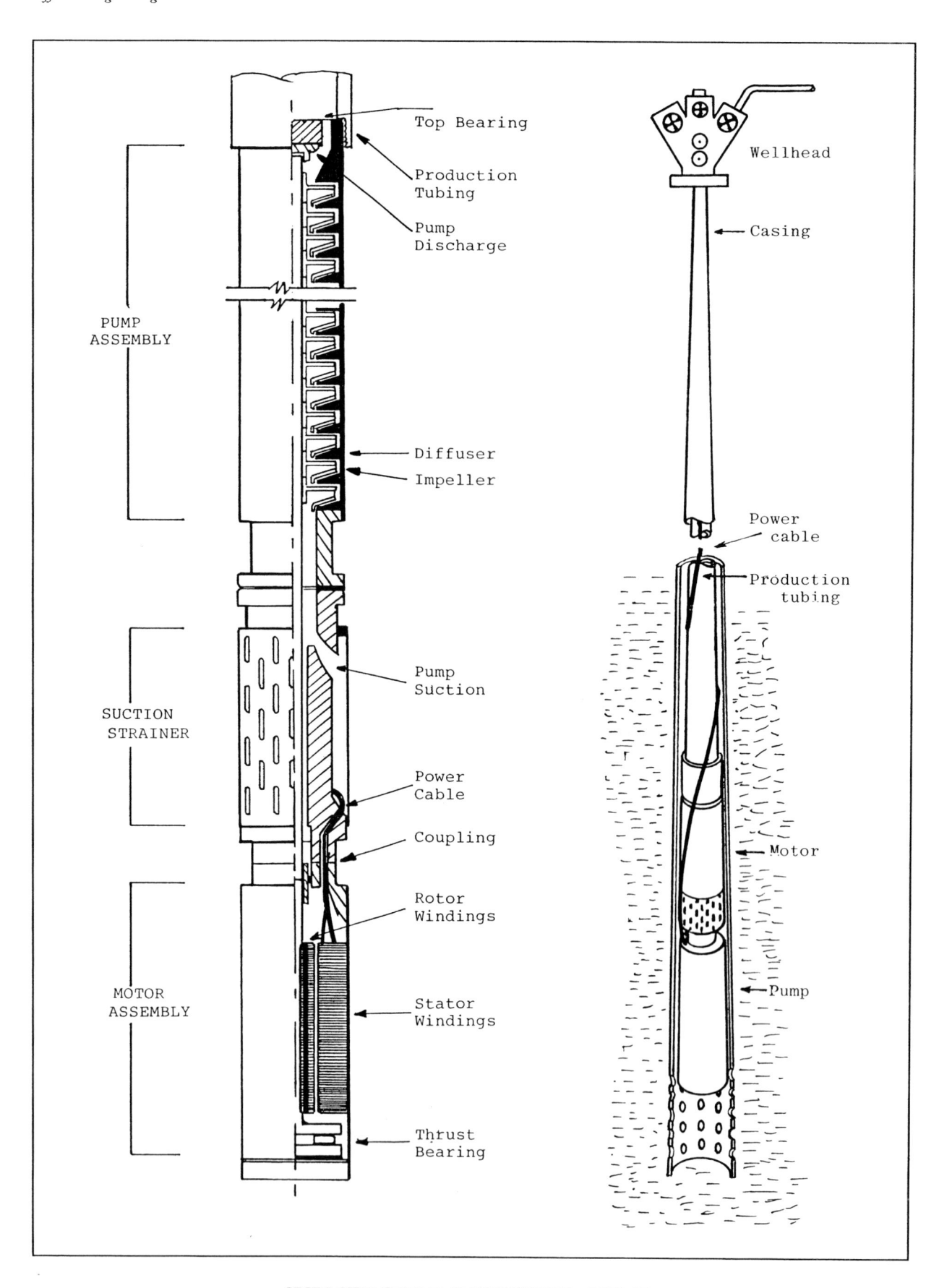

SUBMERGIBLE DEEPWELL PUMP

Downhole pumps have also been developed which will function in wells with extended horizontal sections, a situation perhaps best illustrated by a well on BP's *Wytch Farm* development, an offshore well drilled from an onshore site. The specification is impressive, 5000 metres of power cable providing electricity to a 480 kW motor at a voltage of 4–5 kV, the pump producing 20,000 barrels of oil a day.

6.1.2 HYDRAULIC DRIVE SUBMERGIBLE PUMP (HSP)

In recent years, a hydraulically powered pump has been developed, partly as a competitor to the ESP and partly because the manufacturers saw a hydraulic power source as providing greater potential for dealing with *multiphase* well fluids.

Like the ESP, the hydraulic drive downhole pump utilises a multi-stage centrifugal pump to actually produce the oil but there the similarity ends. A multi-stage, axial flow power turbine is used to drive the pump in preference to an electric motor. The power fluid which drives the turbine is supplied from a charge pump unit located either on the sea bed or on the reception facility. Sea water or produced water may be used as the power fluid, suitably filtered to remove impurities that could cause damage to turbine bearings and seals, delivered at a pressure of 270 bar (4000 psi).

A hydraulically powered pump operates at much higher speeds than an equivalent electrically powered unit, 7000 to 9000 rpm as compared to 3600 rpm, thus they require fewer pump stages, perhaps 20 compared to 90 stages for an ESP of equivalent pumping capacity. This reduction in the number of pump stages results in a more compact, rugged and lighter pump unit.

The hydraulically powered submergible pumps clearly have enormous potential but they are still regarded as *new technology* and sales have only just commenced, there being to-date only a handful of HSPs in service. They should not be confused with the hydraulically powered sea bed booster pumps which are discussed more fully in chapter six.

6.2 SUBMERGIBLE PUMP APPLICATIONS IN SUB-SEA WELLS

Until recently, submergible pumps were not considered suitable for use in sub-sea wells, gas lift being the preferred method of enhanced oil recovery because the equipment is simple, reliable, and can be installed with wireline equipment. The ESPs had a reputation for being unreliable and a service life of as little as 90 days was typical, electrical equipment not being ideally suited to the aggressive operating environment encountered deep within an oil well.

The Achilles Heel of the ESP is the mechanical seal between the motor and the pump but the failure of bearings, electrical connections and the motor itself, all tend to exacerbate the problem. Replacing an ESP is an extremely expensive business even on a fixed platform because it necessitates a well workover to remove the Christmas tree and pull the production tubing. This costs approximately $400,000 and does not include costs attributed to lost production. On a sub-sea well it has been estimated that the cost of replacing an ESP would be in the region of $2,000,000 due to the necessity to mobilise a jack-up or a drill ship to carry out the workover. However, claims of improved reliability and the advent of the *horizontal Christmas tree* has encouraged operators to take a fresh look at the situation.

Pump manufacturers now claim to be in a position to guarantee a service life of two years for their equipment and whilst a workover is still required to replace the pump, horizontal Christmas trees can be left in situ whilst the production tubing is removed which considerably simplifies the situation, the costs being reduced accordingly. These improvements have prompted Shell to install an ESP in the new (1998) *Gannet* field development to tie back sub-sea wells a distance of 30 km (20 miles) from the platform, ultimately looking towards a tie back distance of 50 km (30 miles) for future wells. Texaco have also opted to experiment with a submergible pump in their *Captain field*, a multiphase, hydraulically powered downhole pump they have developed in conjunction with *Weir Pumps* of Glasgow, a 340 kW pump with a capacity of 20,000 barrels/day.

As well as equipment improvements, considerable research has been carried out on installation techniques and there are now both ESPs and HSPs available which can be installed by *coiled tubing* or *wireline equipment*. These are referred to as *through tubing pumps*, wireline installation being used for wells deviated up to 30 degrees from the vertical whilst coil tubing is used beyond these angles and for horizontal wells. This greatly simplifies pump replacement because there is no need to carry out a full well workover which saves both time and money.

6.3 SUBMERGIBLE MULTIPHASE PUMPING

Having resolved the problem of finding a reliable power source for the pump, be it electrical or hydraulic, manufactures have also had to contend with problems with the pump impellers. The conventional *radial* and *mixed flow impellers* fitted to ESPs have difficulty in coping with high gas fraction oils. The gas gravitates to the impeller hub and eventually *gas slugging* or *vapour locking* occurs causing the pump to loose suction and surge, damage often occurring at this point.

Gas slugging is one of the problems that plagues production from horizontal wells when high gas fraction oils are encountered, particularly when producing oil at pressures below the reservoir bubble point where free gas caps may occur. To combat these problems, *axial flow pump impellers* have been developed which can cope with high gas to oil ratios where gas volumes may exceed 90%.

The axial flow impellers have proven to be particularly effective in field trials when powered by a hydraulic turbine, because, being a constant power device rather than a constant speed device, the pump speed adjusts automatically to compensate for the varying product density that occurs with changing void fractions. If a slug of gas enters the pump and the pump looses suction, it accelerates to a point where pumping will resume, the gas slug is dealt with and steady state conditions return. This variable speed capability is a highly desirable feature when dealing with multiphase fluids.

Chapter Six

UNDERWATER ENGINEERING

PART 1. **DIVING — INTRODUCTION**
1. Air Diving
2. Saturation Diving
3. Equipment
4. Vessels

PART 2. **UNDERWATER SURVEYS**
1. Splash Zone Examination
2. Swimaround Survey
3. Non-Destructive Examination
4. Flooded Member Survey
5. Marine Growth Measurement
6. Scour Survey
7. Cathodic Protection Examination
8. Differential Settlement Survey
9. Air Gap Measurement

PART 3. **SUB-SEA WELLS**
1. Sub-sea Development Options
2. Sub-sea Wells

PART 4. **SUB-SEA DEVELOPMENTS**
1. Sub-sea Separation
2. Multiphase Pumping and Metering

Offshore Engineering

System Summary

The vessel is fitted with a saturation diving complex, rated to 450 m depth, for 18 divers in round the clock service, with the possibility of operating two diving teams at different working depths and decompressing a third diving team simultaneously. The decompression chambers and the diving bell are equipped with gas reclaim system.
The diving bell is provided with heave-compensating maincable and guidewires, which can be disconnected when need of sidehauling.
Separate sidehauling-davits with winches are installed. Heave-compensation also provided for, when sidehauling.
The diving bell is driven through the splashzone and moonpool by a cursor-system.
The moonpool abt. 4 x 4 metres is located at the centreline amidships where the motion-amplitude of the vessel is minimum.

1. & 2. Three decompression chambers each for six divers.
- Each chamber with living compartment and ante-chamber.
- Two axial hatches with doors ⌀ 600 mm.
- One lateral hatch with door ⌀ 600 mm to transfer chamber.
- One intermediate hatch with door ⌀ 600 mm between living compartment and ante-chamber.
- Six bunks, eight viewports, one fitted with tv-control.
- Ante-chamber with toilet, shower and washbasin.
- One medical lock.
- One vertical hatch with door ⌀ 600 mm to rescue trunk.

The diving system fulfills classification requirements of Det Norske Veritas.

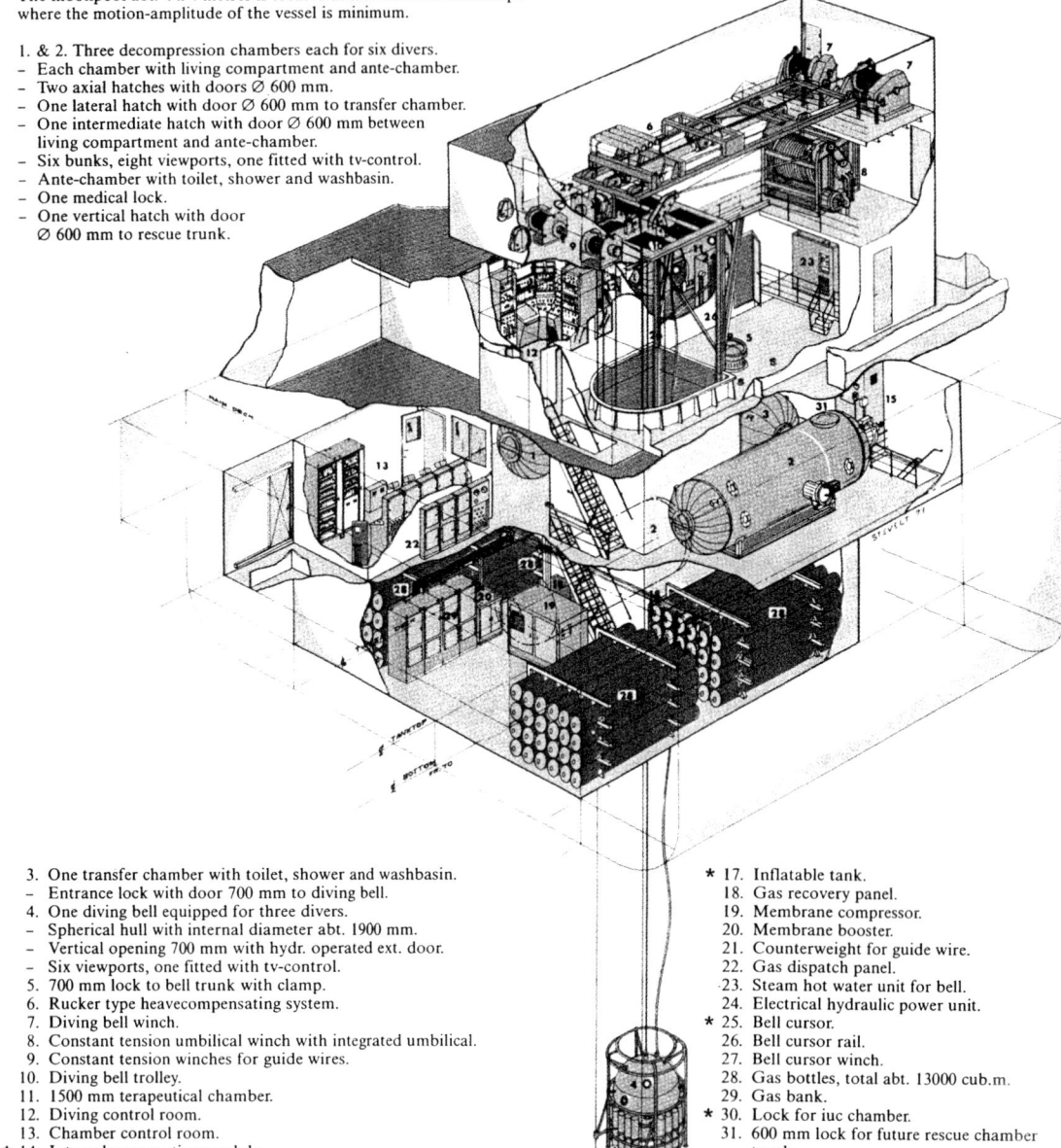

3. One transfer chamber with toilet, shower and washbasin.
- Entrance lock with door 700 mm to diving bell.
4. One diving bell equipped for three divers.
- Spherical hull with internal diameter abt. 1900 mm.
- Vertical opening 700 mm with hydr. operated ext. door.
- Six viewports, one fitted with tv-control.
5. 700 mm lock to bell trunk with clamp.
6. Rucker type heavecompensating system.
7. Diving bell winch.
8. Constant tension umbilical winch with integrated umbilical.
9. Constant tension winches for guide wires.
10. Diving bell trolley.
11. 1500 mm terapeutical chamber.
12. Diving control room.
13. Chamber control room.
* 14. Internal regeneration module.
15. Hot and cold water plant.
* 16. Gas recovery tank.

* 17. Inflatable tank.
18. Gas recovery panel.
19. Membrane compressor.
20. Membrane booster.
21. Counterweight for guide wire.
22. Gas dispatch panel.
23. Steam hot water unit for bell.
24. Electrical hydraulic power unit.
* 25. Bell cursor.
26. Bell cursor rail.
27. Bell cursor winch.
28. Gas bottles, total abt. 13000 cub.m.
29. Gas bank.
* 30. Lock for iuc chamber.
31. 600 mm lock for future rescue chamber trunk.
* 32. 600 mm lock to hyperbaric rescue boat.
* 33. Hyperbaric rescue vessel equipped for 18 divers under pressure.

* Item not shown on this picture.

SATURATION DIVING COMPLEX
(Stena Offshore Limited)

Part 1. DIVING

Offshore installations are analogous to icebergs in as much as the bulk of the structure lies beneath the waves and as such the industry is heavily dependant on the diving fraternity for its continued well-being. The offshore industry in the North Sea provides employment for nearly 1,500 divers who are involved in the installation, repair, maintenance and inspection of structures and pipelines and whilst the current trend is towards a reduction in diver intervention, there will always be a place for the levels of manual dexterity that only man can provide.

Diving operations may be divided into two categories which are dictated by water depth. For depths of less than 50 metres (165 feet) air diving techniques may be employed whilst greater depths necessitate a full *saturation diving* programme. Before discussing the various aspects of air and saturation diving, we should first consider the effects that prolonged immersion in deep water have on the human body, for it is in the sensitivity of the body to environmental change that dictates the type of equipment required for life support.

At normal atmospheric pressure the tissues of the human body are in a state of equilibrium with their surroundings, that is to say they are saturated with the dissolved gases that constitute air, primarily nitrogen and oxygen. This state of equilibrium or saturation is disturbed the instant the diver enters the water.

As the diver descends the hydrostatic pressure on the body increases the gas absorption capacity of the tissues. Given sufficient time at any particular depth they will once again reach a point of equilibrium or saturation. The eventual ascent from saturation depth must be carefully controlled and include in water stops or pauses to permit the natural release of these newly dissolved gases if the diver is to avoid the *bends* (expansion of gas in the blood stream). This is a time consuming process and highlights the major difference between air and saturation diving. The air diver must restrict the depth and duration of his dive to prevent the body reaching the fully saturated condition.

1. AIR DIVING

Air diving is used primarily for dives of short duration in water depths of less than 50 metres (165 feet) where excessive decompression times can be avoided. As the name suggests, the divers breathe compressed air which is supplied through an umbilical.

Air diving is used for the vast majority of underwater inspection and repair programmes carried out in shallow waters such as the southern sector of the North Sea where the strong tides impose a natural restriction on the time the divers can spend immersed, approximately one hour in six. However, it should be appreciated that dive duration is limited dramatically with increase in depth and at 50 metres a stay of only 10 minutes is permitted. The ascent must also include in water stops to facilitate the bodies decompression processes to proceed naturally.

Where conditions permit, a deck decompression chamber (DDC) can be used to increase dive time by removing the need to carry out in water decompression stops. The diver returns to the surface as quickly as possible and enters the DDC which is then pressurised with oxygen to a pressure equivalent to the water depth at the work site. Over a period of about an hour the pressure is gradually reduced to atmospheric, the oxygen assisting in the dispersal of dissolved bodily gases.

The main advantage of air diving is that it does not require the sophisticated equipment associated with saturation diving. Air diving plant may be containerised and located on the deck of a *supply boat*, or a fixed installation, thus saving on the considerable costs incurred when chartering a fully equipped diving support vessel (DSV).

2. SATURATION DIVING

The main attraction of saturation diving is the increased productivity attainable from a diver due to the removal of restrictions on dive time and the need to decompress after each and every dive. Decompression takes place only on completion of the diving programme, or when the diver is due to go on leave, normally after a period of 30 days. The disadvantage of saturation diving is the cost which can be considerable.

A diving programme commences with the transportation of the divers to the work site within the *diving bell*. At the site the divers enter the water through a hatch in the base of the bell having first donned helmets/masks and attached their umbilicals to the distribution manifold on the inside of the bell.

The umbilicals provide the diver with breathing gas and a supply of hot water to heat the diving suit. Subject to favourable tidal conditions, the divers spend approximately 6 to 8 hours at depth during which time the diving bell is used as a habitat where they can eat and rest. On completion of work the divers return to the bell, lock the hatch and are winched back to the diving support vessel (DSV).

Once on board the DSV the bell is locked onto the transfer hatch of the *deck decompression chamber* (DDC) whilst the divers enter the DDC where they will remain until their next shift. Throughout the entire operation the divers remain under pressure which may be considerable, anything from 5 to 25 bar (75 to 375 psi) depending on the saturation depth of the work site. Loss of pressure at any time would cause serious injury or death. This cycle may continue for a period of up to three weeks after which the divers commence *decompression*.

Decompression involves the gradual reduction of the pressure within the DDC to atmospheric to permit the natural dissipation of excess gases from the diver's body. It takes approximately one hour per metre (3.3 feet) of water depth to effect safe decompression thus a saturation depth of 180 metres (600 feet) requires a decompression interval of nearly a week.

3. EQUIPMENT

A saturation diving system consists of a *diving bell* (submersible compression chamber, SCC) and a deck decompression chamber (DCC) and this combination of pressure chambers provide the means by which a diver's body is artificially maintained at a pressure equivalent to the work site for prolonged periods.

i) DIVING BELL (SCC)

The diving bell provides transportation to and from the work site. It is secured to the diving support vessel (DSV) by a steel cable and an umbilical provides the occupants with breathing gas, heated water, electrical power and communication facilities.

The bell is housed within a substantial steel framework designed to provide protection against impact and a location for the ballast tanks and life support systems. An emergency release system permits disconnection of the steel cable and umbilical should the divers encounter a situation which necessitates isolation from the DSV for instance a fire on the DSV or a fouled umbilical. The life support systems contain 96 hours of breathing gas and the facility to de-ballast the bell to permit a return to the surface.

ii) DECK DECOMPRESSION CHAMBER (DDC)

Deck decompression chambers are used for both air and saturation diving operations. The *air diving* DDC is a relatively simple affair consisting essentially of a cylindrical twin chamber pressure vessel. The *saturation diving* DDC is considerably larger and more lavishly equipped and provides an out of water home for up to 6 divers for the duration of the diving programme. It contains beds, toilets, showers, messing and recreational facilities (TV) and small air locks through which hot meals, laundry and medication can be passed to the divers. The occupants of the DDC are monitored 24 hours a day by the diving support crew and a particularly watchful eye is maintained during the decompression process.

iii) DIVING SUITS

Diving suits used for offshore inspection and repair programmes fall into two categories largely dictated by the degree of thermal protection required by the diver.

a. Dry suit

The dry suit consists of a watertight outer layer under which the diver wears thermal clothing to assist in the retention of body heat. In a cold water environment such as the North Sea the dry suit is only suitable for air dives of short duration beyond which a heated wet suit is preferred.

b. Wet Suit

The prolonged periods of submersion associated with saturation diving necessitate the wearing of a heated wet suit. Hot sea water supplied via the umbilical cable is circulated around the inside of the suit prior to exit at the divers extremities.

iv) HELIOX

Air consists primarily of oxygen and nitrogen and both these gases can prove injurious to the human body when ingested under the pressures associated with water depths greater than 50 metres (165 feet). The oxygen becomes toxic whilst the nitrogen leads to *nitrogen narcosis* or *drunkenness of the deep*. Heliox, a specially formulated mixture of helium and oxygen provides a solution to both these problems because it permits the quantity of oxygen to be regulated to suit the body's respiratory needs at a particular depth. Helium like the nitrogen it replaces is an inert gas which fulfils no useful function in the respiratory process other than to bulk out the mixture and support the desired quantity of oxygen.

Whilst heliox may be used by divers who are not in saturation it is a gas primarily associated with saturation diving. The divers breathe the mixture throughout the saturation period, including the time spent in the DDC in between dives. In fact it is the period within the DDC in which another advantage of heliox emerges. The gas is much lighter than compressed air thus requiring less respiratory effort which ensures that the divers obtain restful sleep.

Offshore Engineering

Heliox does have certain disadvantages which are the increase in heat transfer from the body which is considerable, the Donald Duck speech impediment created by the action of the gas on the vocal chords, and the cost. The cost and adverse conductivity can be moderated by incorporating gas heating and recycling equipment whilst a voice synthesizer will minimise speech distortion.

4. VESSELS

i) DIVING SUPPORT VESSEL (DSV)

The primary function of the diving support vessel is to provide a stable platform from which the divers can carry out their duties. The complexity of the vessel will be determined by the type of diving operation for which it was designed, that is to say a saturation diving vessel will be considerably better equipped than one catering solely for air diving.

The DSV represents the diving company's single largest investment and the more divers the vessel can support, the more cost effective it becomes. The latest state of the art vessels employ up to 3 diving systems (3 deck decompression chambers and 2 diving bells) which can accommodate 18 divers working around the clock in water depths ranging from 20 to 450 metres (65 to 1500 feet).

The charter rates for a modern fully equipped saturation diving support vessel are considerable ($80,000 to $160,000 a day) and to minimise lost time due to adverse weather conditions the vessels employ sophisticated active and passive hull stabilisation equipment and dynamic positioning (DP). The DP computer co-relates information pertaining to wind speed, wave height, currents and satellite reference positions and generates command signals for the control of the main propulsion and thruster units so that the vessel can remain on station and continue working in weather conditions approaching Beaufort 8 (gale force) without the need to deploy anchors. The effects of the weather are further negated by the *moonpool*, a vertical shaft which runs through the centre of the ship and is open to the sea. The moonpool provides a sheltered haven for the launch and recovery of the diving bells and represents a considerable improvement on the traditional stern mounted "A" frame launching davit.

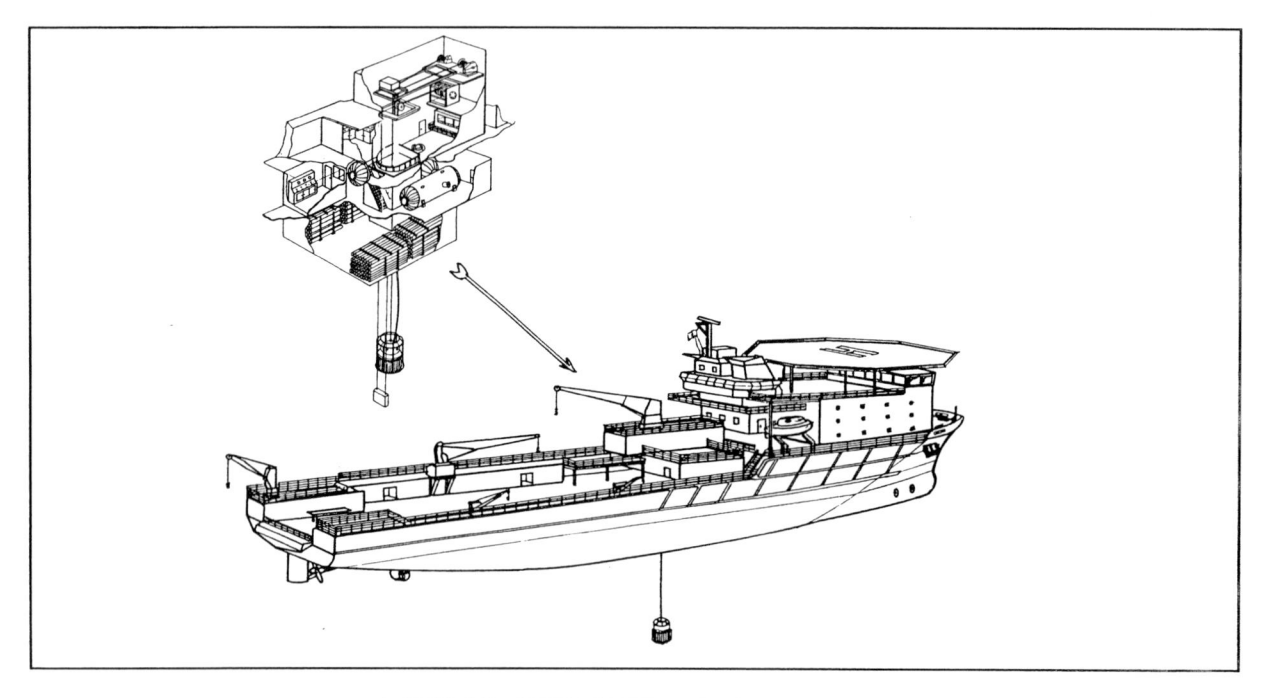

DIVING SUPPORT VESSEL (DSV)

In an effort to increase the operational effectiveness of the DSV a number of vessels have been provided with well service equipment and can carry out intervention and workover programmes for the blossoming sub-sea market. This facility increases the likelihood of obtaining employment beyond the summer months after the routine underwater inspections have been completed.

ii) REMOTELY OPERATED VEHICLES (ROV)

A remotely operated vehicle (ROV) is essentially a mini-submarine that can be controlled from the deck of a ship or an offshore installation via an umbilical. The concept originated in the late 1970's and was developed to provide a cost effective alternative to the deep sea diver for routine inspection duties.

The external appearance of an ROV will depend largely on the type of work it is designed to carry out. The small inspection units often employ a spherical or cylindrical glass reinforced plastic (GRP) or acrylic hull to contain the electronic control and video equipment. The larger work units tend to use aluminium alloy pods to house delicate equipment. A steel frame provides protection against impact, and mounting points for ballast tanks, battery pods, manipulators (if fitted) and attachment of the umbilical. The complete assembly may weigh anything from 0.5 tonnes to 2.0 tonnes. Once in the water the ROV is manoeuvred by a number of motor driven ducted propellers which are electrically powered from self-contained battery packs, or from an external source supplied via the umbilical cable. Buoyancy tanks assist in the maintenance of trim and water depth.

The first generation ROVs were designed purely to *eyeball* underwater activities, being driven around the work site whilst video pictures were relayed back to the command centre. They were, and still are used extensively to carry out the visual inspections on pipelines and jacket structures associated with the annual surveys. Further developments and the addition of articulated manipulators or crabs claws permit operations such as marine growth removal, ultrasonic thickness measurement and bolt tensioning to be carried out. However, it was the offshore industries acceptance of the advantages offered by the sub-sea wellhead that provided impetus to ROV development during the 1980's.

The second generation ROVs were developed hand in hand with the new modular designed diverless sub-sea wellhead packages. These wellheads can be installed, maintained and repaired entirely by ROV and ending the reliance on saturation divers has permitted the exploitation of reserves located in water depths beyond the range of conventional diving techniques.

A discussion on ROVs would not be complete without a brief mention of manned atmospheric diving systems (ADS) such as the Osel Mantis and Slingsby Engineering's Spider. Occasionally referred to as microsubs they share many similarities with an ROV with the obvious exception of the increased hull capacity required to accommodate the pilot and life support systems. In fact the Mantis was constructed as a dual purpose vehicle and could be used in the manned and unmanned ROV mode. The main advantages of manned submersibles are the infinite levels of control afforded by an experienced pilot when compared to the remote operation of an ROV, and the fact that they can be used at depths of up to 1,000 metres (3,300 feet), nearly twice the depth at which a saturation diver can operate. The hulls are maintained at a constant pressure of one atmosphere so there is no need for the pilot to undergo decompression after the dive. Unfortunately, these days manned submersibles tend to be restricted to military service, their use being discouraged by the leading oil companies on the grounds of safety, in spite of their excellent safety record.

Offshore Engineering

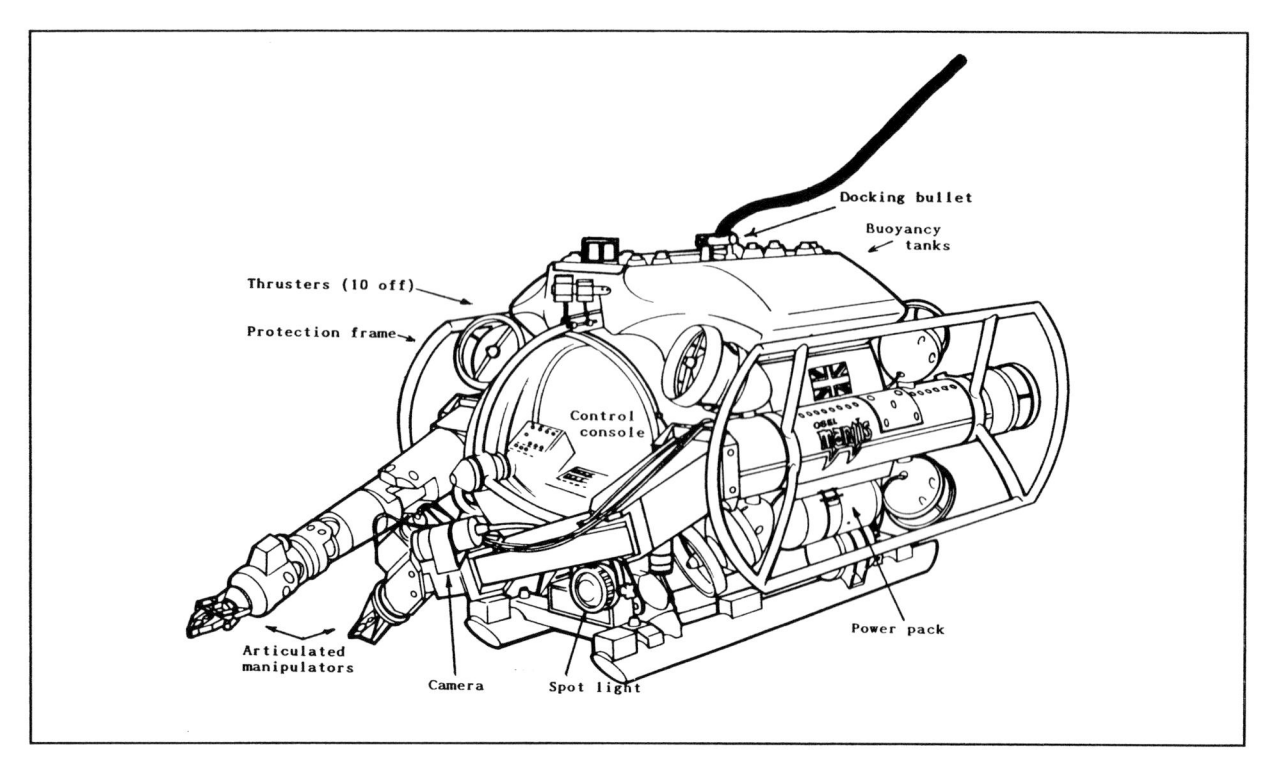

OSEL MANTIS Manned/unmanned ROV
(Hydrovision Ltd)

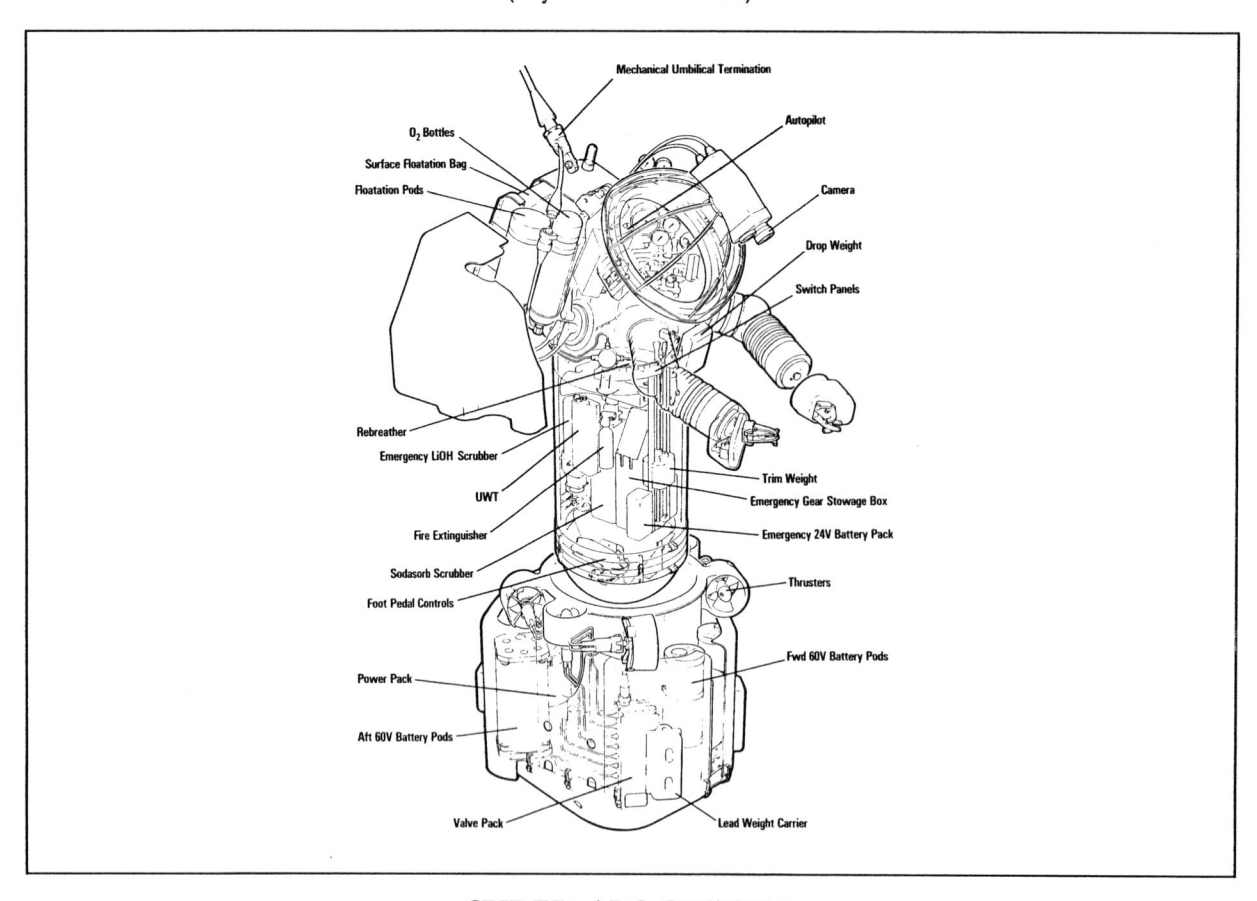

SPIDER ADS SYSTEM
(Slingsby Engineering Ltd)

Part 2. UNDERWATER SURVEYS

Specialised diving contractors are employed by the installation owners to carry out a planned inspection programme of the underwater structure and pipelines designed to detect any potential defects before they develop into major problems.

Up until 1998, in the UK sector of the North Sea, the scope of the planned inspection programme and its frequency were traditionally agreed with the Certifying Authority who issued the Certificate of Fitness for the installation. Under Verification, the role of the Certifying Authority has disappeared having been replaced by the Independent, Competent Person (ICP), and it is the ICP who will now agree the inspection programme. As before it is likely to include the following activities.

1. SPLASH ZONE VISUAL INSPECTION

Definition — *That part of the structure between the crest level of the 50 year (average) wave superimposed on the highest astronomical tide, and 3 metres below the lowest astronomical tide.*

The *splash zone* can loosely be defined as that area of the jacket between the high and low water lines. It is a region which is particularly prone to corrosion and erosion because wave action makes maintenance of the protective paint coatings particularly difficult. It is also the area most likely to sustain damage from impact with supply boats.

Large installations with design life spans in excess of 25 years are occasionally clad with copper-nickel in the vicinity of the splash zone to reduce corrosion and marine growth.

The splash zone survey will include an inspection of spider decks, boat moorings and landings, *caisson* clamps and *risers*.

2. SWIMAROUND SURVEY

The object of the underwater swimaround examination is to detect any obvious signs of damage or distress such as missing structural members, failed joints, impact damage, loose or missing riser and caisson clamps.

The sea bed will also be examined for 'debris' which may have fallen from the rig. This frequently includes gratings, scaffold poles and an assortment of objects dropped during construction projects.

3. NON-DESTRUCTIVE EXAMINATION

An installation jacket consists of a considerable number of tubular members welded together. The point at which one tubular is welded to another is referred to as a joint or node.

The planned inspection programme will identify those joints which due to their design are considered to be particularly highly stressed or to have low fatigue lives. A representative number of these joints will be thoroughly examined throughout the life of the jacket.

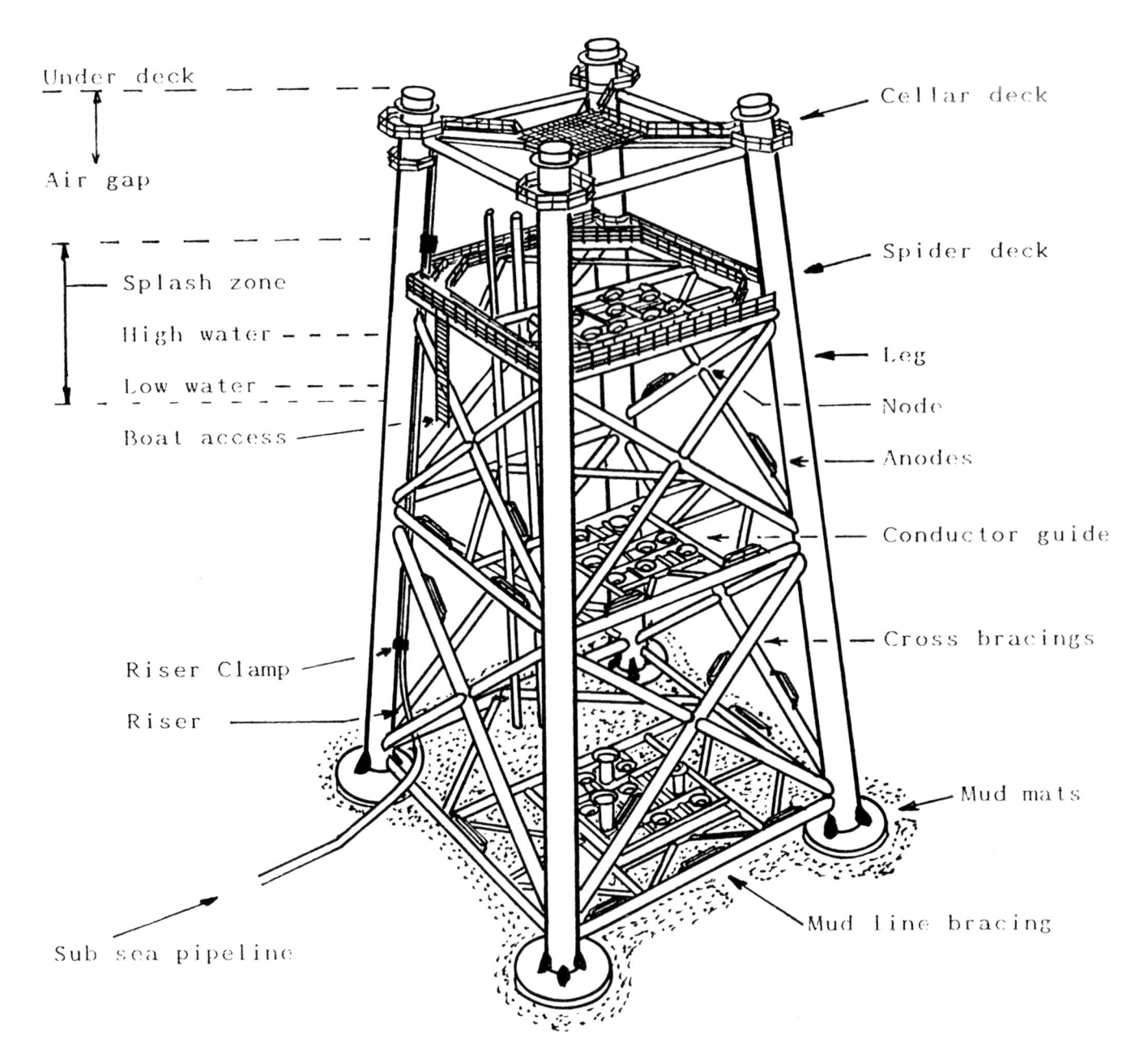

JACKET — COMPONENT PARTS

A thorough examination entails removing marine growth and scale by means of a powered wire brush until bright metal is exposed. The welds will then be subjected to a thorough visual examination as a minimum, supplemented with a non-destructive examination technique such as eddy current or magnetic particle inspection (MPI). Should visual examination reveal a significant defect such as a crack, the suspect area will be subjected to a magnetic particle or ultrasonic inspection. If the crack is confirmed some remedial grinding may be carried out. Should the defect remain after grinding a more extensive repair programme must be considered.

It should be noted that most jacket designs include a certain number of redundant structural members so discovery of a defect should not cause unnecessary concern. An engineering analysis will be carried out to determine the extent and the urgency of the action required.

4. FLOODED MEMBER SURVEY

This form of examination is relatively new to the industry but has gained considerable popularity due to the ease at which it can be carried out. Minimal cleaning of the structural member is required prior to attachment of an ultrasonic measuring device. The device can determine whether or not the member is full or partially full of water. The presence of water would indicate a crack or weld failure. If a flooded member is discovered it must be thoroughly examined using magnetic particle or ultrasonic inspection to determine the extent of the defect.

5. MARINE GROWTH MEASUREMENT

The additional weight imposed on an underwater structure by the presence of marine growth can be quite considerable. It must be carefully monitored if overloading of the structure is to be avoided.

Marine growth can be divided into two categories, hard and soft. The hard growth consists of crustacea such as mussels and the soft growth can be a combination of sea weeds and sponges. The soft weed is measured by wrapping a tape measure around the member to give a compressed growth reading.

A marine growth measurement of 50 mm to 100 mm (2 to 4 inches) in depth can normally be tolerated before a cleaning programme is required. The thickness permitted will depend on the type of growth and the design of the structure.

6. SCOUR SURVEY

The sands of the sea bed have a tendency to drift in the same way as the sands of the desert. Scour is the term used to describe this sub-sea phenomenon. It is measured from a datum point, usually the first horizontal jacket bracing above the sea bed or mud line.

It is not uncommon to find a variation in sea bed levels of up to 4.5 metres (15 feet) between adjacent jacket legs and if left unchecked the foundations of the installation could eventually become exposed and threaten the security of the installation.

Scour can also present problems with sub-sea pipelines by removal of the sea bed under the pipeline. The pipeline can be left suspended for considerable distances. These *free spans* can induce excessive strains in the steel and leave the pipeline in a position where damage may be sustained from ships anchors or fisherman's nets.

The problem is normally rectified by dumping rocks, gravel or laying concrete mats to build up the sea bed and prevent further removal of sand. Another alternative is to pin nylon scour mats to the sea bed. These fibrous mats are constructed to encourage sand retention.

7. CATHODIC PROTECTION EXAMINATION

The bulk of the underwater structure consists of bare steel and a cathode protection system is employed to inhibit corrosion. This involves the attachment of sacrificial anodes, typically zinc or an aluminium/zinc/indium alloy at strategic positions on the jacket legs and bracings. Galvanic action promoted by the presence of sea water decomposes the anodes in preference to the steel structure.

Sufficient anodes (anything from 7 to 70 tons) are attached during construction of the jacket to last the life of the installation, normally 25 years. The underwater survey consists of monitoring anode condition and measurement of the potential difference generated. The potential should exceed 0.84 volts and can be measured directly using a *Bathy corrometer*.

There are two further surveys associated with monitoring the performance of the installations support structure. Whilst not part of the underwater survey they have been included in this chapter because they are structure and foundation related.

8. DIFFERENTIAL SETTLEMENT SURVEY

Differential settlement readings provide an indication as to the continuing satisfactory performance of the jacket foundation piles.

The measurements may be obtained using an instrument similar to a building surveyors theodolite, or laser lines of sight directed at a fixed target on the shore line. In effect it entails measurement of the height of each corner of the installation relative to the other corners to see if any tilting has occurred.

Minor variations may be attributed to changes in deck loadings as may occur when any major items of equipment are added, or removed.

9. AIR GAP MEASUREMENT

An offshore platform must be designed to resist the severest weather conditions. To withstand wave action there must be sufficient, or air clearance under the lowest deck on the installation to accommodate what is known as a design extreme wave crest based on a 50 year return period, and still retain 1.5 metres (5 feet) of clearance gap. Occasional wave heights of 25 metres (85 feet) have been measured in the North Sea. The air gap must be monitored on a regular basis because it gives an indication of any settlement or sinkage that may have occurred due to reservoir depletion and sea bed subsidence.

Severe problems were encountered in the Norwegian *Ekofisk field* when several installations showed signs of sinking. The problem was addressed by cutting the jacket legs, jacking up the topside structures a distance of 6 metres and installing spacer spools.

Part 3. SUB-SEA WELLS

INTRODUCTION

Sub-sea wells have been around for over 30 years and there are over 600 sub-sea wells operating in the UK sector of the North Sea alone. Basically sub-sea wells have two applications. They may be used in support of a fixed installation as an alternative to a satellite platform for recovering reserves located beyond the reach of the drill string, or used in conjunction with a *floating production system* (FPS). The combination of sub-sea wells and floating production system provide a cost effective partnership for the recovery of oil from marginal fields, those containing recoverable reserves of less than 50 million barrels of oil or gas equivalent, and often the only means of developing very deep water fields.

The Rob Roy development illustrated is a good example of a field development using a floating production system. Commissioned in 1989, Rob Roy is still actively producing oil and likely to remain so as new wells were added in 1998. The *British Petroleum Eastern Trough Area Project* (BP-ETAP) is also illustrated, one of the largest and most innovative North Sea developments of the decade. The ETAP development is based around a conventional manned platform, the *Central Processing Facility* (CPF) which processes oil and gas from the Marnock reservoir and the six satellite reservoirs. A *not normally manned installation* (NNMI) is located over the Mungo reservoir whilst Machar and Monan are produced from sub-sea manifolds, as are Heron, Egret and Skua (owned by Shell) which are developed jointly as three separate sub-sea wellhead clusters.

The individual reservoirs which comprise the ETAP development are relatively small, but collectively they constitute some 435 million barrels of oil and NGL (natural gas liquids — propane and butane), and 1.1 trillion cubic feet of gas. At peak production the development will produce 216,000 barrels of oil a day and 553 million standard cubic feet of gas. In the North Sea, which is now considered to be a mature province, there are a considerable number of these *marginal fields* and extensive research has been carried out on the various options for the recovery of the oil and gas as the number of new, large field discoveries have diminished.

1. SUB-SEA DEVELOPMENT OPTIONS

A sub-sea development may consist of any number of wells. Where the development consists of more than one well, the wells may either be located on a *template*, in clusters feeding back to a *manifold*, or a sub-sea *flowbase* may be used.

IVANHOE, ROB ROY & HAMISH FIELDS SCHEMATIC

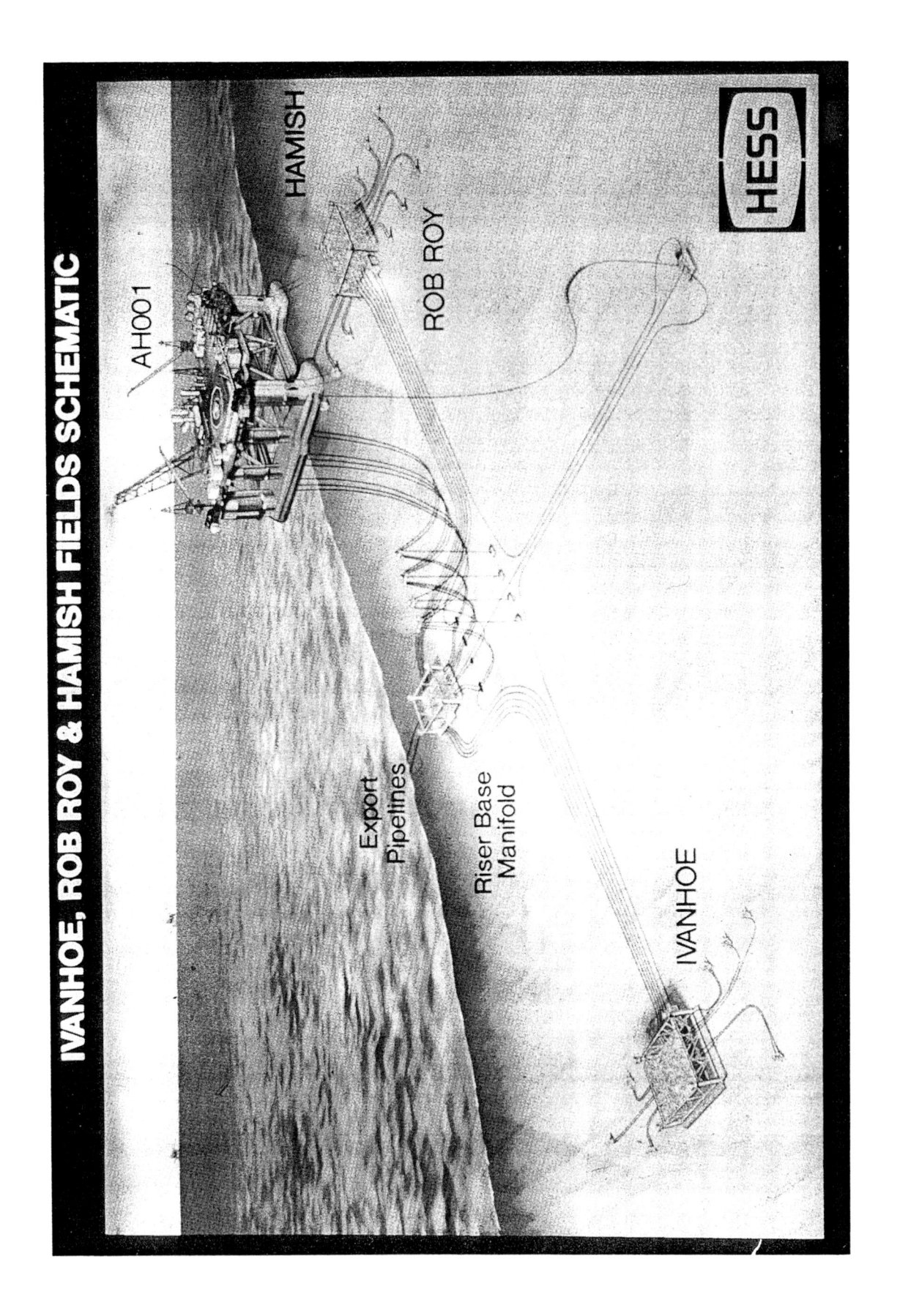

SUB-SEA FIELD DEVELOPMENT

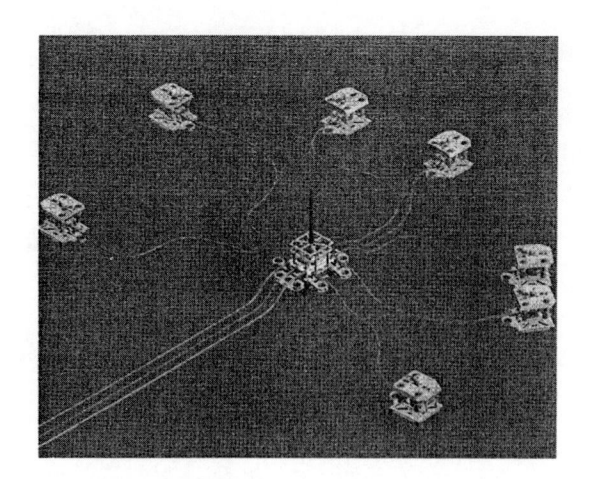

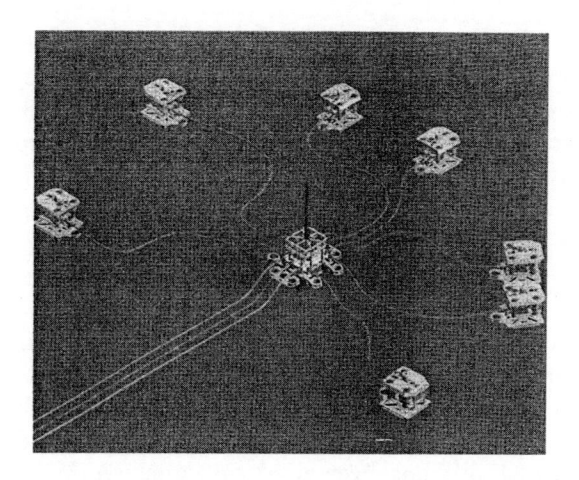

HOIST MANIFOLD CENTER WITH SATELLITE WELLS   **HYBRID HOST APPLICATION**

(Reproduced with permission of Konsberg Offshore as)

1.1 TEMPLATES

A sub-sea template is essentially a steel guideframe which is secured to the sea bed by piles or suction anchors and through which the wells are drilled. The templates may be extremely large, heavy structures, up to 800 tonnes, which can accommodate a considerable number of wells, or they may be small and light with provision for just two to four wells. The template provides guides to assist in the location of the Christmas trees and a manifold arrangement to connect the trees with the sub-sea pipeline. A removable structure sits over the structure where over trawl protection is required.

1.2 MANIFOLDS

At first glance, a sub-sea manifold looks very similar to a sub-sea template, the difference being that the structure contains just a manifold, the Christmas trees being located around the structure rather than within it, often at locations several hundred metres distant. The manifold is essentially a marshalling point for the flowlines and control umbilicals.

There are advantages and disadvantages associated with both manifolds and templates. Manifolds may reduce the amount of extended reach drilling required because the wells can be sited at the optimum field location. Also, being relatively light, the manifold structure can be installed from the drilling rig rather than a crane vessel which considerably reduces installation costs. However, the cost of the additional sub-sea flowlines and control umbilicals associated with manifold structures can be prohibitive as they are extremely expensive to manufacture, install and test. Ultimately the decision as to whether manifolds will be used in preference to templates or vice versa will be governed partly by the reservoir characteristics and partly by installation and material costs. Some large developments may even use a combination of the two.

Offshore Engineering

1.3 SUB-SEA FLOWBASE

A flowbase allows one well to be flowed and then commingled with the output from a second well in what is generally described as a *daisy-chain* arrangement. The use of flowbases allows the operator to drill and tie back wells over an extended period of time, the only drawback being that it reduces the flexibility for individual well testing. When reliable sub-sea *multiphase meters* become available the need to individually test wells will no longer be necessary.

1.4 PIPELINES AND RISERS

A sub-sea pipeline is generally used for the transportation of the reservoir products from the manifold or template back to the host installation and there are a number of issues to be dealt with if transportation problems are to be avoided, particularly on some of the new developments.

Due to the waxy nature of some crude oils, a system for the circulation of a warm fluid (diesel oil) has been used on some recent developments to preheat the flexible production risers, pipelines and sub-sea manifold in order to minimise the deposition of wax from the oil on start up. Other developments have dealt with similar situations by installing one pipeline within another to provide an outer jacket through which hot water can be circulated, or in which insulating materials may be housed.

Another difficulty that has to be overcome is the *pigging* of the pipeline, a subject explained more fully in chapter five. Pigging is facilitated by either installing two pipelines to the manifold or template so that a *round trip* pigging operation can be carried out to clean the pipeline, or by installing a loop at the blind end of the pipeline with a series of remotely operated valves, the pig can be re-directed within the loop prior to its return to the installation. Alternatively, sub-sea *pig launcher-receivers* may also be installed but all these options are extremely expensive to put into effect.

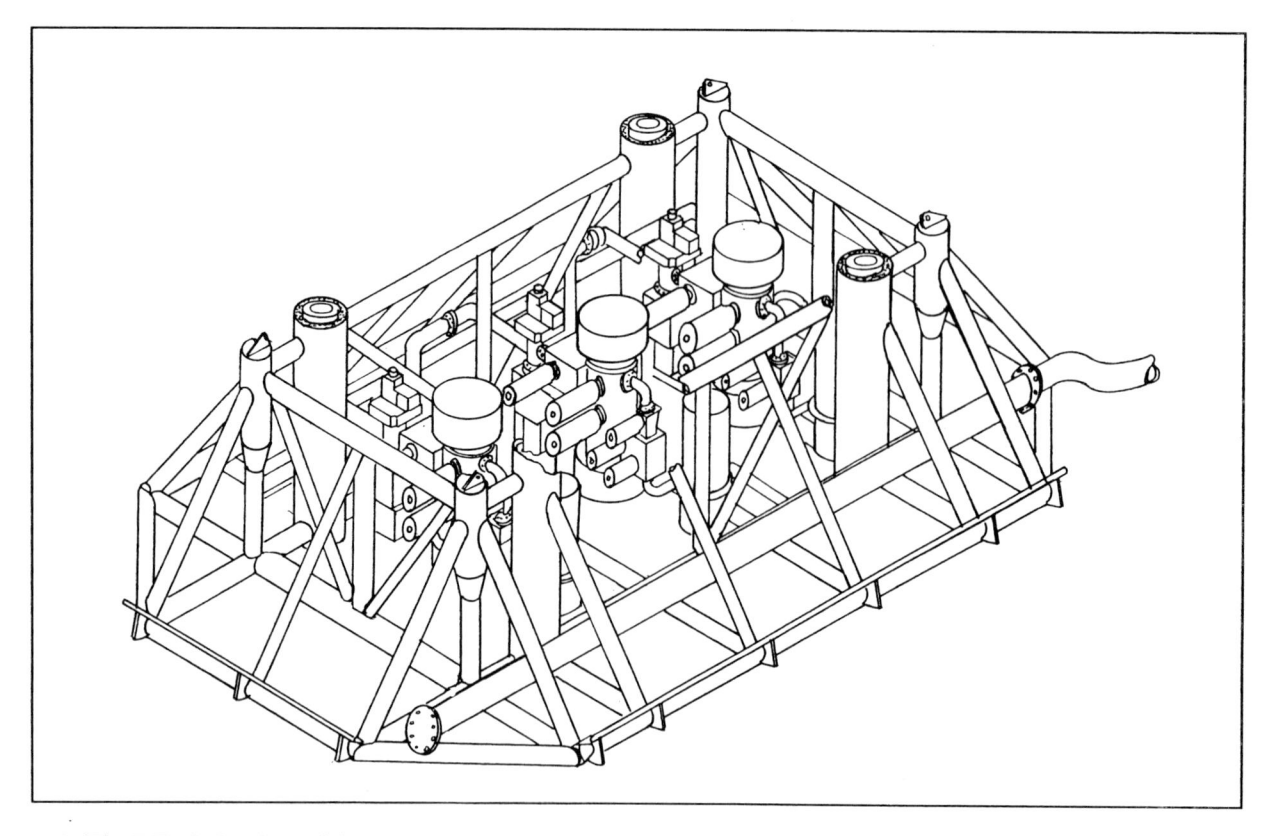

3 TREE SUB-SEA TEMPLATE AND WELLHEAD PROTECTION STRUCTURE

2.0 SUB-SEA WELLS

A sub-sea well consists essentially of a Christmas tree, wellhead assembly, guide frame, and protective structure. The wells are drilled from a jack-up or a drill ship in a manner identical to conventional wells which terminate on a fixed platform.

2.1 CHRISTMAS TREES AND WELLHEADS

The most basic sub-sea wells employ a simple diver assist Christmas tree mounted over a *mudline suspension system*, the Christmas tree being bolted to a hydraulically activated *collet connector* prior to immersion to facilitate attachment to the wellhead adaptor. However, as the need has arisen to produce oil from deeper waters, more specialised sub-sea wellheads and Christmas trees have been developed.

From an installation point of view, there are basically four types of sub-sea wellhead and Christmas tree and it is the water depth that ultimately governs the type of system that will be used.

i) **MUDLINE/SIMPLE**

In waters that are regarded as shallow, less than 90 metres (300 feet) in depth, very basic systems are used which are totally reliant on divers to connect the flowline and control systems and, to assist in well workover operations.

ii) **DIVER ASSIST**

Systems installed in waters regarded as moderate in depth, 60–210 metres (200–700 feet) will have some diverless features. Diver assistance will be required for the installation of the flowline and control systems but workover operations will be carried out without the assistance of divers.

iii) **DIVERLESS**

Whilst divers can operate at water depths of up to 540 metres (1800 feet), the realistic limit for effective work is nearer 180 metres (600 feet). Consequently, for systems installed in deep waters in the range of 180–900 metres (600–3000 feet), totally diverless designs are required and there are two very different approaches to solving the problems associated with developing fields in very deep waters.

The two systems are categorised as either with, or without guidelines, the latter being referred to as *guidelineless*. The first system employs four guidelines which run from the drill ship or semi-submersible down to guide posts mounted on the sub-sea guidebase. The tree, flowline and control systems are designed so that they can be guided into position by the wires and as the component parts are stacked, one on top of the other, they latch together and seal.

iv) **GUIDELINELESS**

Deepwater guidelineless systems are designed to be installed from vessels which are dynamically positioned, often in water depths where conventional anchors cannot be deployed. All functions, including well workovers are performed remotely, that is without mechanical links between the ship and wellhead. The location of each component is achieved using either *funnel up* or *funnel down* designs, the operation of which is self explanatory as shown by the illustrations. The deepest sub-sea well installed to-date is in the *Campos Basin* of Brazil in a water depth of 1709 metres (5700 feet).

Offshore Engineering

Sub Sea Christmas Tree

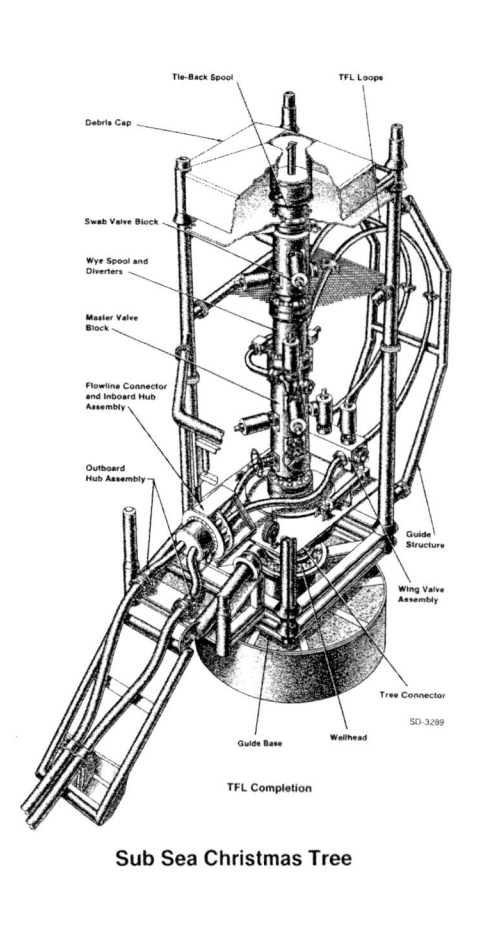

Sub Sea Wellhead

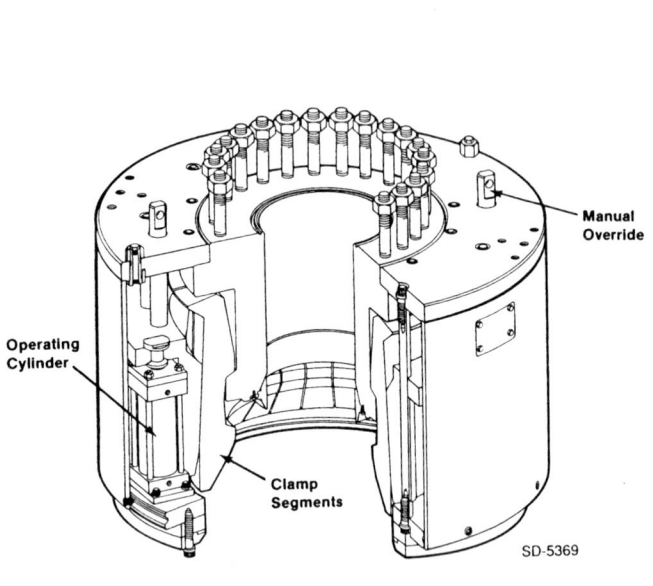

Model 70 Hydraulically Operated Tree Connector

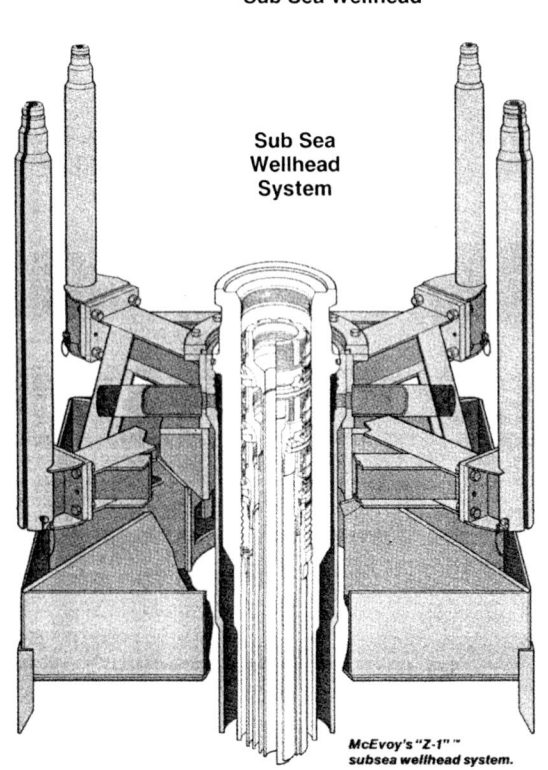

Sub Sea Wellhead System

McEvoy's "Z-1"™ subsea wellhead system.

SUB-SEA WELLHEAD EQUIPMENT
(Reproduced with permission of Cooper Oil Tools, Houston, Texas)

Ending the reliance on *saturation divers* has opened up vast reserves up to and beyond the thousand metre (3,450 feet) depth range that were previously thought to be unattainable. Whilst still referred to as new technology the *diverless wellhead* has proven extremely reliable and is being considered for use in more moderate water depths where the high initial cost would still appear to be justified when compared with the financial savings to be made on conventional diver intervention.

2.2 GUIDEFRAMES

The sub-sea tree is surrounded by a guideframe which acts as an installation device for the component parts of the tree and for the wellhead protection framework.

From an operational viewpoint the guideframes provide three basic functions:

i) **INTERFACE**

The guideframes are designed specifically for a particular type of wellhead system and provide an interface which assists in the alignment of the tree during installation.

ii) **MOUNTING POINTS**

Attachments may be provided on the guideframe for the mounting of the flowline connector, the ROV docking station, and the control umbilical through which the electrical or hydraulic signals are conveyed to operate the Christmas tree valves.

iii) **RE-ENTRY GUIDANCE**

The top of the guideframe may be fitted with guidance posts or funnels to assist in the alignment of the tree running tool and tree cap which are used to install, retrieve and workover the tree.

2.3 WELLHEAD PROTECTION STRUCTURE (WPS)

Sub-sea trees are prone to damage, the most common causes being impact damage from ships anchors or from fishermens trawlboards. Protective structures are normally fitted to provide some degree of protection and these will be installed over the well bay where the trees are installed on a template or built directly over a satellite or cluster tree. The protection structure may be rigidly attached to the well or it may alternatively be a gravity device. Either way it must be of a design which can be removed with the minimum of effort to facilitate drilling, well completion and workover operations to proceed.

Offshore Engineering

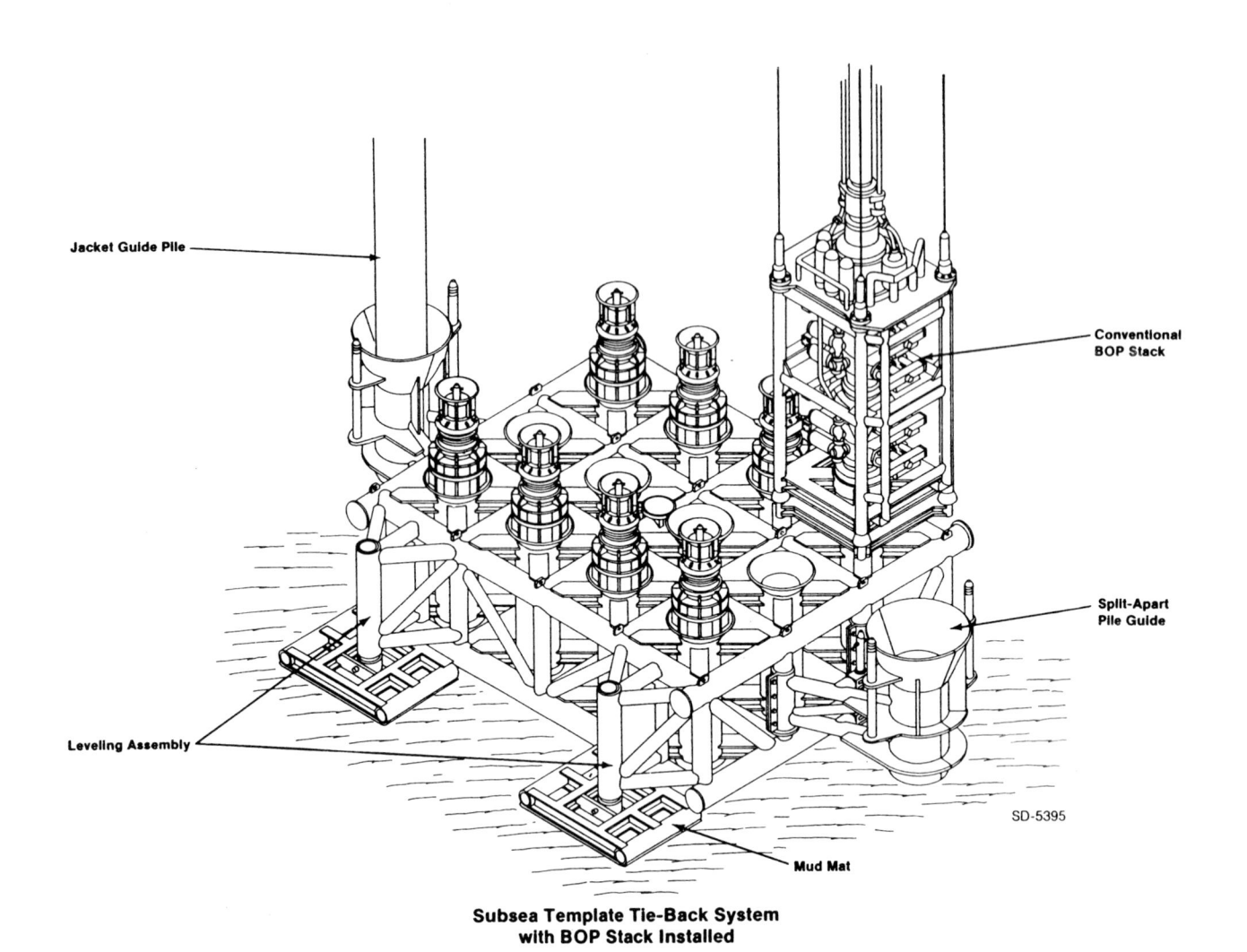

**Subsea Template Tie-Back System
with BOP Stack Installed**

SUB-SEA TEMPLATE
(Reproduced with permission of Cooper Oil Tools, Houston, Texas)

17 – *The jack-up Galveston Key testing an appraisal well. The vent boom can be seen flaring the gas. The V door and pipe draw can be seen at the base of the derrick.*

18 – The Ocean Benarmin jack-up with its drilling derrick cantilevered over the Mobil Camelot 53/1a unmanned gas platform. The Benarmin is completing the wells following installation of the topside structure.

19 – Drill crew hard at work on the drill floor. Note the drill pipe stands standing vertically to the left of the picture. Inset – the drill crew are dwarfed by the travelling block hook, the gooseneck and rotary hose, the swivel and drive motor.

*20 – Varco BJ top drive drilling motor undergoing an examination.
The motor guide rails and drill pipe stands are clearly visible.*

*21 – Blowout preventer (BOP) stack piped up and ready for action.
The stack consists of an annular BOP twin, and single ram BOP's.
Inset – the tool pushers console from where all the drilling activities are controlled.*

22 – Drilling programme complete the Arch Rowan is lowered to the water and towed to its new destination. The Wrestler is taking the tow.

*23 – The Big Orange XVIII well service vessel alongside a gas production installation.
Inset — sphere launcher, an interlocking door prevents inadvertent opening whilst the chamber is under pressure.*

24 – Cameron Iron Works (Cooper Oil Tools) emergency shutdown valve (ESDV) installed on a subsea pipeline riser as part of the new regulations introduced after the Piper Alpha disaster. The fail safe actuator can clearly be seen as can the heat resistant cementations coating on the pipeline below the valve.
Inset – a solid block Vetco Gray Christmas tree. On this particular tree the lower master and swab valves are manually operated whilst the upper master and wing valves are fitted with remote control actuators.

Part 4. SUB-SEA DEVELOPMENTS

The use of sub-sea wells to recover reserves in the immediate vicinity of a host installation, be it fixed or floating, is a relatively straight forward operation. However, as the distance from the installation increases, artificial means may be required to augment the natural reservoir pressure if the flow of hydrocarbons, and oil in particular is to continue. Whilst water injection is often used as a means of boosting reservoir pressure, this option is only viable on relatively large oil fields, the only other alternative being the installation of gas lift equipment.

Strictly speaking, water injection and gas lift are means of improving the flow of oil from the reservoir, a process referred to as *enhanced oil recovery* (EOR), and what we are about to discuss relates more to the problems caused by the distance the wells are located from the host facility, rather than the condition of the reservoir. Indeed, the equipment we are about to describe may well be provided in addition to one of the previously mentioned forms of enhanced oil recovery, a subject more fully discussed in chapter five.

The transportation of unstabilised well fluids, liquids or gases, is fraught with difficulties due to the volatile nature of the component parts of the hydrocarbon mixture, a problem exacerbated by pressure drop as pipeline length increases. The flow of oil in a pipeline may be interrupted by slugs of vaporised gases whilst gas pipelines are likely to suffer from slugs of condensed liquids which drop out of the gas during pressure fluctuations. Large volumes of sand and water create additional problems, particularly as the reservoir ages. Providing the equipment that can solve these problems has to-date, presented equipment designers with a number of extremely challenging tasks.

The present horizontal range for directional drilling from a fixed structure is about 8 km (5 miles) whereas sub-sea wells may be located up to 15 km (10 miles) distant from the mother platform. Extending the distance that wells may be located from the host installation has attracted considerable funding in recent years with the research proceeding along two quite different routes, namely *sub-sea separation*, and *multiphase* or *sub-sea pressure boosting*, the ultimate aim being to provide a reliable system for the exploitation of oil reserves up to 50 km (30 miles) from the host facility.

Before embarking on a description of sub-sea separation and pressure boosting systems it should be noted that these options are both regarded as *new technology* and are still undergoing development. Whilst field trials have been carried out and prototype systems ordered, these systems are unlikely to see widespread use until the start of the new millennium.

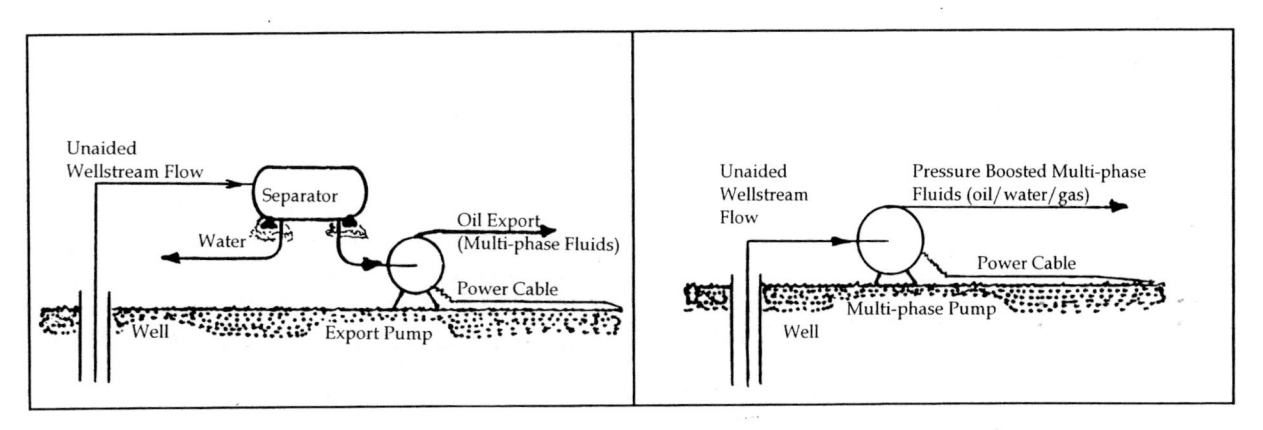

FIGURE 1: THE MAIN COMPONENTS OF A SUB-SEA SEPARATION SCHEME

FIGURE 2: THE MAIN COMPONENTS OF A SEA BED MULTIPHASE PUMPING SCHEME

1. SUB-SEA SEPARATION

INTRODUCTION

Sub-sea separation or sea bed processing involves the separation of the gas from the liquids on the sea bed in the vicinity of the sub-sea wells so that they may be transported back to the host installation through separate pipelines. This approach virtually eliminates the problems associated with the pumping and the conveyance of unstabilised reservoir products (referred to as *multiphase fluids*) over long distances through a single pipeline, although the injection of a hydrate inhibitor such as *methanol* or *glycol* into the gas pipeline may still be required in some instances.

Whilst their use is not widespread, sub-sea separation systems have been available for a number of years. Recent developments however have centred around the potential benefits to be obtained by disposing of the separated water at source, or at least near the wellhead, the water either being re-injected into the formation, or dumped into the sea, rather than transporting the combined products back to the platform or FPSO for separation. With some new developments producing well fluids which initially contain 50–65% water, often rising to 95% water as the field ages, this option has considerable benefits and as such is one of the emerging technologies receiving considerable industry funding at the present moment in time.

A prototype *sub-sea separation and re-injection system* is under development by *ABB Offshore Technology*, referred to as the SUBSIS concept, field trials are scheduled for the spring of 1999.

1.1 SUB-SEA SEPARATION AND RE-INJECTION — SUBSIS

ESSENTIAL FEATURES

The sub-sea separation system sits on the sea bed in the general vicinity of the template or manifold and relies primarily on a conventional gravitational separation vessel, a *production separator*, for the separation of the oil from the water, prior to the water either being re-injected into the formation or dumped into the sea. Single phase, constant speed, electrical powered centrifugal pumps are employed to boost the pressure of the liquids as they exit the separator and provided that the pipeline is suitably sized to maintain the pressure above the bubble point of the oil, the subsequent transportation should proceed smoothly.

Whilst the advantages of a sub-sea separation system are considerable, there are some very demanding challenges still to be overcome such as the development of high voltage electrical connections and a reliable *oil-in-water* monitoring device. A water purity of 1,000 ppm (parts per million) is required for water destined for re-injection whilst 40 ppm is the maximum permissible value for oil-in-water if the produced water is to be discharged directly into the sea.

The demand for electrical power will be considerable because the system will require as a minimum, sufficient power for the pumps used to increase the oil pressure for transportation back to the host installation. Power will also be required if gas compressors and water re-injection equipment are specified. Electrical power will be fed through an umbilical which will also contain fibre-optic control lines, chemical injection and hydraulic oil supply lines. Electrical connectors which can cope with 24 to 36 kV have been specified and the equipment will have to be extremely rugged if it is to function reliably in the harsh environment sub-sea, for anything from 10 to 25 years.

Sub-sea Separation and Injection System (SUBSIS)
Reproduced with permission of ABB Engineering as

OPERATIONAL BENEFITS

The two main benefits of a sea bed separation system relate primarily to the saving of weight on the reception facility and the avoidance of problems associated with pumping multiphase well fluids.

Reducing the volumes of liquids being transported back to the reception facility can result in a considerable reduction in the process equipment required and whilst the reduction of weight is highly desirable on a fixed installation, it can have a dramatic effect on the sea keeping properties of a *floating production and off-loading installation* (FPSO).

The removal of water virtually at source, will also permit the use of much smaller diameter pipelines, flowlines and risers which will result in a sizeable reduction in material and installation costs, particularly on fields being developed by FPSOs where there is also a limit to the number, and diameter of the pipelines and flexible risers that can be installed. The pipelines and risers often appear to consume a disproportionate part of the field development budget.

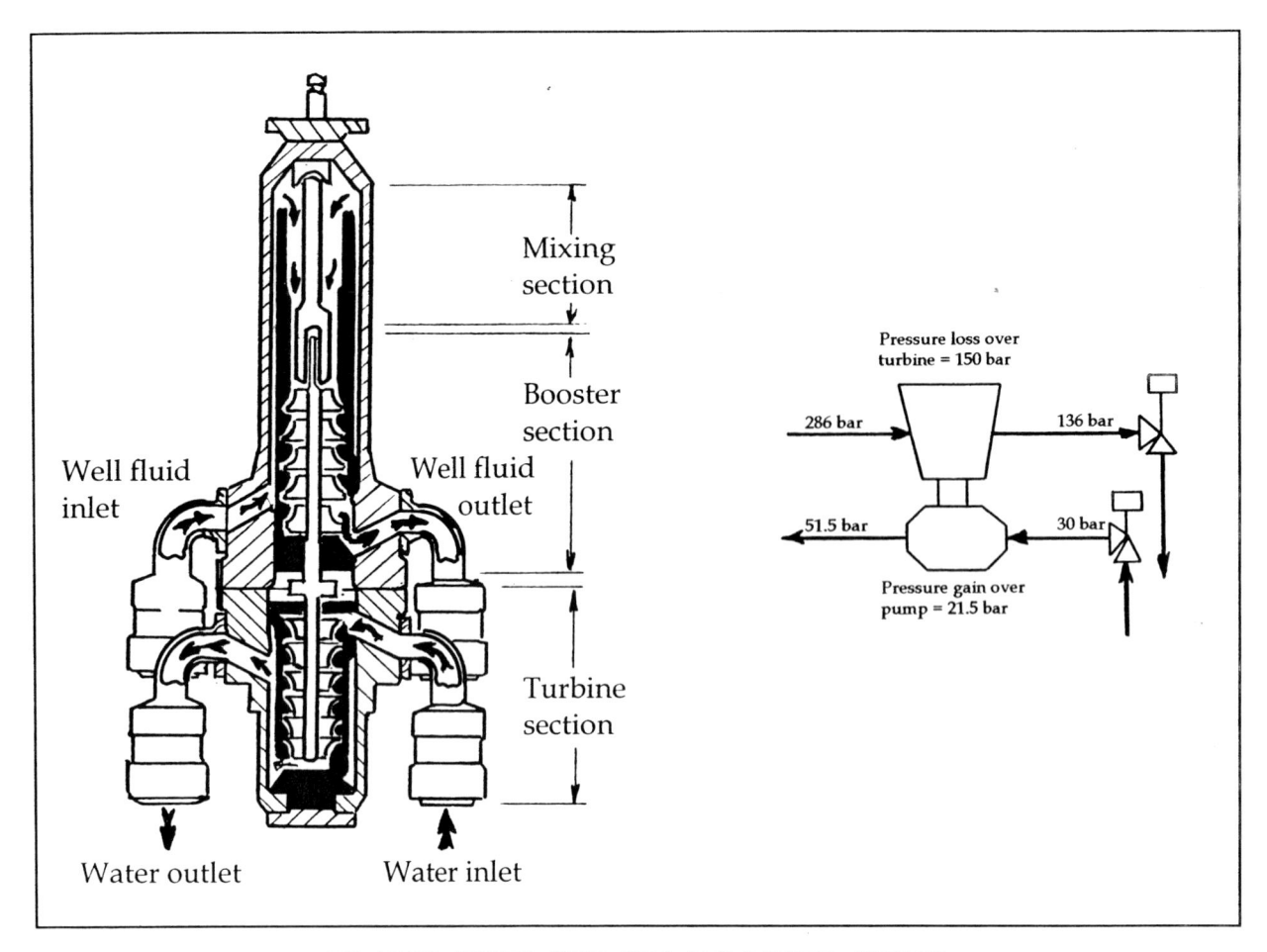

FRAMO TYPE SUB-SEA BOOSTER PUMP

2. MULTIPHASE PUMPING AND METERING

A colossal amount of time and money has been expended on research into *multiphase pumping*, and *multiphase metering* in recent years for the potential savings are enormous if reliable equipment can be developed. Multiphase pumps and meters have been used successfully on land based applications for a number of years and their development for use offshore has been motivated by the desire to simplify equipment and thus further increase the potential of unmanned installations and sub-sea developments.

2.1 MULTIPHASE PUMPING AND SUB-SEA PRESSURE BOOSTING

INTRODUCTION

Sub-sea pressure boosting is the process by which additional pressure energy is imparted to an unprocessed well stream as it exits the wellhead in order to prevent the formation of gas slugs in the sub-sea pipeline whilst the well fluids are transported back to the central processing installation, often up to 35 km distant. This has entailed the development of a pump that can cope with well fluids containing high percentages of gas without vapour locking, that is a multiphase pump.

The main advantage of multiphase pumping over equipment which relies on phase separation prior to pumping is the fact that the reservoir products are transported through a single pipeline. This represents a considerable saving on costs when compared with installing both oil and gas pipelines over extended distances.

In 1994, the worlds first commercial sub-sea multiphase booster pump unit, referred to as the SMUBS concept, was installed on a spare slot on the water injection template of Shells Draugen field development and production was enhanced by 5,000 bbl/d. Following the success of this system, which was supplied by *Framo Engineering AS* of Norway, the BP ETAP consortium contracted Framo to provide two sub-sea booster pump units for the *Machar* field development for installation in 1999.

The sub-sea pressure boosting equipment is located on the sea bed in the vicinity of the sub-sea wells and should not be confused with the *submergible pumps* which are installed *downhole*, in the wells. The submergible pumps are used for *enhanced oil recovery* and whilst the sub-sea booster pumps do improve the flow of oil from the reservoir, they are installed primarily to improve the transportation of the oil in the sub-sea pipeline.

ESSENTIAL FEATURES

The pumps are installed on a separate structure as part of the water injection valve module located approximately 30 metres downstream of the production manifold. They are based on the self regulating, *helico-axial* (Poseidon) principle and are driven by an hydraulic turbine powered by high pressure water supplied from the central processing platform. The water provides the motive power for the turbine prior to being re-injected into the formation for enhanced oil recovery.

Prior to the well fluids entering the pump inlet they are thoroughly homogenized by a *common flow mixer assembly* which is located immediately upstream of the pumps. Optimising the inlet conditions of the well fluids prevents *slugging* and broadens the operating range of the pumps. However, it should be noted that the pumps are designed to operate safely in both pure gas, and pure liquid conditions.

Under normal operational conditions the full injection water flow rate of approximately 32,000 bbl/d at a pressure of 286 bar (4,147 psi) will be routed through each pump turbine and a pressure drop of 150 bar (1,972 psi) is expected. The pumps have been designed with reliability being of paramount importance and consequently, no mechanical seals are used and all the bearings are lubricated by the drive water.

An alternative to the water turbine powered pump, the ELMSUBS is also being developed by Framo. A direct drive, high speed, oil filled electric motor will be powered from a 15 kV power supply fed from a sub-sea step-down transformer located on the sea bed where the voltage will be reduced to 1 kV. This arrangement provides an alternative to the hydraulic drive when water injection is not required and has been designed for tie-back distances of up to 50 km.

2.2 MULTIPHASE METERING

INTRODUCTION

Accurate measurement of the quantities of oil, gas, water and solids exiting a well is an essential prerequisite to successful reservoir management. At present, it is a time consuming and labour intensive operation which involves the use of test separators to separate the flow into its constituent components, that is oil, water and gas prior to measuring them separately using straightforward, single phase orifice metering techniques. While this approach is effective, the equipment required is considerable and somewhat impractical for sub-sea applications. The alternative is to provide equipment which can measure the flow of multiphase well fluids on line, that is without phase separation, a *multiphase flow meter*.

There are at present a number of multiphase meters on the market, most of which are claimed as being suitable for both topside and sub-sea applications. In reality, sub-sea multiphase metering is still in its infancy with only a handful of instruments installed to-date. The technology varies considerably from manufacturer to manufacturer and electronic sensors, or a combination of mechanical separators and electronic sensors are employed in various arrangements to analyse and measure the flow of hydrocarbons. The diagram shows a *Framo* sub-sea multiphase meter which has been selected for the *BP-ETAP* development and a brief description of the principle features will now be given.

2.2.1 FRAMO MULTIPHASE FLOW METER

ESSENTIAL FEATURES

The Framo multiphase flow meter is described as an *inert* design. The main body or receiver barrel is permanently installed on the sub-sea manifold structure and acts as the inlet and outlet housing, and as a guide and support during the installation of the *insert cartridge* which contains the actual measuring equipment. This arrangement facilitates the removal and replacement of the operational components as one unit either by diverless guidewire, or guidelineless equipment, a feature great importance given the relatively unproven nature of the equipment.

Output from the flow meter is processed by a computer which provides the actual flow measurements. The equipment may be electrically interfaced to a control pod on the sub-sea production system with separate wires being provided for power supply and data transmission, or the flow meter maybe fed from a dedicated control umbilical. Typical supply voltage will be 24 volts DC with a power consumption of approximately 20 watts.

OPERATING PRINCIPLES

The heart of the Framo multiphase meter is the *flow mixer body* which is designed to provide an homogeneous flow to the measuring sensors located within the venturi. The capacity of the mixer body is such that it can accommodate sufficient fluids to ensure that the transient conditions associated with multiphase fluids are smoothed out, that is, the measuring sensors are never exposed to unexpected slug flow conditions.

Well fluids entering the mixer body separate with the heavier liquid phases gravitating to the base of the chamber whilst the gases migrate upwards and into the ejector via the injection pipe. The liquid and gases are re-united as they enter the ejector nozzle which generates a turbulent shear layer and provides the venturi with an homogeneous mixture, a known flow regime which can be accurately measured.

As the homogenised mixture passes through the venturi, the fluid velocity is measured as if it were a single-phase fluid with equivalent mixture properties and standard venturi calculations can be applied. The fractions of oil, water and gas are determined by a dual gamma fraction meter, the calculations being based on the attenuation of two different gamma energy levels emitted from a radioactive Barium 133 isotope. Measurement of the detector hit rate of photons from the two different energy levels can, by physical equations, be expressed as a function of oil, water and gas volume fraction. The advantages claimed for a dual energy gamma meter is that the system is completely independent of whether the liquid is dominated by oil, water or even emulsions.

FUTURE DEVELOPMENTS

The use of radioactive isotopes is becoming more of a common feature in measurement analysis and the latest technology under development, *Nuclear Magnetic Resonance* (NMR) and *Pulsed Neutron Activation* (PNA) devices appear to offer direct measurement of individual phase fractions and velocities. Both developments are technically challenging and expensive to produce, particularly for sub-sea applications and a great deal of further prototype testing is required before units become commercially available.

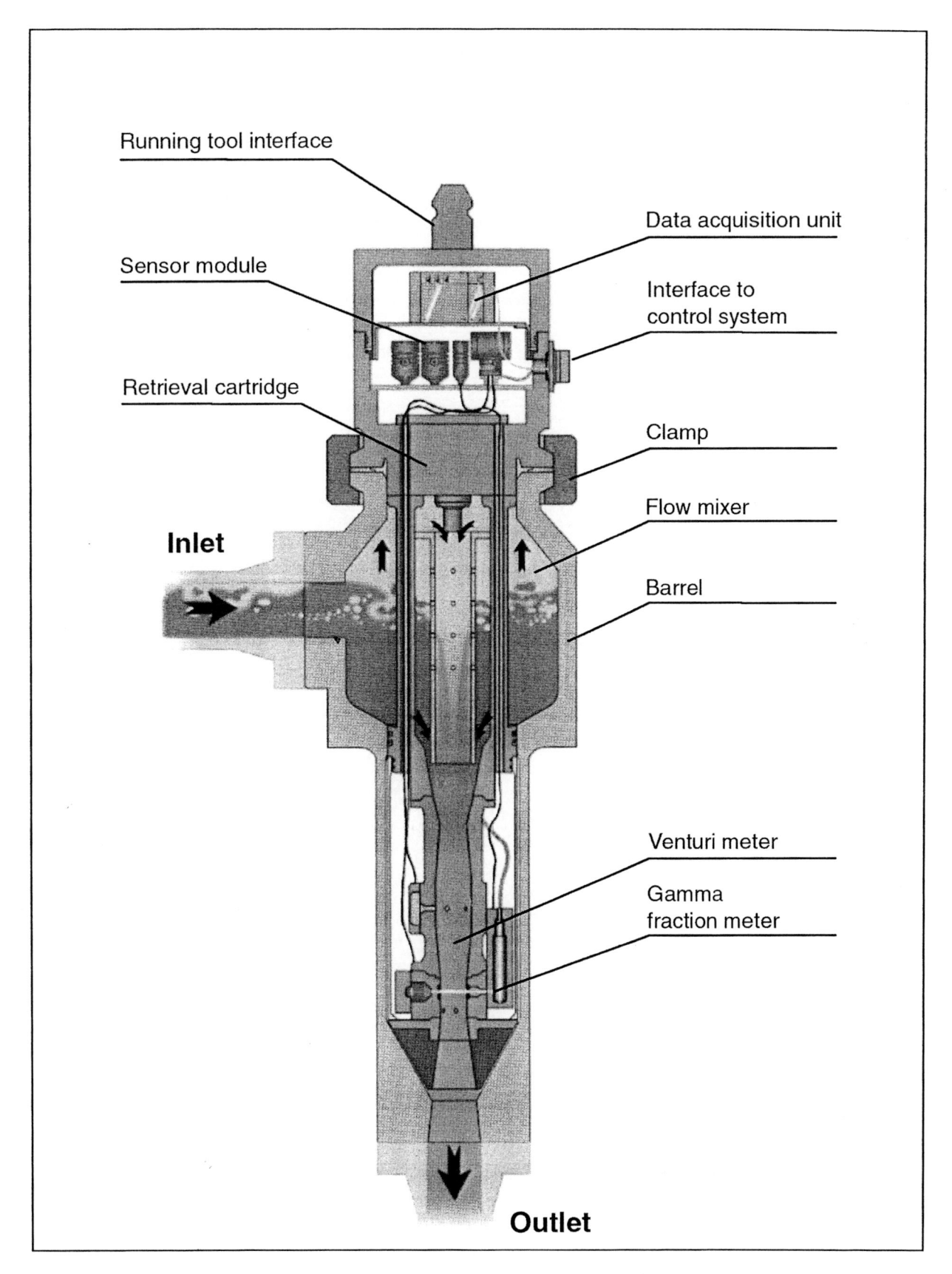

FRAMO MULTIPHASE FLOW METER
(Reproduced with kind permission of Framo Engineering AS)

Chapter Seven

DRILLING

The drilling industry is unique and to obtain an understanding of the drilling process can prove to be quite difficult, access to the drill floor being restricted to key personnel and reading material tending to be highly specialised, often presented in the form of highly technical papers prepared by the specialist service support companies.

In an effort to unlock some of the mysteries of the drill floor, the following text has been divided into 7 parts and further sub-divided where required to assist in the explanation of what is a very involved subject.

PART 1.   INTRODUCTION

      1.   The Formation Of Oil And Gas

      2.   Exploration

PART 2.   THE WELL

      1.   Well Construction

      2.   Cement Job

PART 3.   EQUIPMENT

      1.   Drilling Derrick

      2.   Drill String

      3.   Drilling Mud

PART 4.   OPERATIONS

      1.   Drill Crew

      2.   Making Hole

PART 5. WELL CONTROL EQUIPMENT

    1. Diverter

    2. Blowout Preventer (BOP)

    3. BOP Operations

PART 6. DRILL SHIP EQUIPMENT

PART 7. DEVIATED WELLS

    1. Directional Drilling

    2. Horizontal Wells

    3. Geosteering

    4. Multi-Lateral Wells

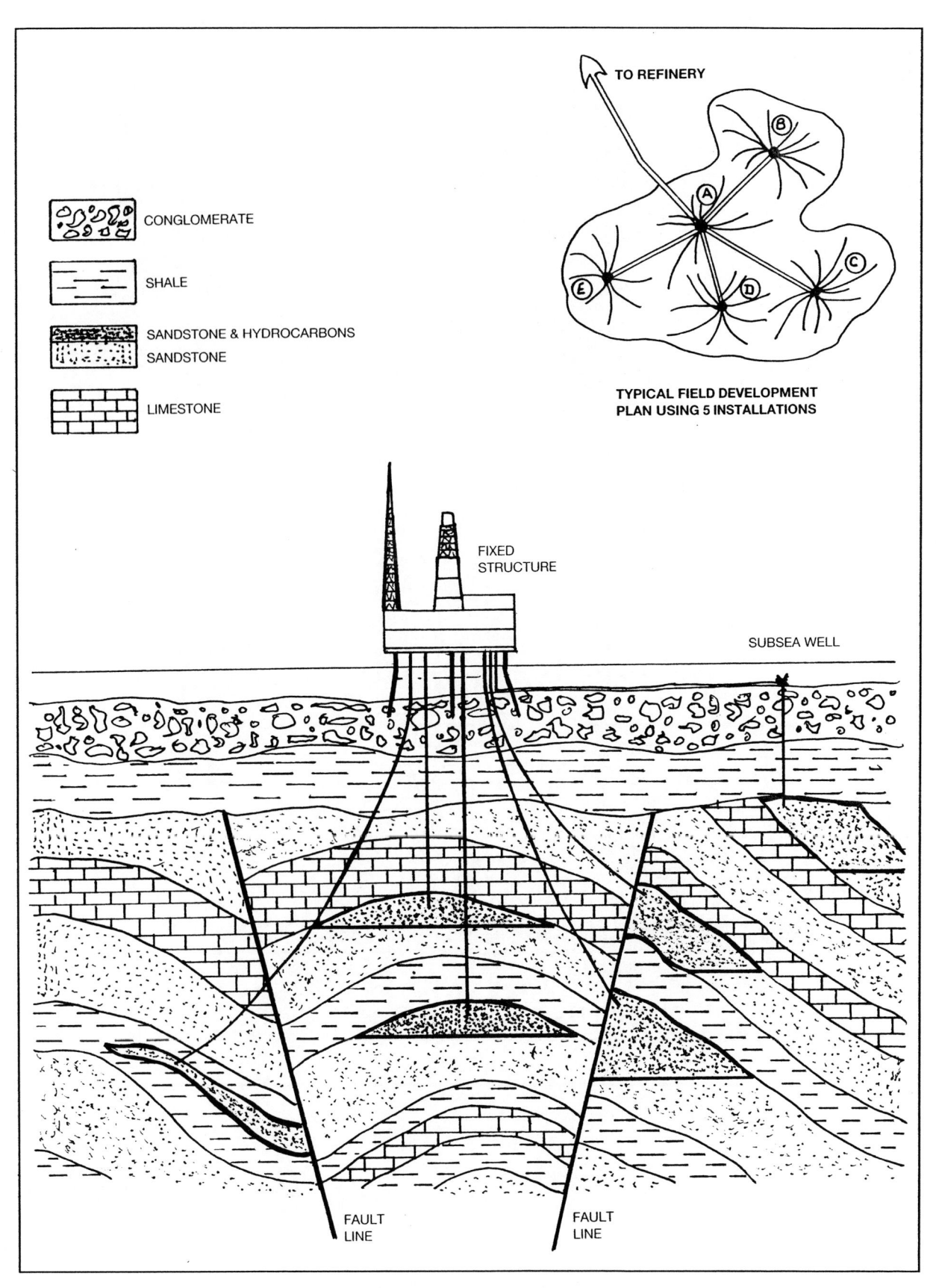

FORMATION OF OIL AND GAS
OIL/GAS FIELD SCHEMATIC

Offshore Engineering

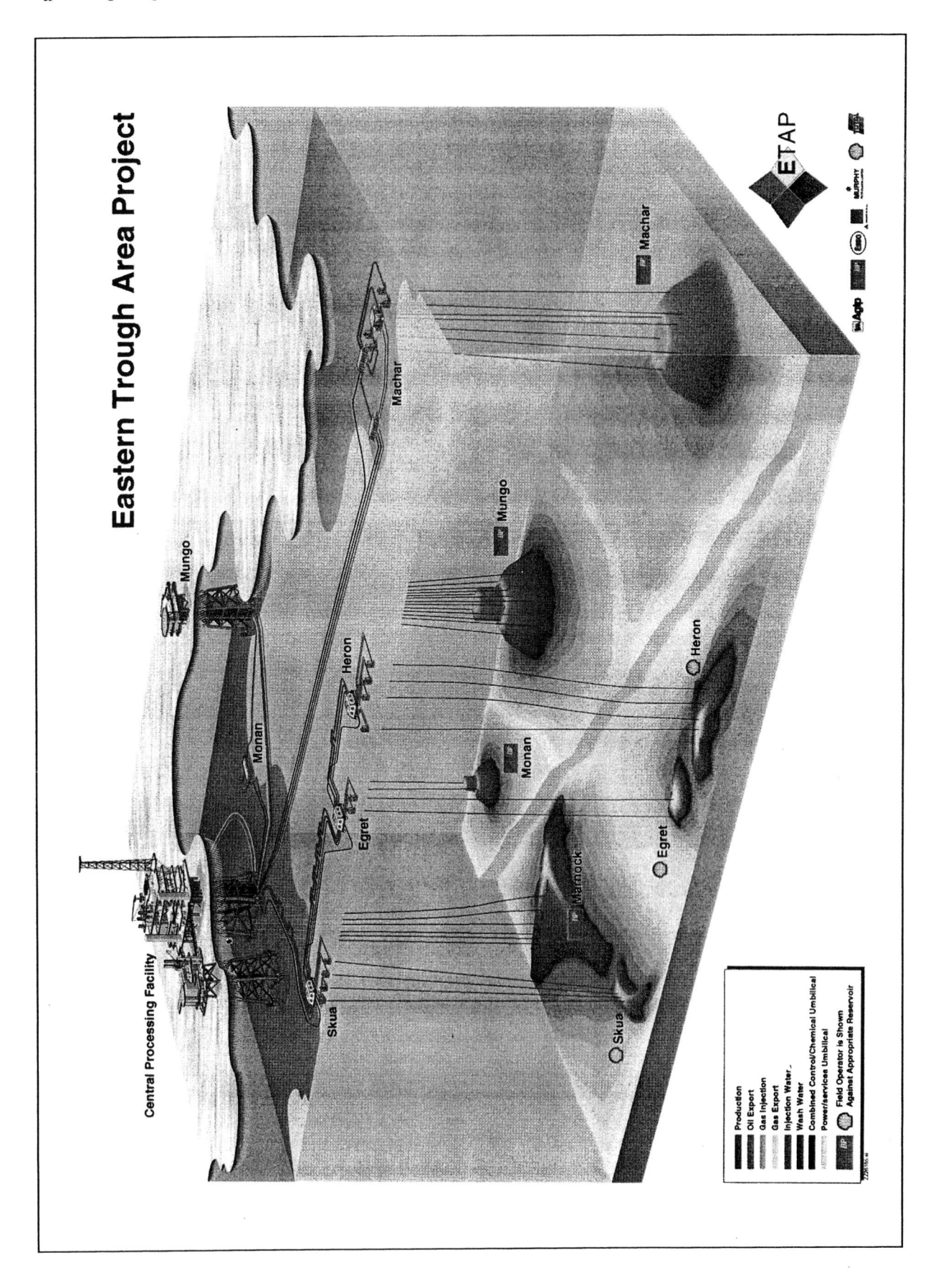

Part 1. INTRODUCTION

1. THE FORMATION OF OIL AND GAS

It is thought that oil and gas are derived almost entirely from decayed plants, algae and bacteria in what is essentially an energy regeneration process in which energy from the sun, that originally fuelled the growth of the plants and bacteria, is recycled into a useful energy source in the form of coal, oil or gas.

Only a very small percentage of the vast quantities of vegetation and bacteria that are produced by nature are converted into hydrocarbon compounds because in most cases, they decompose naturally and are absorbed by their surroundings. It is only where there is insufficient oxygen to feed the bacteria which would normally break down this material, that it remains intact and stagnates, the first stage in a unique set of circumstances required for the formation of an oil or gas field.

Where swamps, flooded forests, sheltered lakes and river estuaries occur, huge volumes of organic material can accumulate, and trapped under layers of silt and mud, the materials may be preserved in sufficient concentration to form a *source rock*. As the trapped deposits are further buried by younger sediments and pushed deeper towards the earths crust, the temperature increases and a *cooking* process commences.

With increasing heat and pressure, the fats, waxes and oils from the algae, bacteria, spores and cuticle (leaf skin) breakdown into their chemical component parts which link to form *Kerogen*. The cellulose and woody parts of land plants are converted to coal and woody kerogen and as the source rock matures and is further heated, long chains of hydrogen and carbon atoms break from the kerogen and form waxy and viscous heavy oils. At higher temperatures the shorter hydrocarbon chains break away and form lighter oils and then gas, predominantly methane.

The generation of hydrocarbons continues as the source rock matures and as the pressure increases, migration of the oil and gas commences through cracks and fissures in *permeable rocks* to areas of lower pressure. This migration to areas of lower pressure continues in an upward direction until, in many cases, the oil or gas escapes into the atmosphere as natural gas or tar seepages. Clearly, the provision of suitable organic materials and sedimentary deposits does not guarantee the formation of an oil or gas field. There are a number of further natural circumstances which must coincide with the deposition and breakdown of the base materials.

The upward migration of oil or gas through permeable or *reservoir rocks* can be halted only when they are overlain with an impermeable *cap rock* such as cemented sandstones, clays or salt, all of which act as a sealing medium, salt being particularly effective as it is dense and heals itself when cracks appear due to movements in the earths crust. Only where the reservoir and cap rocks occur in the correct formation, shapes and relative positions, will oil and gas accumulate to form oil and gas reservoirs. These reservoirs should not be thought of as huge underground caverns, the hydrocarbons actually exist within the structure of porous rocks in a condition analogous to a sponge full of water. All hydrocarbon fields are thus formed by a chance combination of events that produce the raw materials, suitable rock formations and structures, together with the right timing.

Offshore Engineering

In the North Sea, the formation of gas commenced approximately 300 million years ago during the *Carboniferous* period, when the area which is now the southern North Sea, was covered in lush, tropical rain forests. It is thought that *methane*, the primary constituent of natural gas, was formed from the vegetation after it had been converted to coal bearing rock. The oil, which is found in a deep rift valley in the northern North Sea, was formed from the remains of planktonic algae and bacteria that flourished in what were the tropical waters of the *Jurassic* period, approximately 140 million years. In some parts of the North Sea, trap structures existed 125 million years ago but were not filled with oil until 100 million years later. These processes continue today. The floor of the North Sea is still sagging and new areas of source rock are reaching depths where they too are generating oil and gas.

2. EXPLORATION

Prior to the commencement of drilling activities, a Government licence must be obtained which will allocate an area or block of territorial waters designated as suitable for offshore exploration. Having obtained an exploration block, the first operation entails enlisting the services of a survey vessel to carry out a *seismic* investigation.

2.1. SEISMIC SURVEYS

Seismic surveys provide the means by which underground rock formations are examined to determine if potential hydrocarbon bearing reservoirs exist.

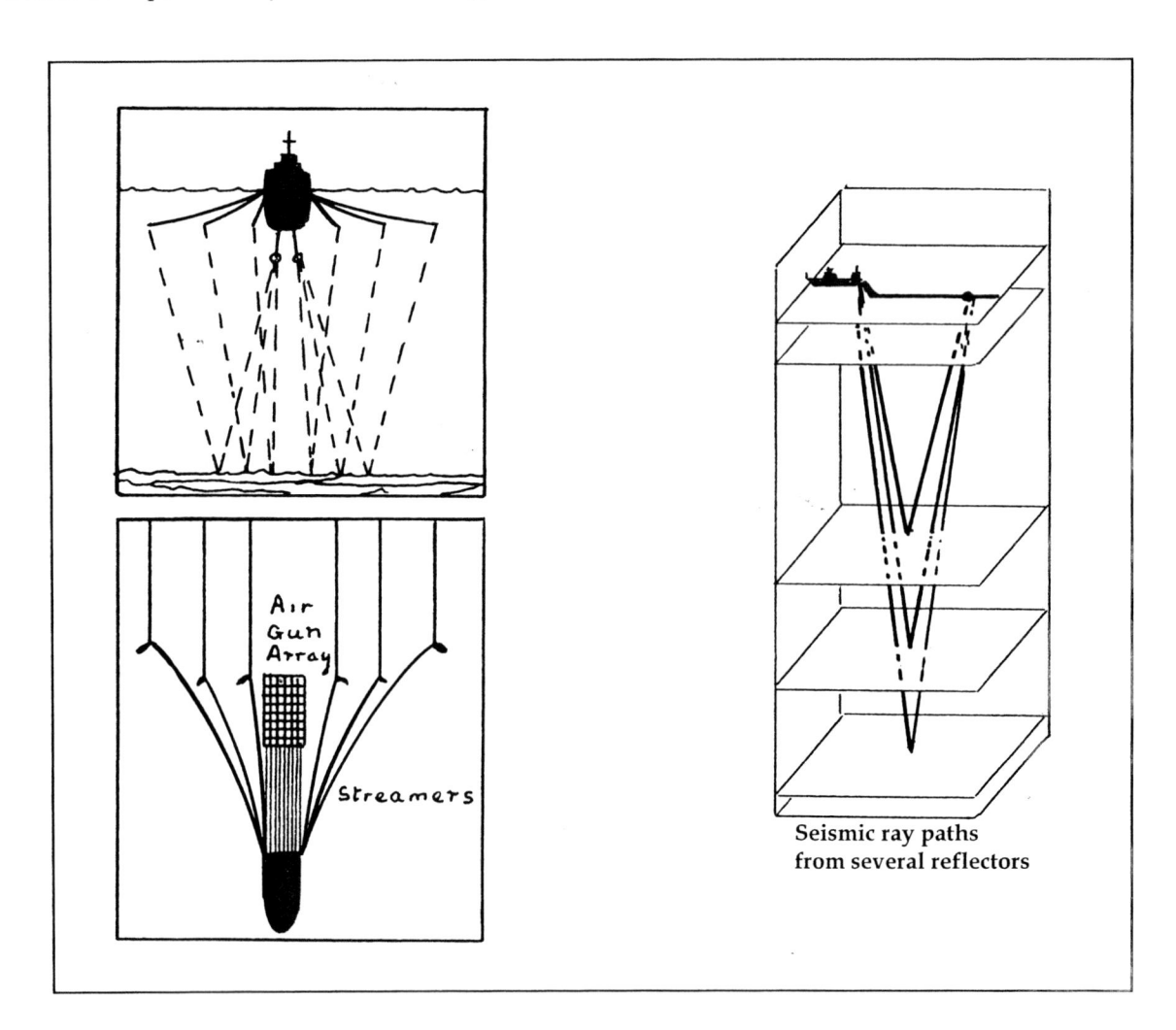

Seismic ray paths from several reflectors

242

The surveys are carried out by specially equipped survey vessels which traverse the area under investigation in a systematic pattern, emitting controlled burst of sound energy from a *submerged air or water gun array* towed behind the vessel. The sound waves radiate through the water and into the rock formations under the sea bed where they are refracted and reflected by the different layers of rock. The return signals are absorbed by hydrophones towed behind the vessel, up to 240 contained within a *streamer*, several kilometres in length.

Interpreting seismic survey information takes many months. The data received by the hydrophones is used to plot graphs of sound intensity against elapsed time. Vertical *seismic slices* are produced which are then *stacked* side by side to form a *seismic cross section*, a two dimensional (2-D) view of the underground structure where the reflective and refractive surfaces correspond with the boundaries between rock layers. Fault lines, stratigraphic and structural traps can be identified on the seismic cross sections and because the velocity of sound is sharply reduced in gas bearing rock, they provide some indication of where hydrocarbon deposits may be located.

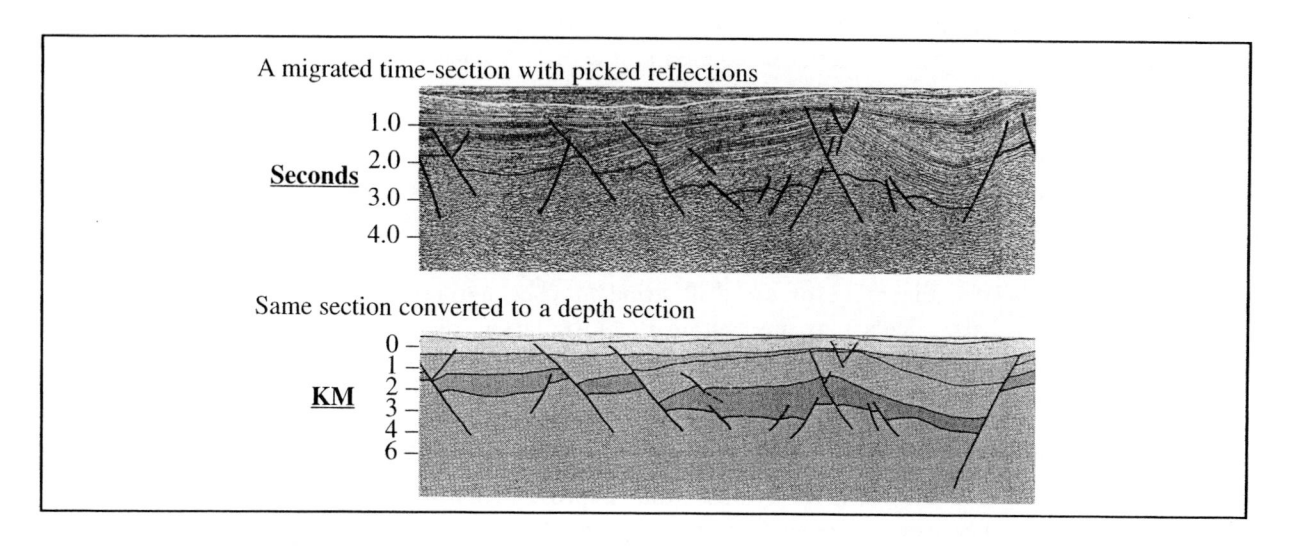

A migrated time-section with picked reflections

Seconds 1.0 2.0 3.0 4.0

Same section converted to a depth section

KM 0 1 2 3 4 6

Whilst 2-D seismic surveying has formed the backbone of the exploration business for many years, the recent advances in the power and speed of computers, combined with more sophisticated sound transmitting equipment deployed at much denser spacings, has permitted the development of three dimensional (3-D) seismic images. These images comprising both *vertical* and *horizontal sections* or *time slices* allow for the delineation of deeper, more subtle stratigraphic plays (the study of the depositional interrelationships of sedimentary rocks), a great advantage, particularly in geologically challenging areas such as West Africa.

The end result of the seismic survey investigations are the formulation of contour maps and 3-D colour and shade enhanced images which show very detailed geological structures. Visualisation software further permits the interpreters to view the 3-D data as a cube which can be rotated and cut at any angle in order to view the subsurface geometry of the rock structures more closely. The results can be used to determine where further seismic surveys would prove beneficial or wildcat drilling commenced, for it is only by drilling a well that the presence of hydrocarbon deposits can be confirmed.

Recently, BP installed an array of hydrophones on cables buried just beneath the sea-floor of their Foinaven field development, the object being to carry out what is referred to as a 4-D survey, essentially a time elapsed 3-D survey. By shooting over the array with a seismic vessel once a year, BP will be able to monitor any changes to the seismic wave character throughout the field life and to actually *see seismically*, the oil draining from the rock, thus facilitating an accurate study to indicate the areas of the field where oil is not being produced efficiently.

2.2 DRILLING PROGRAMME

As previously stated, the drilling of a *wildcat* or exploration test well is still the only fool proof means of confirming the existence of a hydrocarbon reservoir and a success rate of only 1 in 4 wells gives some indication as to the difficulties encountered in determining the existence of hydrocarbon deposits. It takes approximately 8 weeks and cost up to $10,000,000 ($60,000,000 in Arctic waters) to drill a test well so it doesn't pay to have too many failures.

On the occasions where drilling operations prove successful, a well test programme will be instigated to evaluate reservoir conditions and ascertain the exact nature of the hydrocarbon deposits. This entails flowing the well through a test facility located on the drilling vessel and the results will provide information as to the type of equipment that will eventually be required to produce the oil/gas and give an indication of any problems that may be encountered.

A successful well test programme will be followed by the drilling of a series of *appraisal* wells at locations considered by the geologists to represent the boundaries of the field. This will enable the physical dimensions of the field to be calculated and give an indication as to its development potential.

Where exploration surveys indicate that a field represents a commercially viable proposition, a permanent installation will be located at a position where it can reach the more favourable hydrocarbon bearing formations. A field can cover several square miles and consists of several geological traps or reservoirs, all of which must be tapped individually by separate wells. Deviation drilling techniques permit the exploitation of reservoirs situated up to 8 km horizontally distant from the rig whilst reserves located beyond the range of the drill string may be harnessed by sub-sea wells linked to the Mother platform. Large fields may necessitate the installation of additional satellite platforms. The Leman gas field located off the coast of East Anglia supports more than 30 such platforms which are operated by Amoco and Shell.

Frequently fields are discovered which are considered to be marginal or borderline in terms of prospective profitability and often these must await improvements in the economic climate, or available technology before they can be considered for development. As an example, the advances in sub-sea equipment and drilling technology have recently permitted the exploitation of fields which were discovered 10 or 15 years previously and which were considered uneconomic at that time.

Part 2. THE WELL

The majority of existing hydrocarbon reservoirs are located between the depths of 7,000 to 12,000 feet (2,100 to 3,500 metres) although discoveries of reserves located at depths of up to 15,000 feet (4,550 metres) are becoming more frequent. As one might imagine, if a hole is to be drilled to such a depth it must be drilled in several stages and reinforced to provide resistance against external pressure created by the rock formation. The process may be compared with the digging of a tunnel as much as they both require additional support when passing through soft rock or sand. However, unlike a tunnel, the hole also requires reinforcing to withstand internal forces generated by reservoir pressure.

During the drilling operation a column of dense cutting fluid (drilling mud) stabilises the hole and prevents the ingress of loose formation products. On completion of each intermediate hole, permanent support is provided by the insertion of a steel pipe referred to as a casing string. The casing strings are assembled from 30 foot (9 metre) lengths of pipe which are screwed into tubular couplings or "joints" as they are lowered into the hole. The casing string is then cemented into the rock formation prior to commencement of the next stage of drilling.

The process by which a hydrocarbon well evolves from a drilled hole will now be described.

1. WELL CONSTRUCTION

i) The drilling operation commences with the installation of a large diameter steel pipe known as a conductor. Typically of 30 to 36 inch (0.75 metre) diameter the conductor provides the foundation for subsequent drilling operations and a support mechanism for the intermediate casing strings.

The conductor may be installed into a pre-drilled hole and secured with cement, or it may be piled into the sea bed. The latter approach involves driving the conductor to refusal, the depth at which it will go no further. The formation products are then drilled from the centre of the pipe and piling recommenced. This process is repeated until the conductor has been driven anything from 200 to 800 feet (60 to 240 metres) into the sea bed. A starting head is then fixed to the top of the conductor and a diverter installed.

ii) Drilling commences in earnest with the second hole which will be sunk to a depth of approximately 2,000 feet (600 metres). On completion of the hole, the drill string must be removed whilst the first intermediate casing string is lowered into the hole and cemented in place. A casing head spool is mounted above the starting head to support and seal the top of the casing and provide a temporary home for the BOP stack.

iii) The number of subsequent holes drilled, their depth and diameter will be governed by the type of rock formation and the overall depth of the hydrocarbon reservoir. Very broadly speaking the drill bit will be reduced in diameter every 3,000 to 6,000 feet (900 to 1,800 metres) and the hole *cased off* (casing inserted). However, on occasions hole collapse or loss of drilling mud into the formation will necessitate suspension of the drilling programme and premature installation of the intermediate casing string. Having thus reinforced the hole, drilling may be resumed using a smaller diameter drill bit.

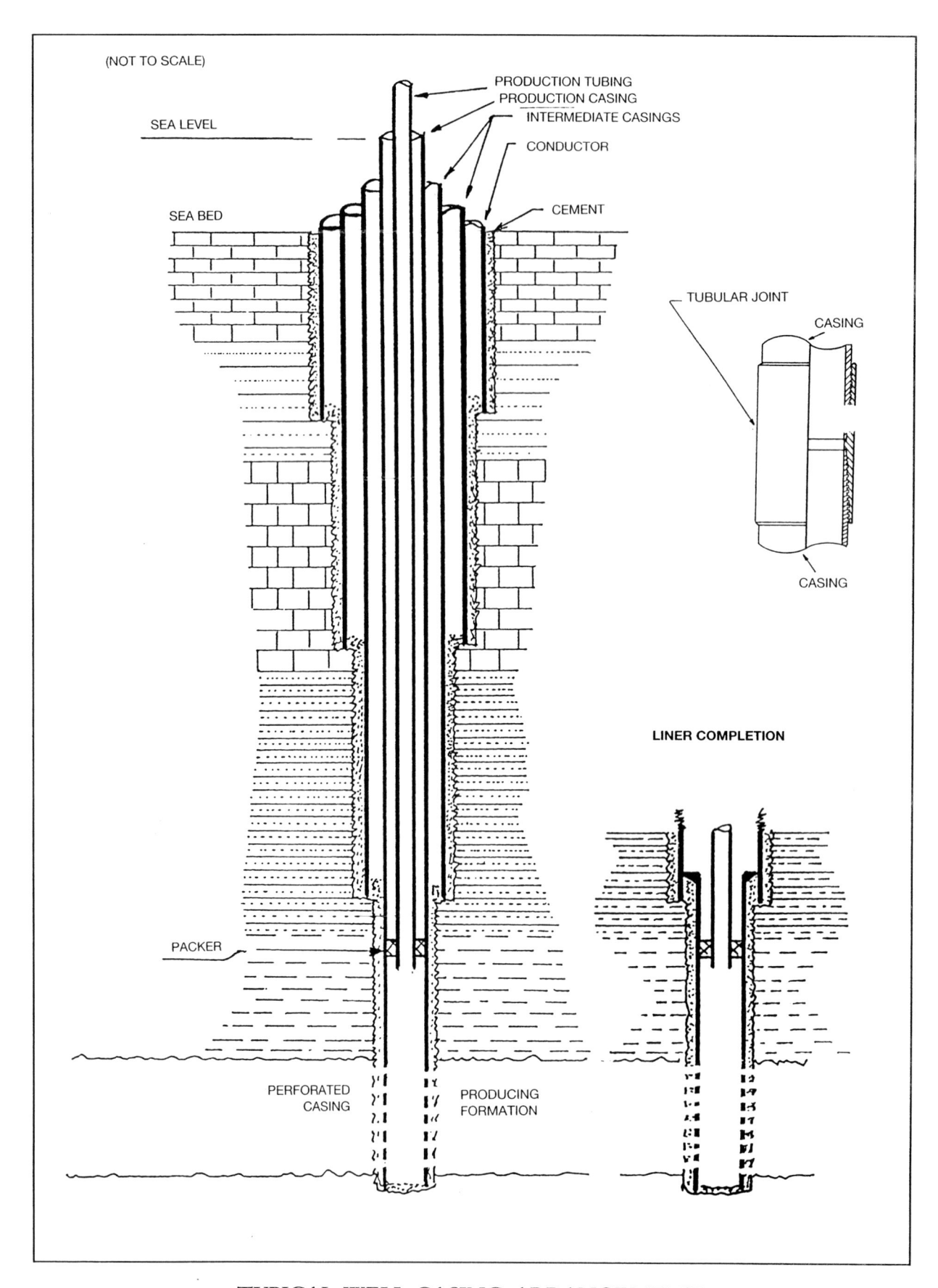

TYPICAL WELL CASING ARRANGEMENT

iv) Eventually the drill bit enters the hydrocarbon bearing formation. When the hole has been drilled to the required depth the drill string must be carefully removed to permit completion of the well, the hydrostatic head of drilling mud preventing hydrocarbon products entering the hole. Completion of the well involves the insertion of the final casing string which may be terminated using any one of three configurations.

a) Casing completion

As shown in the main sketch, the final casing string is run to the bottom of the hole and cemented in an identical manner to the preceding casing strings.

b) Liner completion

As shown in the detail sketch, the casing or liner extends from the base of the previous casing and is not returned to the surface. It is installed on the end of a drill string attachment and secured by a packing device prior to cementing. Liner completions are used extensively and permit a considerable saving on casing, in excess of 3,650 metres (12,000 feet) in a typical North Sea well.

c) Open hole completion

The open hole completion refers to the practise of leaving the hole uncased in the vicinity of the hydrocarbon formation, the casing or liner being terminated at the top of the hole.

v) The penultimate operation entails installation of the production tubing and Christmas tree in preparation for perforation and the commencement of production.

The production tubing is essentially a small diameter casing string which is suspended from a tubing hanger located under the Christmas tree and held within the lower reaches of the preceding casing by a packing device. This arrangement facilitates removal of the tubing should well modification work be required at a later date. The production tubing contains a number of machined recesses or nipples designed to accommodate the mudline safety valve, and steel plugs that are used to isolate the well during repair and maintenance operations.

With the exception of the casing located below the packer, the production tubing is the only component within the well which receives direct exposure to the reservoir pressure and it is through the production tubing that the oil and gas will flow to the surface. Once the Christmas tree has been *nippled up* (bolted up) to the production tubing hanger and hydrostatically tested, the perforation operation can commence.

vi) It should be remembered that prior to perforation, the well bore is isolated from the hydrocarbon bearing formation by the casing string and is still full of conditioning or well preservation fluid (normally brine or diesel oil). Perforation involves the puncturing of the final length of casing string in the vicinity of the hydrocarbon bearing formation, a distance which can extend over several hundred feet. It is achieved by lowering shaped explosive charges into the well on a wireline tool string and detonating them by remote control, either electronically or by hydraulic pressure signals. The charges blast through the steel casing and cement and permit the passage of reservoir products in to the well bore.

Initially entry of the oil or gas into the production tubing is prevented by the hydrostatic head of preservation fluid and there are two main methods by which the hydrocarbon flow can be initiated and the well brought *on line*:–

DRILLING PROGRAMME
1. Conductor hole drilled.
2. Conductor installed, cemented and starting head fitted.
3. First intermediate hole drilled, annular BOP in situ.
4. First intermediate casing string inserted, cemented and casing hanger set.
5. Drilling of second intermediate casing hole in progress.

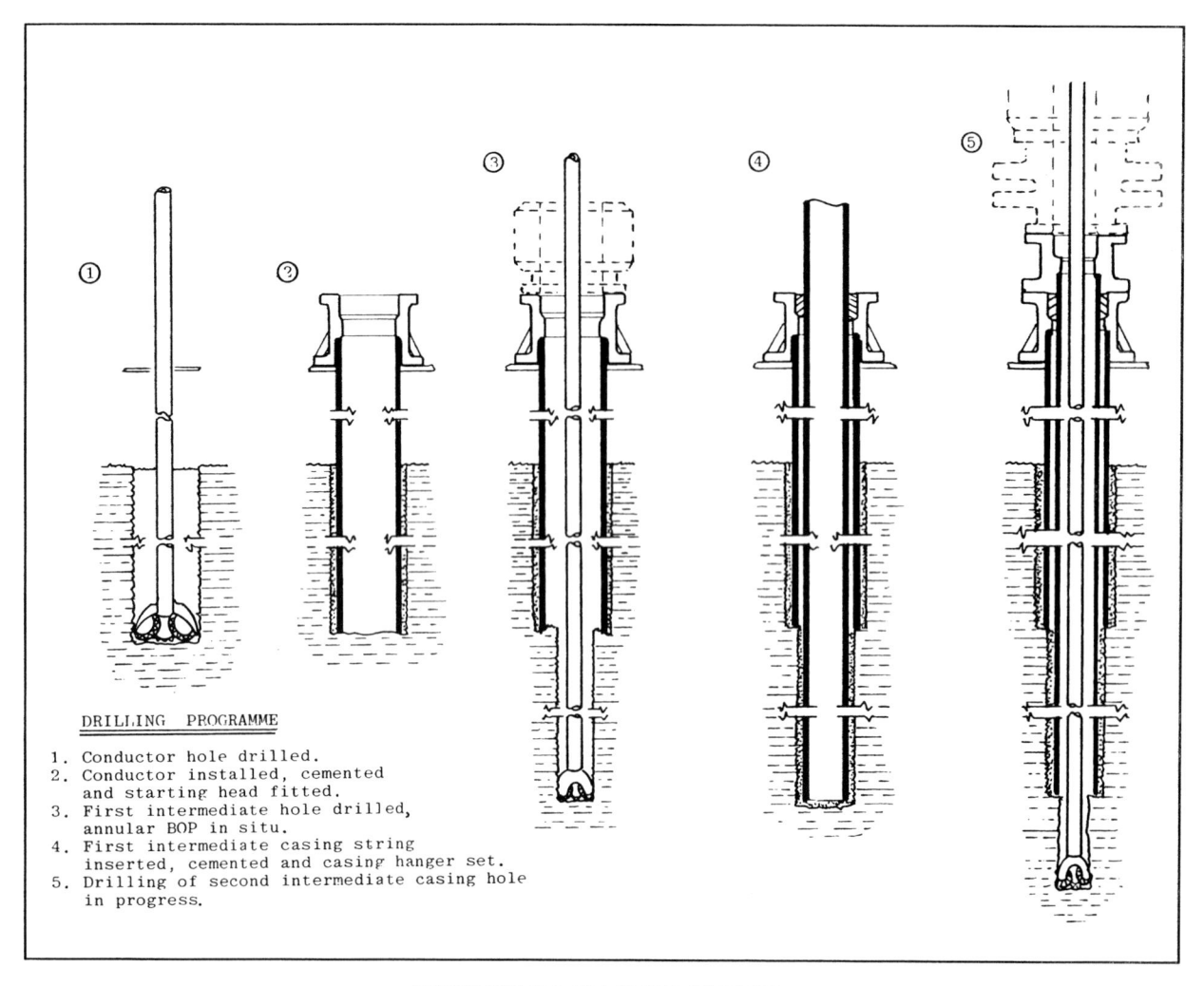

DRILLING PROGRAMME

a) **Coil tubing**

Coil tubing involves the insertion of a small diameter (1" or 25.4 mm) steel tube through the Christmas tree and down to the base of the well. High pressure nitrogen gas admitted through the tube helps displace the preservation fluid so that the flow of hydrocarbons can commence.

b) **Blanket gas**

Prior to perforation, the hydrostatic head of preservation fluid is reduced to a value less than the reservoir pressure and a charge of nitrogen gas is then admitted to the well bore. The combined pressure created by the brine and nitrogen exceed reservoir pressure and provide the necessary safety barrier. On completion of perforation, the nitrogen gas can be vented to atmosphere and the reduction in pressure will permit the reservoir products to displace the remaining column of brine and let flow commence.

Having dealt with the drilling of the hole and the installation of the intermediate casing strings into the rock formation, we should consider how the casings and production tubing are secured on the production platform. The securing system is referred to as the wellhead and has been dealt with in Chapter 8.

2. THE CEMENT JOB

As explained in the preceding section, a well is drilled in stages all of which are individually lined with a steel *casing string* that is secured into the formation by cement. Cementing the casing strings produces an extremely rigid assembly and provides protection against corrosion products released by the formation.

A sketch has been prepared to assist in the explanation of the cementing process or *cement job*.

On completion of each stage of drilling and prior to the introduction of the casing string the hole must be *conditioned*. Conditioning involves the replacement of the drilling mud with a conditioning fluid, frequently a brine solution suitably weighted to prevent the ingress of hydrocarbons into the hole, should they occur. The conditioning process removes any *oilyness* left by the drilling mud and improves the chances of obtaining a secure bond between cement, casing and hole. With the hole suitably conditioned the installation of the casing string can commence.

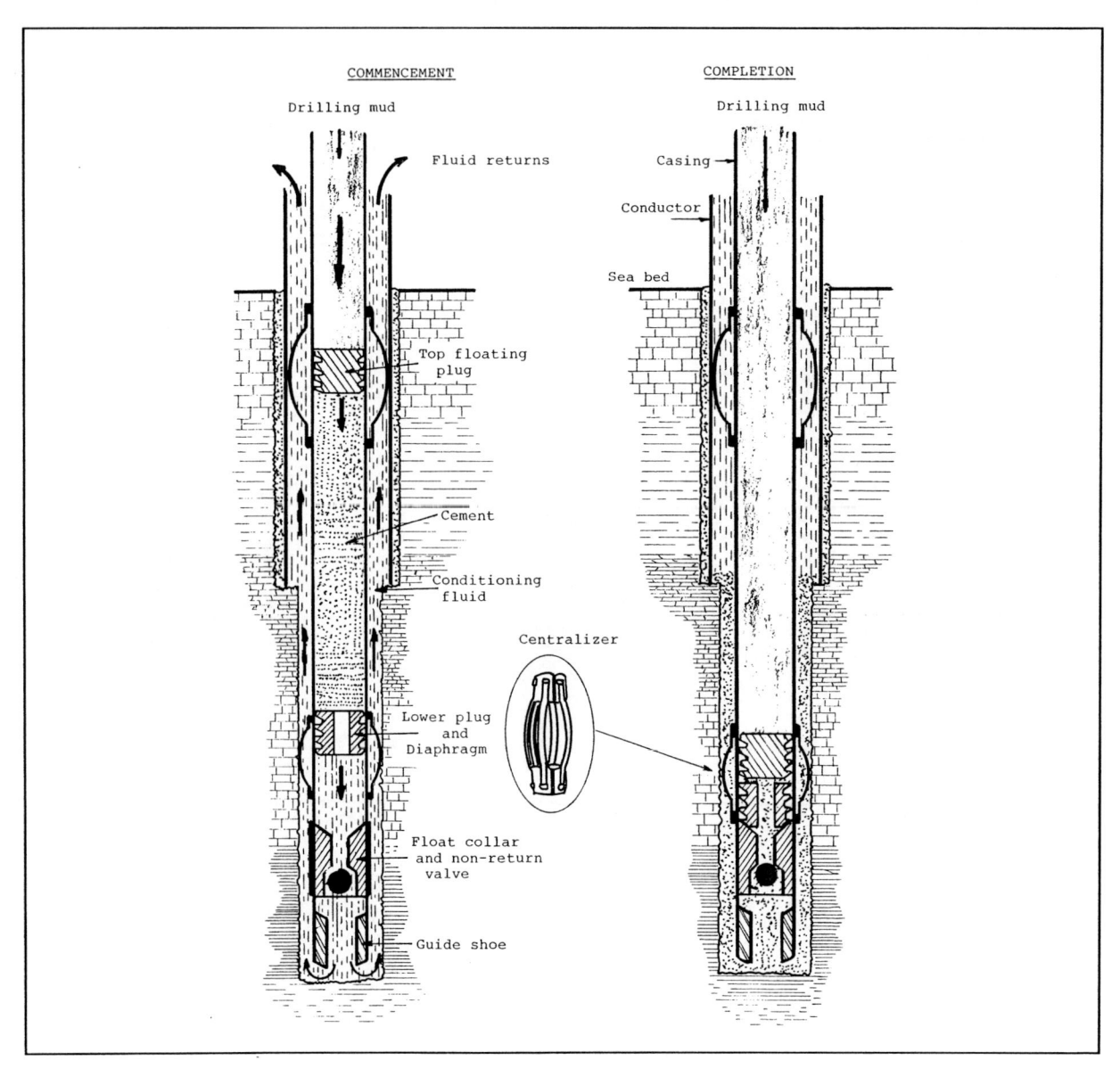

THE CEMENT JOB

Offshore Engineering

The first section of each casing string lowered into the hole is fitted with an internal *guide shoe* and *float collar* which contains a non-return valve. Subsequent casings are fitted with external sprung steel *centralizers* at regular intervals to ensure that they locate centrally within the hole and provide an unrestricted channel for the passage of cement. The cementing process can proceed as soon as the final length of casing enters the hole.

The first operation entails insertion of a floating plug into the casing that will provide a barrier between the preservation fluid and the cement. A measured quantity of cement is then pumped into the casing, the volume calculated from the size of the hole and the external diameter of the casing string. A second floating plug is then installed above the cement and the drilling mud system re-connected.

The casing now resembles a giant syringe and as pressure is applied to the drilling mud, the column of cement moves slowly down the inside of the casing until it reaches the non-return valve located in the float collar at the base of the casing. When the bottom plug contacts the non-return valve, the diaphragm within the plug ruptures permitting the passage of cement through the plug and into the space created by the drilled hole and the external surface of the casing. Drilling mud pressure is maintained until a rapid increase in resistance indicates that the top plug has contacted the bottom plug and the cementing process is complete.

The cement takes from 6 to 24 hours to cure depending on the downhole temperature and during this period it is prevented from flowing back into the casing by the non-return valve in the float collar.

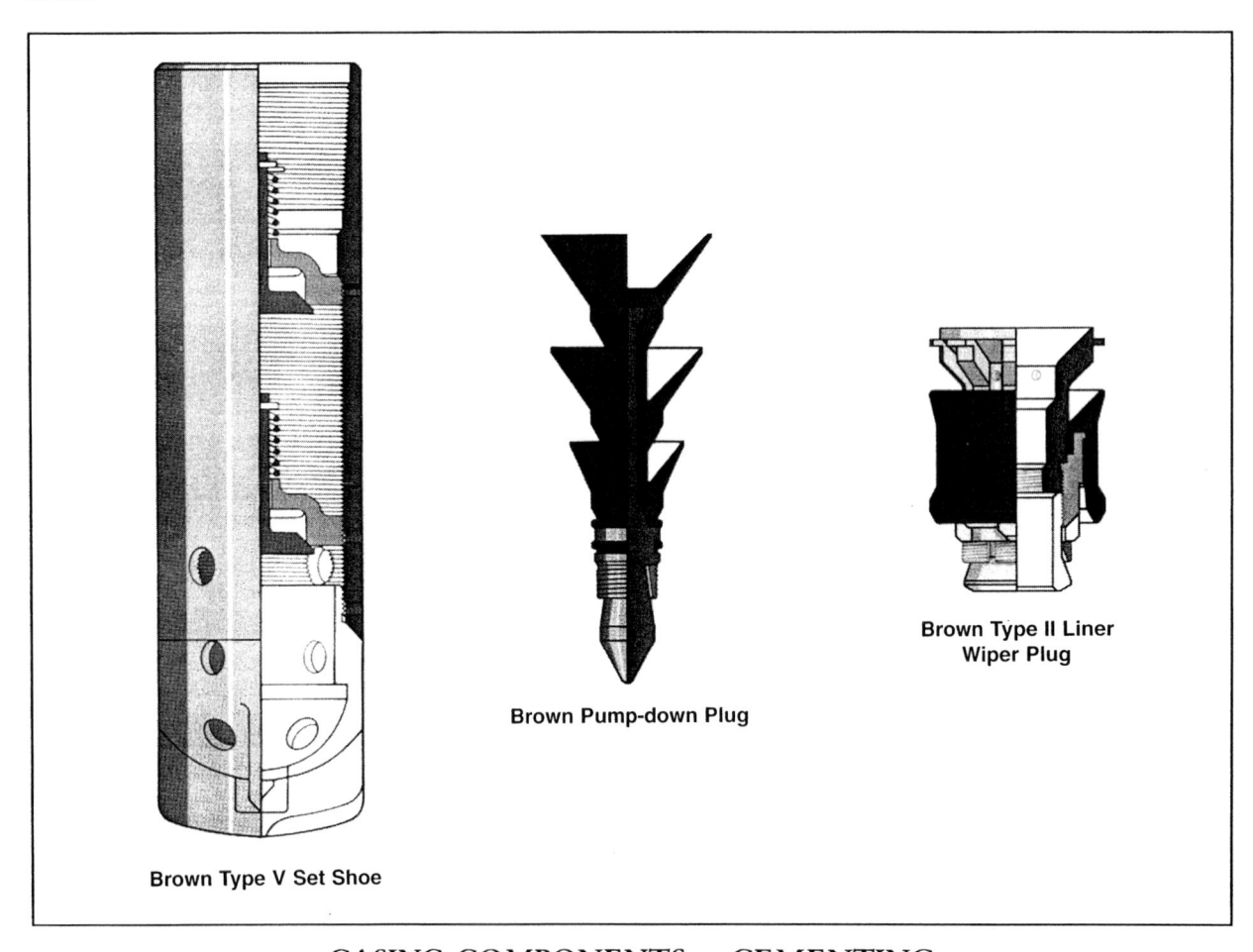

Brown Type V Set Shoe

Brown Pump-down Plug

Brown Type II Liner Wiper Plug

CASING COMPONENTS — CEMENTING
(Reproduced with kind permission of Baker Service Tools)

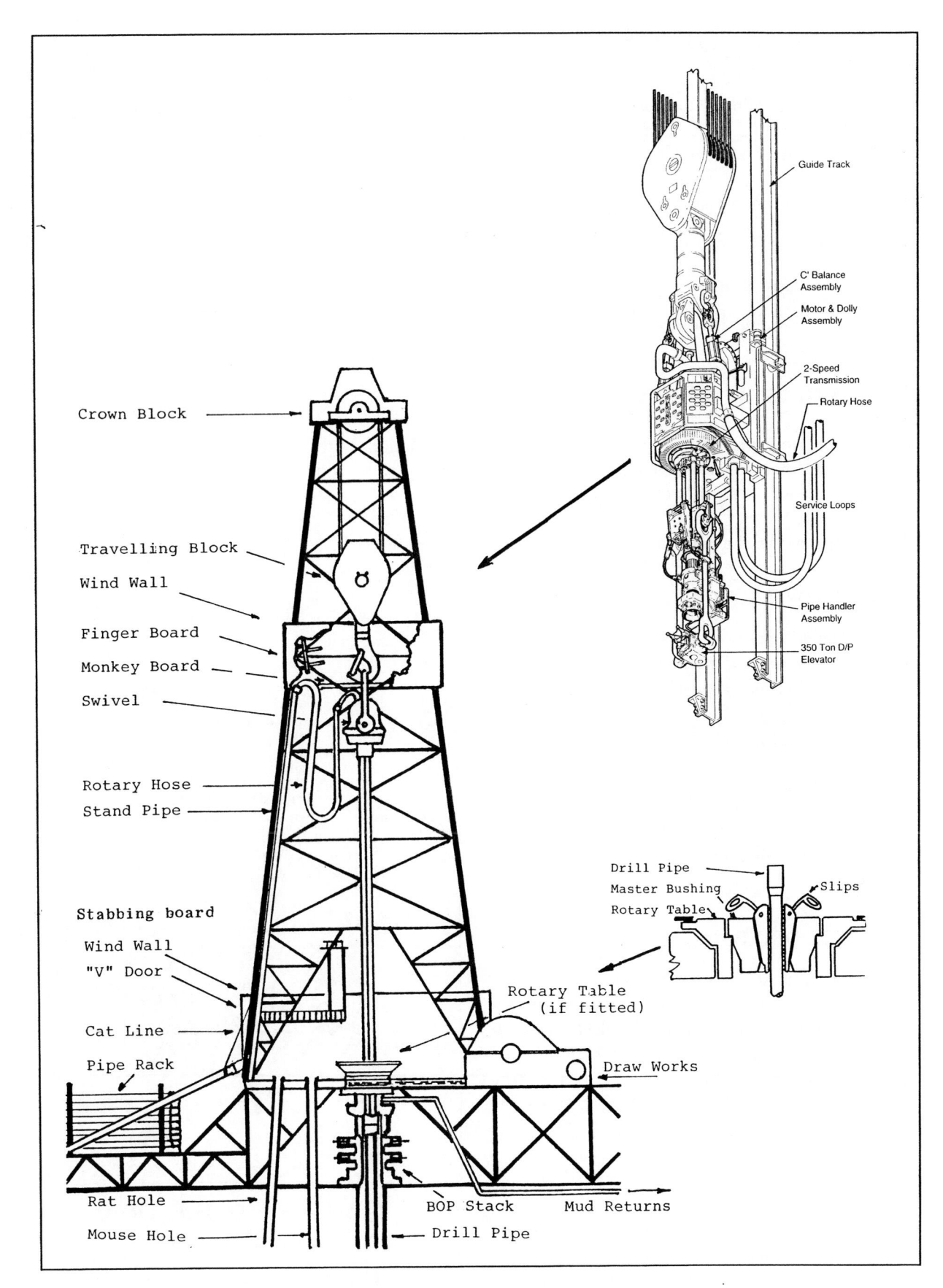

DRILLING

DRILLING DERRICK

Offshore Engineering

Part 3. EQUIPMENT

Having outlined the basic process by which a well is drilled and completed we can now discuss in greater detail the major items of equipment required to perform successful drilling operations from a sea bed supported structure such as a jack-up or a fixed installation.

The main components are:

1. Drilling Derrick.
2. Drill String.
3. Drilling Mud System.

1. DRILLING DERRICK

The most prominent feature of an offshore installation is the drilling derrick. It consists of a steel lattice tower approximately 50 metres (165 feet) in height which supports the Crown Block and provides temporary storage facilities for drill pipe stands.

The sketch shows the main component parts of the drilling derrick.

i) CROWN BLOCK

The stationary sheaves mounted at the top of the derrick over which the wire ropes attached to the travelling block pass.

ii) TRAVELLING BLOCK

The large heavy duty, multi-sheave lifting block which is used primarily to support the weight of the drill string during drilling operations and to hoist drill pipe and casing into and out of the hole.

iii) TOP DRIVE

The top drive assembly has all but superseded the rotary table as the means by which rotary motion is imparted to the drill string. It consists of a large electrically or hydraulically powered motor which is suspended from the travelling block and connected directly to the drill string.

iv) MONKEY BOARD

The small platform located in the top half of the drilling derrick from where the monkey, or derrick man can manoeuvre pipe stands into and out of the finger board during a trip.

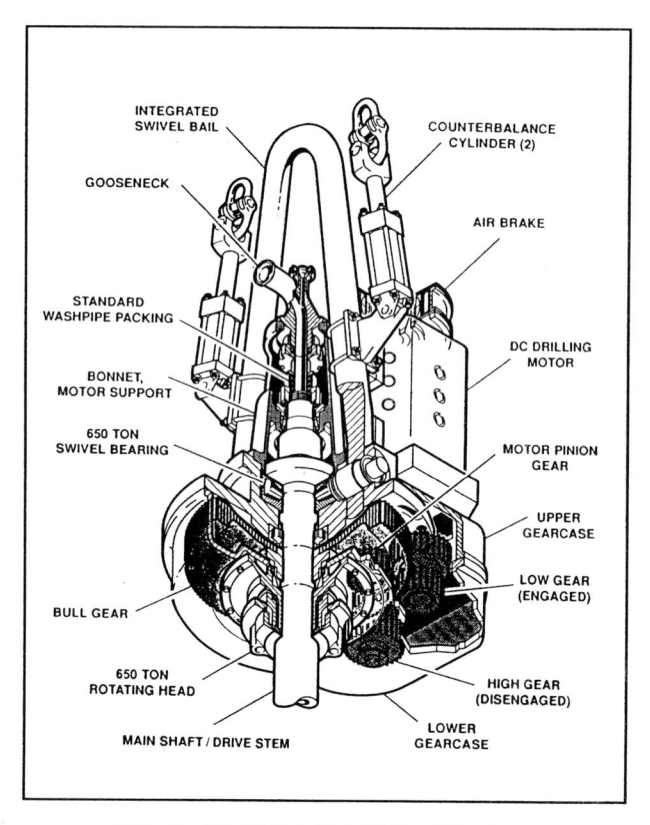

TDS-4S TOP DRIVE TRAIN
(Dual Speed)

(Reproduced with permission of Varco BJ, Orange, California)

252

v) FINGER BOARD

The finger board resembles a large peg board and is located alongside the monkey board. The finger board provides a location for the drill pipe stands.

vi) DRAW-WORKS

The collective name given to the power control centre on the drill floor. It consists essentially of a large hydraulically or electrically powered winch which provides the motive power to operate the travelling block and the rotary table (if fitted).

vii) ROTARY TABLE

Where fitted, the rotary table consists of a large casting located centrally within the drill floor and it is used to impart rotary motion to the drill string, being chain driven from the draw-works. It contains removable bushings which permit the passage of the drill string components whilst also providing a drive mechanism for the rotation of the square section kelly.

viii) DOG HOUSE

Tin hut or shed located at the side of the drill floor and used as a shelter by the members of the drill crew.

ix) MOUSEHOLE

A pipe let into the drill floor in which a new section of drill pipe is inserted prior to its connection to the kelly and subsequent inclusion in the drill string.

x) RATHOLE

A pipe let into the drill floor to accommodate the kelly during the making of a trip.

xi) PIPE RACK

Area of deck in front of the drilling derrick where drill pipe and casing are stored prior to use.

xii) PIPE DRAW

Ramp leading from the pipe rack to the drill floor over which drill pipe and casing are dragged by the catline.

xiii) "V" DOOR

The V shaped opening in the windwall between the pipe rack and the drill floor.

xiv) CAT LINE (or Tugger)

Small winch used to transport drill pipe and casing from pipe rack to drill floor.

Offshore Engineering

2. THE DRILL STRING

The drill string is the collective name which describes the assembly of components used to drill a hole. They are:

i. Drill bit

ii. Drill collars

iii. Drill pipe

iv. Kelly

v. Saver-Sub Assembly

vi. Swivel

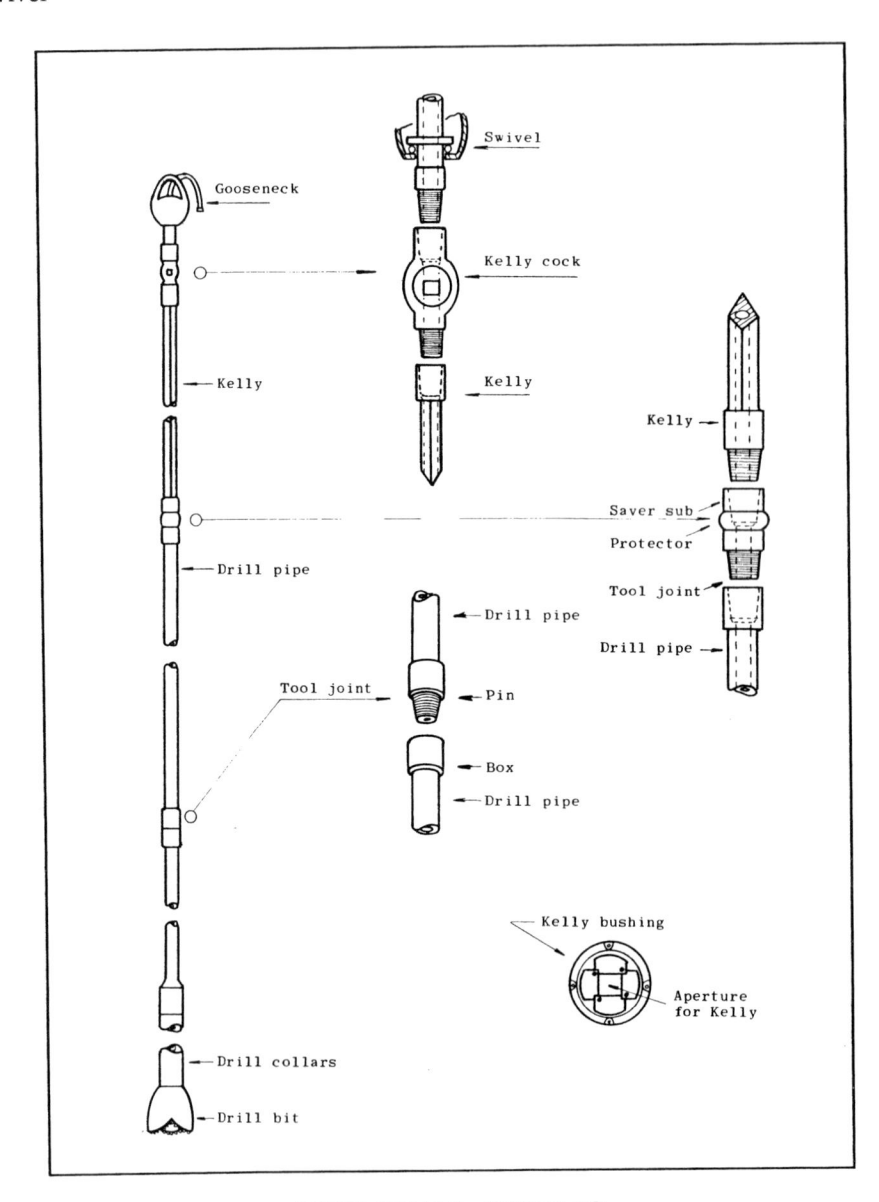

THE DRILL STRING

254

i)   **DRILL BIT**

The drill bit, roller cone or rock bit consists of three rotating cones which are fitted with hardened steel, carbide tipped or diamond edged teeth. The choice of tooth material will depend on the type of rock formation through which the drill bit must pass.

ii)   **DRILL COLLARS**

The drill collars are heavy, thick walled tubular couplings which connect the drill bit to the drill pipe. Anything from 2 to 30 collars may be used to provide concentrated weight in the vicinity of the drill bit to assist in the rock breaking process. The collars also provide rigidity to the drill string which prevents the bit from wandering from the vertical during drilling operations.

iii)   **DRILL PIPE**

Drill pipe is supplied in 30 foot (9 metre) lengths and consists of 3½ or 5 inch diameter heavy wall pipe with male and female couplings welded to each end. The couplings are referred to as pins (male) or boxes (female) and permit assembly and disassembly of the drill pipe during drilling operations.

iv)   **KELLY** (required only for rotary table drive)

The Kelly consists of a long hollow, square section forging which screws into the top section of a drill pipe. It is driven by the rotary table and provides the drill pipe with both rotational movement and drilling mud.

v)   **SAVER-SUB ASSEMBLY**

The saver-sub assembly consists of a simple male/female threaded spacer designed to protect the Kelly or power take off shaft threads from damage during the repeated make up or break out of the sections of drill pipe.

vi)   **SWIVEL**

The swivel permits the drill string to revolve freely whilst suspended from the travelling block and provides for the passage of drilling mud. The mud is supplied via the rotary hoses which are connected to the goose neck on the swivel.

3.   DRILLING MUD

Drilling fluid or *mud* as it is universally referred to is the lifeblood of the drilling operation. It lubricates the drill bit, removes drilling debris, stabilises the hole and provides the principal safety barrier against the ingress of hydrocarbons into the well bore.

In actual fact, drilling mud is quite a sophisticated fluid consisting essentially of *bentonite*, a colloidal clay dissolved in either water or fuel oil. *Baryte* (Barium Sulphate) is added as a weighting medium to permit variation of the specific gravity and the only resemblance the final product has to *mud* is its consistency and the dreadful mess it creates.

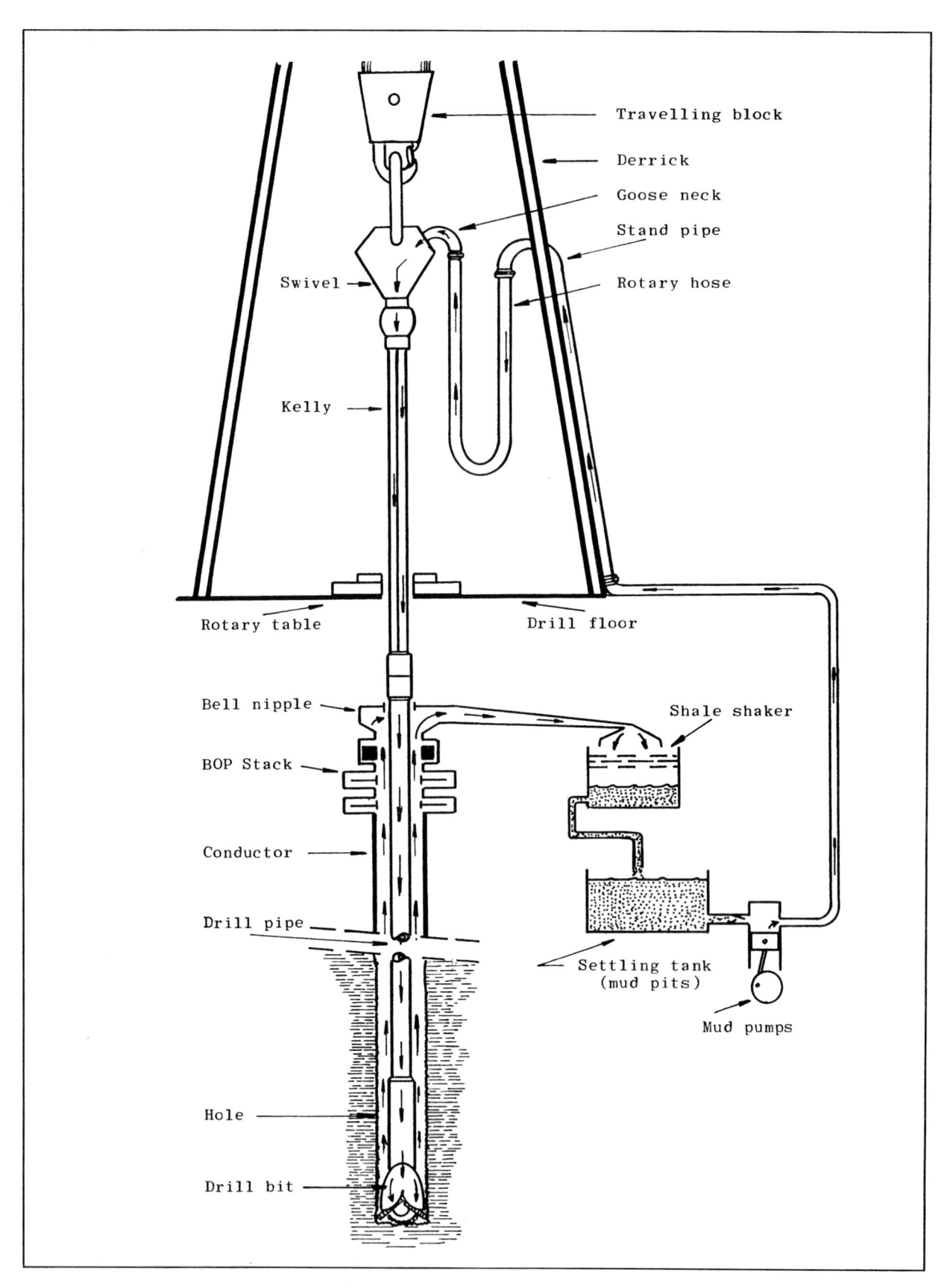

DRILLING MUD SYSTEM
(rotary table)

Considerable care must be exercised in the preparation of the drilling mud to ensure that drilling operations progress in a smooth safe manner. The density of the mud must be continuously monitored and adjusted to ensure that the hydrostatic head it creates within the hole is always greater than the pressure likely to be encountered should the drill bit strike an unexpected pocket of oil or gas. The formation pressure is primarily a function of hole depth and can be estimated with a fair degree of accuracy. An adequate head of mud will prevent a *blow out*, that is a blow back of hydrocarbons into the hole and up to the rig floor. However, an excess of mud pressure must be avoided, particularly when drilling through soft formations where the mud will simply force its way into the structure of the rock and circulation will be lost. When this happens drilling must be suspended and a casing string inserted to reinforce the hole.

The majority of drilling operations are performed using a *water based mud* (WBM) which is both cheaper and kinder on the environment than the oil based varieties. However, *oil based muds* (OBM) are preferred for drilling operations through water sensitive shale formations and for the drilling of highly deviated wells which benefit from the additional hole stability provided by the denser mud as it cakes the well bore.

The disposal of drilling debris can create tremendous problems, particularly when one considers that the drilling of one well can liberate up to 2,000 tons of cuttings. Cuttings produced using water based muds can simply be flushed into the sea, whereas most countries have now prohibited the disposal of oil based mud cuttings in this manner on environmental grounds.

Operating Principles

The schematic illustration shows the basic layout of the drilling mud circulation system employed on a *rotary table* powered drilling rig. The system is identical to that used on the more modern *top drive* systems with the exception of the Kelly which is no longer required.

i) The cycle commences as high pressure reciprocating pumps deliver mud to the rotating drill string via the *standpipe*, flexible *rotary hoses* and *gooseneck* attached to the swivel.

ii) The mud passes down the centre of the Kelly and into the bore of the drill pipe where it eventually exits around the teeth of the drill bit. The mud acts as a cutting fluid, cooling the drill bit and retaining rock chippings produced by the drilling process in a suspension which can be channelled back up through the hole and on to the rig.

iii) The mud returns are collected within the *bell housing* situated on top of the diverter or BOP stack and are directed into a *shale shaker* prior to entry into the settling and storage tanks (or *mud pits*). The shale shaker consists of a series of vibrating gratings or screens which sieve the mud and remove the particles of drilling debris. Particles removed by the shale shaker are examined to provide the drill team with information pertaining to the physical characteristics and water content of the formation through which the drill bit is passing.

Where an improved level of filtration is required *centrifuges* or *hydro-cyclones* may be used in addition to the shale shakers to clean up the mud.

iv) Prior to re-circulation of the drilling mud it is allowed to stand in the settling and storage tanks which are open to the atmosphere. This permits the release of any gaseous hydrocarbon products dissolved during the drilling process as will be the case when the drill bit enters a hydrocarbon bearing formation.

Recent Developments

Drilling muds are being continually improved and *synthetic-based drilling muds* (SBM) are now available as an alternative to oil based muds (OBM). The synthetic based muds are claimed to reduce drilling torque, the risk of the drill string sticking in the hole, and to improve penetration rates. Unlike oil based muds the cuttings produced with synthetic based muds can be discharged directly into the sea but one of the disadvantages is that it is extremely expensive, between three and five times the price of oil based muds. Another disadvantage is that the synthetic based muds must be displaced with a completion fluid prior to cementing if a paste with the consistency of peanut butter is to be avoided. Clearly no one mud system provides the ideal solution to all the technical, environmental and commercial challenges which must be considered when embarking on a new project.

Underbalanced

The preceding text outlines the significance that drilling mud plays in the drilling programme, acting as the primary safety barrier against the ingress of hydrocarbons into the well bore. The process is described as drilling in the overbalanced condition. However, in recent years some companies, mostly operating onshore in Canada, have exploited the advantages of drilling in the underbalanced condition, that is where the column of drilling mud is lightened by the injection of foams or gases, the object being to keep the pressure exerted by the mud in the well above the collapse pressure of the hole but below the pressure in the formation. The well will thus flow as soon as the hydrocarbon bearing formation is entered.

Underbalanced drilling is generally used when modifying an existing well such as in the drilling of a new horizontal section, or to finish a new well in an existing reservoir where the reservoir pressure has subsided significantly. The well is drilled in the conventional overbalanced mode until the hydrocarbon bearing formation is reached and the remainder of the well is then drilled underbalanced. The main advantage of underbalanced drilling is that by preventing the well bore and surrounding formation from becoming impregnated with drilling mud, the permeability of the rock is maintained which saves on well clean up operations. It also greatly reduces the tendency of the drill string getting stuck in tight horizontal holes.

Whilst there are numerous means of achieving underbalance, offshore, nitrogen gas is used, injected into the mud in the standpipe at a pre-calculated pressure to achieve the desired level of underbalance. The nitrogen may be supplied in tanks as a liquid, or a nitrogen generator may be used. Conventional drilling equipment is used with the addition of a rotary BOP (RBOP) which is installed on top of the conventional BOP. The inner element of the RBOP rotates with and seals the drill string and ensures that the mud returns which contain rock cuttings, nitrogen and hydrocarbons can be directed in to a four phase separator. The gases are flared, the hydrocarbon liquids are deposited into storage tanks and the cuttings disposed off.

Part 4. OPERATIONS

No account of offshore drilling would be complete without introducing the drill team and providing a brief description of the more routine operations they perform during drilling or *making hole*. The terminology reflects the American origins of the industry for whilst it may have been a Scotsman, James Young, who patented the first process for the production of refined petroleum products, it was an American, Colonel Drake, who the same year, 1850 drilled the first well specifically for the production of oil. Colonel Drake's well in Titusville, Pennsylvania produced oil from a depth of 68 feet (20 metres).

Making hole is a particularly time consuming operation. Drilling can proceed at speeds of up to 360 feet (108 metres) an hour or down to 1 foot (0.3 metres) depending on the type of rock formation. A typical well takes approximately 8 weeks to drill and consumes 5 miles of casing and tubing.

Whilst automation is slowly creeping onto the drill floor the environment still dictates that drill crews work long hours in physically demanding conditions.

1. THE DRILL CREW

The drill team consists of the following personnel:

i) **Toolpusher.** The toolpusher or drilling supervisor is in overall charge of the drilling programme.

ii) **Driller.** The driller is in charge of all operations carried out on the drill floor, in effect the drill floor foreman.

iii) **Assistant driller.** The assistant drillers main duties involve the preparation of drill floor machinery.

iv) **Derrick man.** The derrick man works from the monkey board and assists in the installation and removal of drill pipe during the making of a trip, guiding the stands into and out of the finger board.

v) **Roughnecks.** Drill floor labourers.

vi) **Roustabouts.** General rig labourers.

2. MAKING HOLE

Whilst the rotary table has largely been superseded by the top drive assembly during the last 10 years, an account of drilling operations using both types of equipment will be given as the rotary table may still be encountered on older installations and as part of drilling history will frequently be referred to offshore.

Offshore Engineering

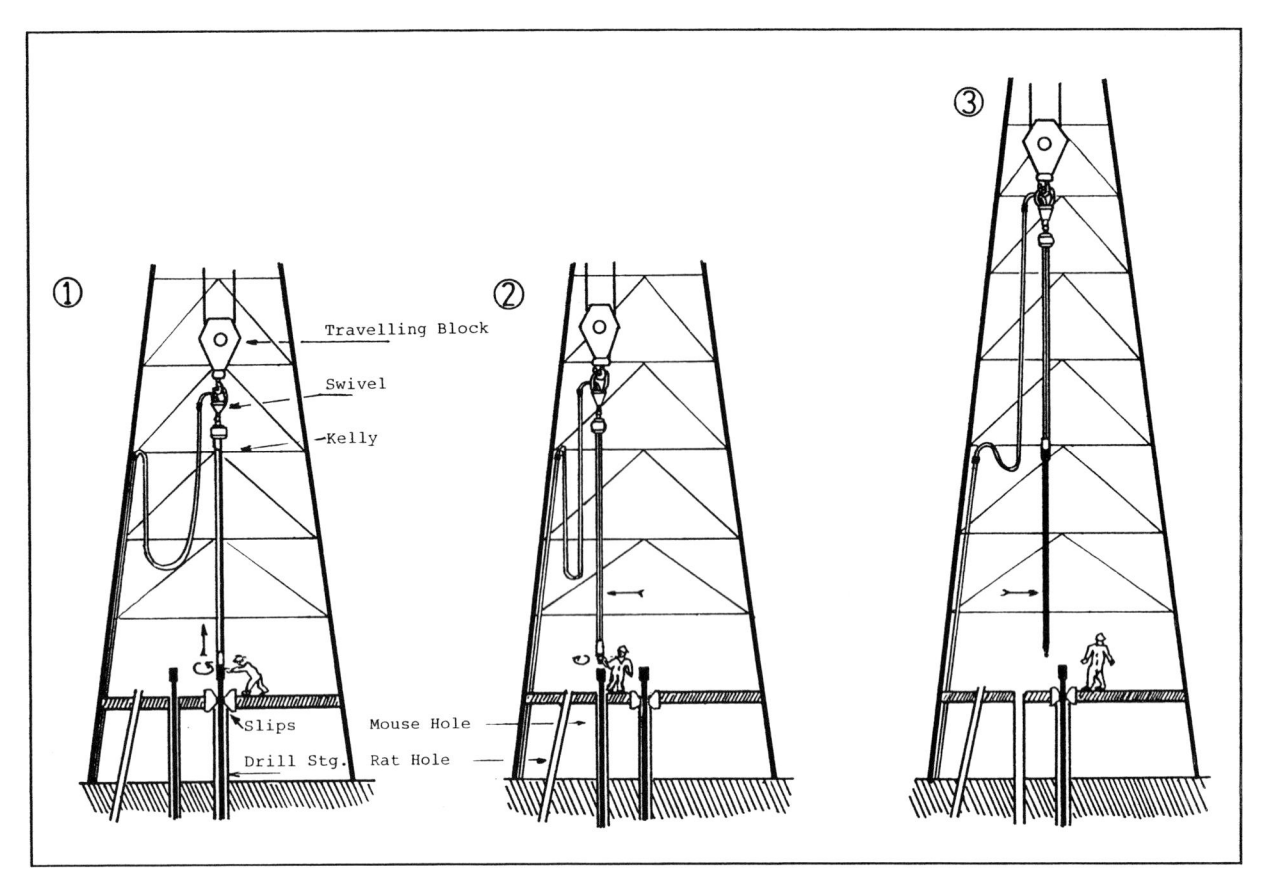

MAKING A CONNECTION

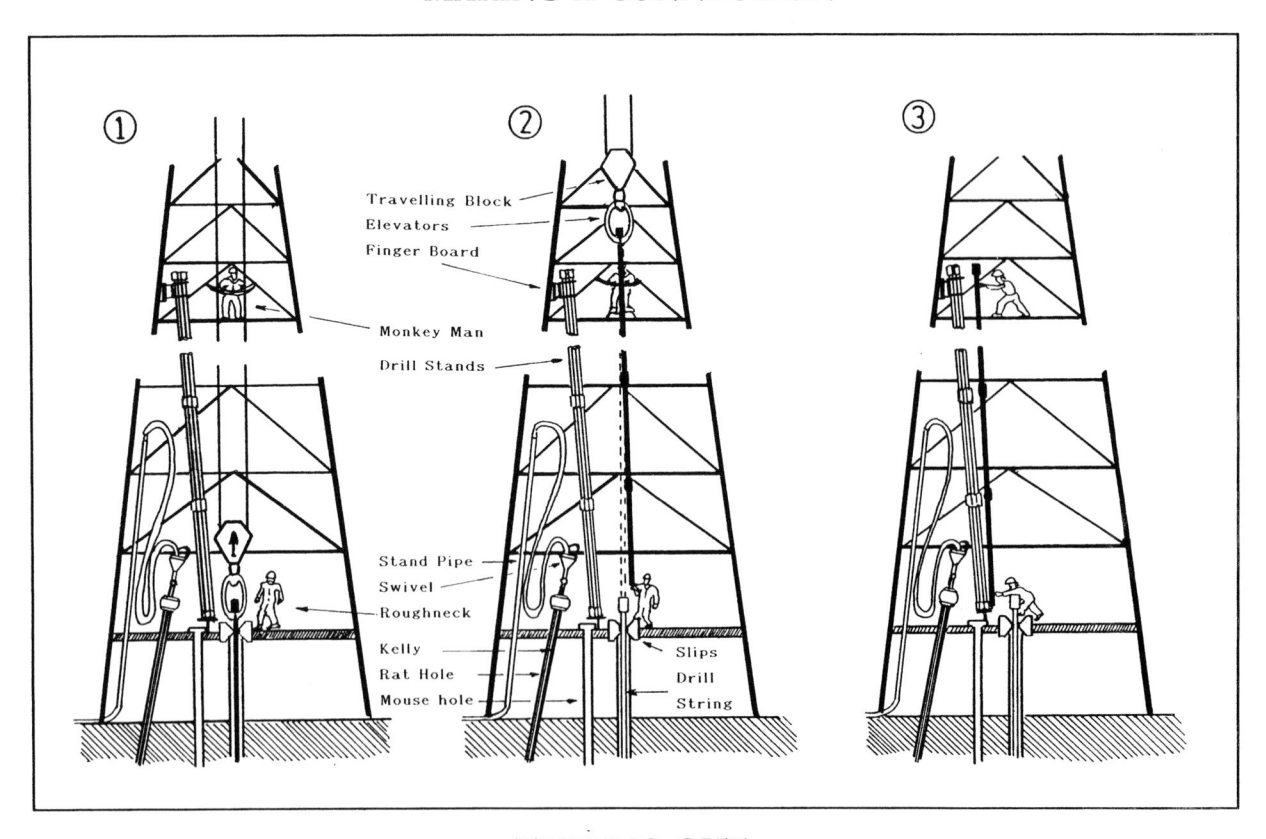

TRIPPING OUT

260

i) ROTARY TABLE DRIVE

Rotational movement is imparted to the drill string by the rotary table, via the Kelly bushings. As the hole progresses the Kelly slides through the bushings until approximately 30 feet (9 metres) of hole have been drilled and a new section of drill pipe is required. The addition of drill pipe is referred to as *making a connection* and the process may be compared with the way a chimney sweep increases the length of his brush as it proceeds up the chimney.

To make a connection the rotary table must be stopped and the flow of drilling mud interrupted. The drill string is then lifted by the travelling block until the Kelly is clear of the rotary table. The Kelly bushings are then replaced by *slips*, a pair of serrated steel wedges which prevent the drill string falling back down the hole when the base of the Kelly is disconnected. The Kelly is connected to a new section of drill pipe located in the mouse hole and the complete assembly re-positioned over the drill string. The drill string joints are tightened using a pair of hydraulically powered pipe grips or *tongs* after which the drill string is ready for re-entry into the hole. The drill string weight is once again taken by the travelling block whilst the slips are removed and the drill string is lowered back into the hole. Once the Kelly bushings have been replaced, drilling may continue.

ii) TOP DRIVE

The top drive assembly employs a large hydraulically or electrically powered motor to impart rotary motion to the drill string. The motor is suspended from the travelling block and 90 feet (27 metres) of hole can be drilled before additional drill pipe is required. The facility to drill three times as much hole as the rotary table between connections represents a considerable saving in time and is the main reason for the success of top drive systems.

To make a connection with a top drive assembly the motor must first be stopped and the flow of drilling mud suspended. The travelling block briefly supports the entire weight of the drill string whilst the slips are inserted into the drill floor. The motor can then be disconnected from the drill string and hoisted into position over a new section of drill pipe.

Drill pipe awaiting inclusion into the drill string is re-assembled into *stands* consisting of three sections of pipe stood vertically inside the derrick *finger board*. The drive motor is connected to the top of a stand and the travelling block is used to lift it into position over the drill string whilst the two are coupled together. Drilling may be resumed once the slips have been removed and a further 90 feet of hole drilled.

Whilst the preceding text provides an outline of the way in which a hole is progressed, it will be appreciated that occasions arise which require the removal of the drill string in it's entirety, the operation being referred to as *making a trip or tripping out*. It is a time consuming operation which may require the removal of up to 12,000 feet (3,650 metres) of drill pipe during the latter stages of drilling.

The making of a trip will be required prior to the installation of a casing string and whenever the drill bit requires replacement. Drill bit life will be dependant on the type of formation being drilled but a life of 10 to 72 hours can be expected unless specialised diamond tipped cutters are used.

The tripping operation commences with the suspension of the drill string from the slips in the drill floor. The procedure is identical for top drive and rotary table powered equipment once the Kelly and swivel have been removed. The swivel is replaced with *drill pipe elevators*, a large clamp arrangement which grips the drill string below the uppermost drill joint. The travelling block is then raised until three lengths of drill pipe (a *stand*) are exposed and the slips are re-set in the drill floor. The stand is then disconnected and stood vertically inside the derrick, and held in the finger board until required for further use. The process is repeated until all the drill pipe has been retrieved from the hole. The procedure is reversed once the new drill bit or casing has been installed.

Varco BJ Drilling Systems

Description:
The Varco Pick-up/Lay-down System PLS, is a new generation pick-up/lay-down system that together with a rig floor pipe racking system drastically reduces the manual handling of pipe on the rig floor and pipedeck. The PLS automatically picks up pipe from special pipe containers which may be loaded onshore or at the pipe yard. The pipe is horizontally transferred to the rig floor where a pipe pick-up boom rotates the pipe vertically and presents it to the arm of the pipe racking system. This eliminates the need to manually latch elevators on the pipe as well as removing the tugger or crane handling of each joint. The danger associated with tailing pipe into the V door and the handling of the pipe on the pipe deck is completely eliminated.

The PLS system consists of two major components, a pipe Elevator and a Pick-up/Lay-down boom. The pipe elevator is designed to raise pipe from of the pipe deck to the rig floor level. The elevator assembly sits on top of the pipedeck next to the catwalk. The pipe containers are positioned on either side of the elevator and automatically feed the drill pipe, collars or casing to the lifting clamps on either side of the elevator. On top of the elevator is a series of V rollers that feed pipe to and from the pick-up boom. The pipe boom is designed to pick-up the pipe from its horizontal position by rotating it 90 degrees permitting the upper end of the pipe to pass through the existing V door opening in the derrick. The PLS is controlled from the same control system used to control the pipe racking system. A control station for manual operation is located on a work platform near the rig floor. Semi automatic sequencing is controlled by the pipe racking system operator from his console.

Benefits:
- Eliminates the need to feed each joint of drill pipe, collars or casing into the V door by crane
- Can be combined with a PHM racking system for fully automatic operation
- Can be operated in manual mode if required
- Reduces the number of Roustabouts required
- Improves pipe deck safety

Specifications:
Elevator High Speed: 2 ft per sec
Elevator Low Speed: 6 inches per sec
Boom Elevation Time: 18 sec
Pipe Sizes High Range: 9-3/4 to 20 inches
Pipe Sizes Low Range: 3-1/2 to 9-3/4 inches

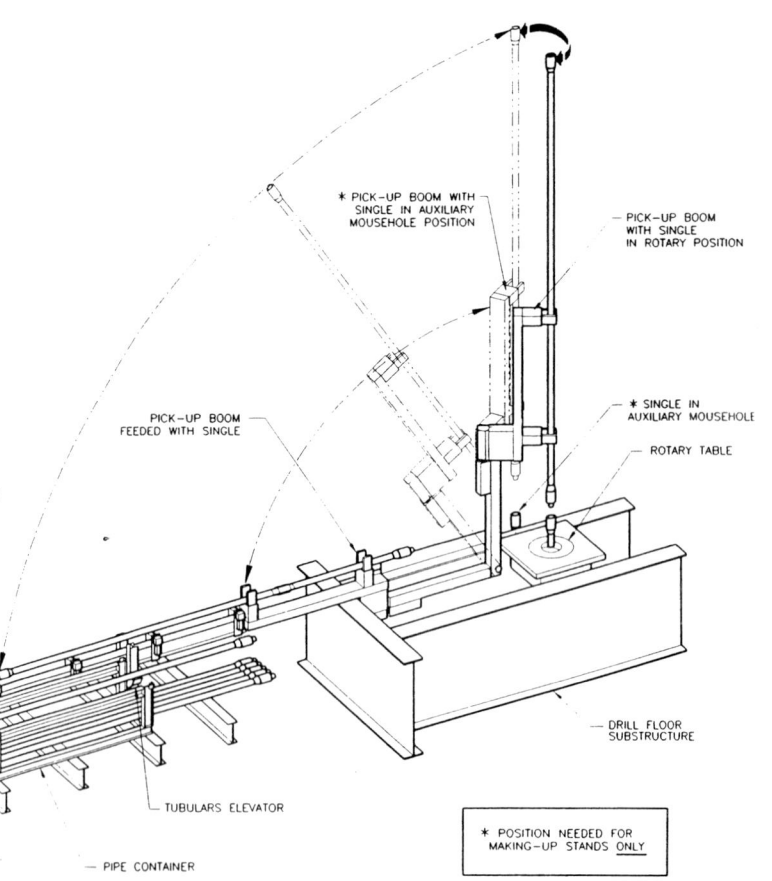

Varco BJ Pick-up/Lay-down System PLS

AUTOMATED DRILL PIPE HANDLING SYSTEMS
(Reproduced with permission of Varco BJ, Orange, California)

Part 5. WELL CONTROL EQUIPMENT

The *diverter* and *blowout preventer* (BOP) are probably the two most important items of equipment on a drilling rig and their description has been left until last because to understand their function one must first have a basic appreciation of drilling operations. Basically, diverters and BOPs are the last line of defence which provide the drill crew with a breathing space to take corrective action, or abandon the installation when control of the well is lost.

The initial stages of drilling a well are without doubt the most dangerous due to the ever present risk of penetrating shallow gas bearing sands, a hazard which has resulted in numerous fatalities and the complete destruction of both drill ships and jack-ups. The diverter is used in preference to a BOP to provide a degree of protection against these hazards because attempting to confine hydrocarbons within a shallow conductor may lead to a blowout of the formation around the outside of the conductor which would create an unrecoverable situation. Once the hole depth has been substantially increased and the first intermediate casing string cemented in place, the diverter can be replaced with a BOP stack.

1. THE DIVERTER

As the name suggests, the diverter provides a means of diverting an unexpected release of well fluids, primarily gas and occasionally solids, to a location at the extremities of the rig where they can be discharged safely.

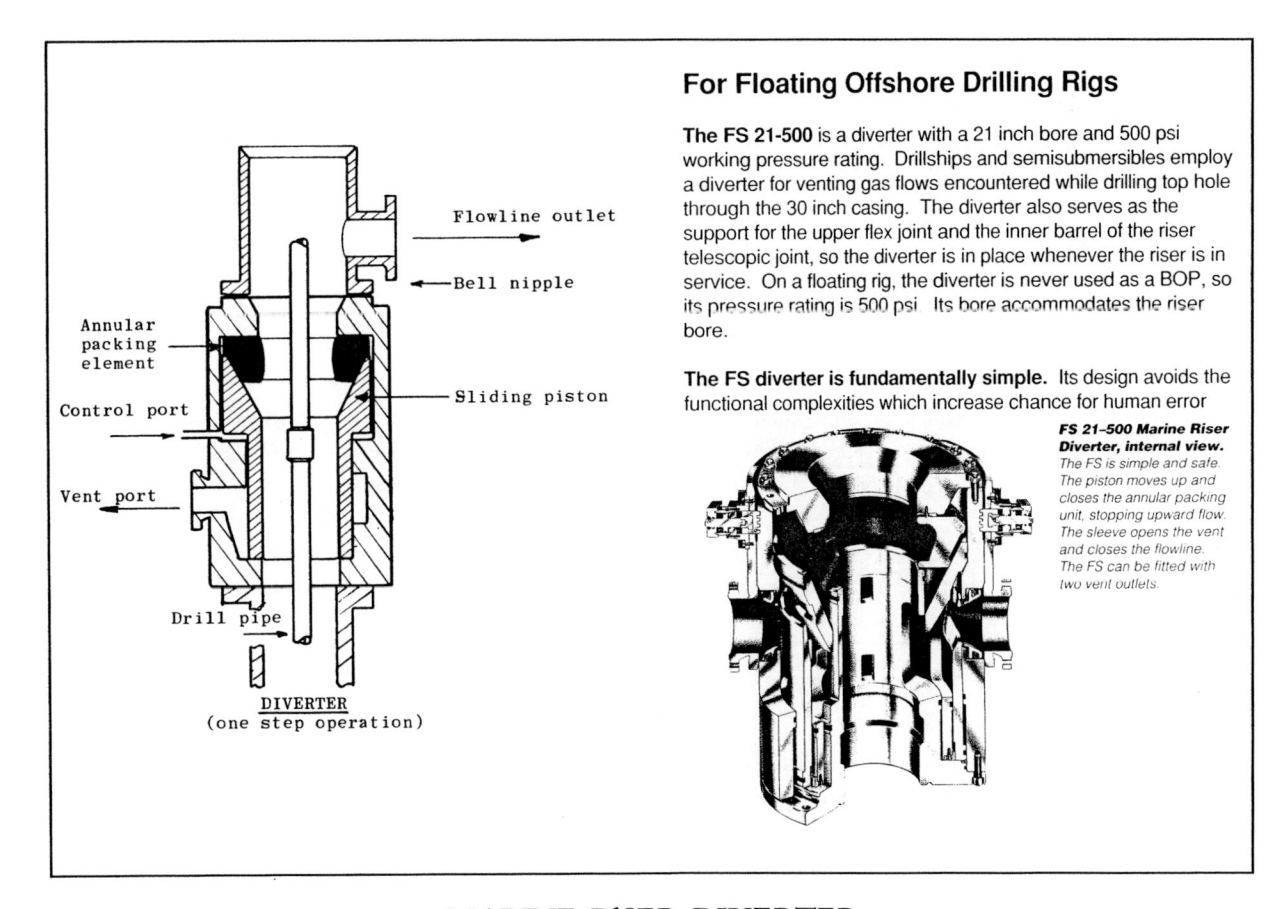

For Floating Offshore Drilling Rigs

The FS 21-500 is a diverter with a 21 inch bore and 500 psi working pressure rating. Drillships and semisubmersibles employ a diverter for venting gas flows encountered while drilling top hole through the 30 inch casing. The diverter also serves as the support for the upper flex joint and the inner barrel of the riser telescopic joint, so the diverter is in place whenever the riser is in service. On a floating rig, the diverter is never used as a BOP, so its pressure rating is 500 psi. Its bore accommodates the riser bore.

The FS diverter is fundamentally simple. Its design avoids the functional complexities which increase chance for human error

FS 21-500 Marine Riser Diverter, internal view.
The FS is simple and safe. The piston moves up and closes the annular packing unit, stopping upward flow. The sleeve opens the vent and closes the flowline. The FS can be fitted with two vent outlets.

MARINE RISER DIVERTER
(Reproduced with permission of Hydril, Houston, Texas)

The diverter is situated on top of the *conductor* and must permit the passage of the *drill bit* whilst still being capable of effecting a seal around the *drill pipe* or *Kelly*. During normal drilling operations the diverter vents are closed and the drilling mud returns flow upwards and into the *bell housing* from whence they are channelled into the *shale shaker*. Activation of the diverter results in the closure of the packing element around the drill string and the sequential opening of the vents, thus providing an unrestricted passage to atmosphere for well fluids.

The diverter may consist of a proprietary item or it may be assembled from a fabricated manifold and a conventional *annular blowout preventer* (BOP). The illustrations show a diverter in which the constriction of the packing element, the opening of the vent port, and the closure of the flowline port are combined into a single operation controlled by the annular piston, an inherently more reliable arrangement than can be achieved with built up diverters.

The drilling of wells from a floating rig (semi-sub or drill ship) necessitates location of the BOP stack on the ocean floor. In these cases a diverter is normally mounted on top of the *marine riser* as a precautionary measure throughout the drilling programme.

The most noteworthy development in diverter technology in recent years has been the concept of diverting at the wellhead rather than on the rig. This is accomplished by locating a diverter on the sea bed in addition to the diverter on top of the drilling riser and permits the venting of well fluids into the sea. The risks associated with a possible release of gas on the drill floor during the early stages of drilling are thus avoided.

2. THE BLOWOUT PREVENTER

The primary function of the blowout preventer is to confine well fluids within the well bore. It operates as part of a BOP system which includes the BOP stack, choke and kill valves, choke manifold and a hydraulically powered control unit.

There are two basic types of blowout preventer, that is the annular type and the ram type, examples of which are shown in the illustrations.

i) ANNULAR BOP'S

The annular BOP relies on the constriction of a reinforced elastomeric packing element to effect a seal around the drill pipe or Kelly. The flexible packing element can accommodate considerable changes in diameter and cross section but it typically cannot withstand pressures as high as the ram type BOPs. Consequently, it is used in conjunction with ram type BOPs which are better suited to retaining extremes of reservoir pressure. In addition to containing well bore pressure, annular BOPs are also used for *stripping* and *snubbing* operations which involve the vertical movement of drill pipe and casing into or out of the well under well pressure.

ii) RAM TYPE BOP

As previously stated, ram type BOPs are used in combination with an annular BOP and may feature any one of four ram configurations:

a) **Pipe rams**

These are supplied in various sizes and are designed to close and seal around the drill pipe.

DRILLING

ANNULAR BOP IN DIVERTER MODE
(shown closed)

Hydraulic Control System

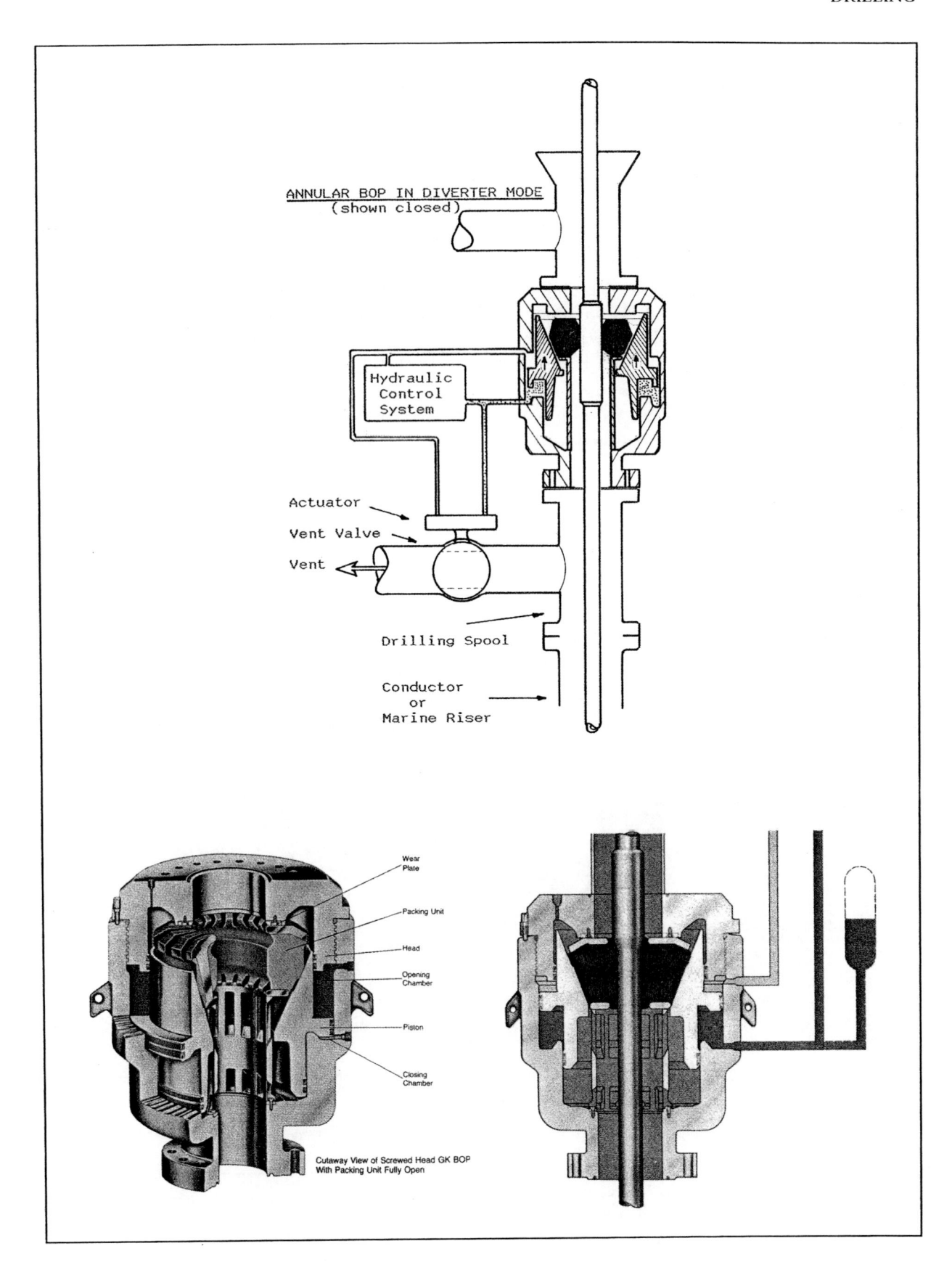

Actuator
Vent Valve
Vent

Drilling Spool

Conductor or Marine Riser

Wear Plate
Packing Unit
Head
Opening Chamber
Piston
Closing Chamber

Cutaway View of Screwed Head GK BOP
With Packing Unit Fully Open

ANNULAR BLOWOUT PREVENTER
(Reproduced with permission of Hydril, Houston, Texas)

Offshore Engineering

b) **Variable rams**

These are capable of closing and sealing on a limited range of pipe sizes.

c) **Blind rams**

These seal against each other and are designed to seal the well bore in the absence of the drill pipe.

d) **Blind/shear rams**

These are primarily used during sub-sea drilling operations where weather conditions may necessitate the rapid disconnection of the rig from the well. The blind rams incorporate a cutting edge capable of severing the drill string and sealing high pressure, high temperature well fluids within the well bore for an extended period of time.

An hydraulically controlled, fail safe power unit typically provides the fluid power for operation of the rams, and the annular piston used in the annular BOPs.

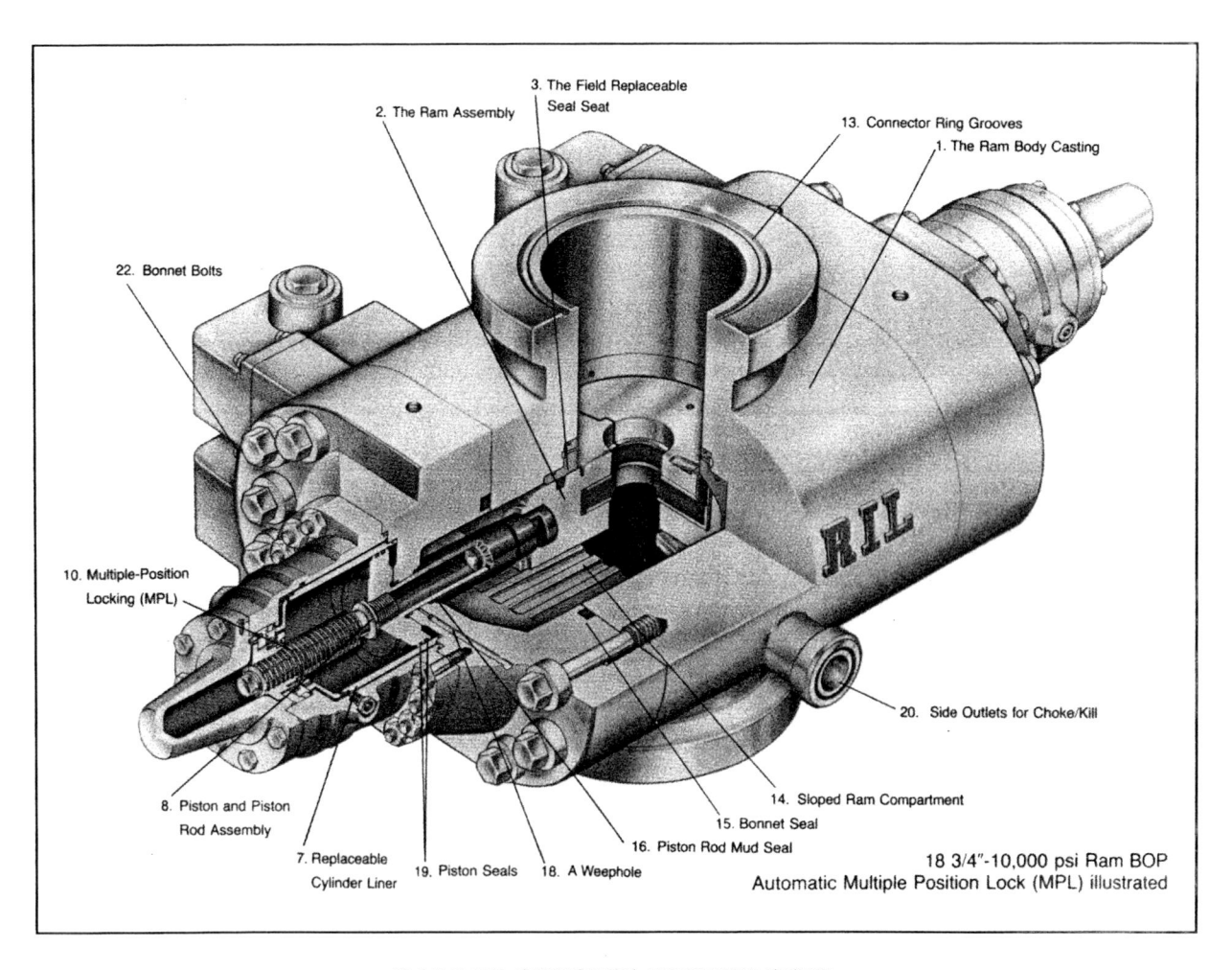

RAM BLOWOUT PREVENTER
(Reproduced with permission of Hydril, Houston, Texas)

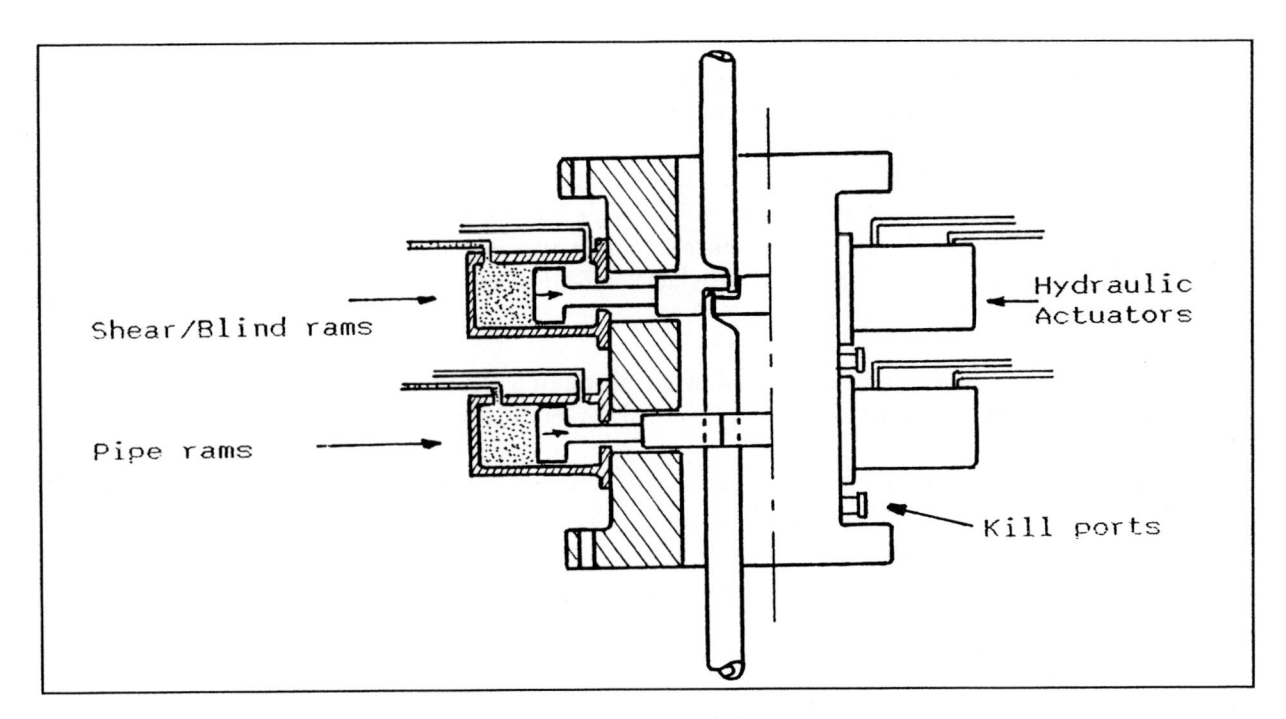

RAM TYPE BOP SCHEMATIC

3. BOP OPERATIONS

Drilling operations proceed through the centre of a BOP stack which is mounted on top of the *casing head*. The stack consists of a *drilling spool* incorporating *choke and kill* connections, an annular and at least one ram type BOP. As anticipated well pressures increase, additional sets of rams are installed up to a maximum of three sets. A typical surface mounted, high pressure well BOP stack would consist of one annular BOP, two sets of pipe rams and one set of blind/shear rams. A sub-sea BOP stack normally contains an additional set of pipe rams and two annular BOPs.

During normal drilling operations both ram and annular BOPs are held in the open position there being no hydrocarbon pressure within the hole. Should the drill bit penetrate an unexpected pocket of oil or gas the hydrostatic head created by the drilling mud should prevent well fluids from entering the hole.

When the drill bit enters a hydrocarbon bearing formation and the hydrostatic head exerted by the column of drilling mud is insufficient to cope with the formation pressure, the well fluids will commence displacement of the drilling mud which will be indicated by an increase in the volume of mud returning to the mud tanks. This situation is referred to as a *kick* and must be dealt with immediately if it is not to escalate into a full blown *blowout*.

To control a kick the drill crew must first close the BOP pipe rams and the Kelly cock in order to contain the formation pressure within the well bore. Eventually the pressure will stabilise at which time a specially prepared mixture of dense *drilling mud* can be pumped into the well via the kill connections located on the drilling spool or BOP. The well pressure is thus neutralised by circulating the mud back through the choke line to the choke manifold. Having successfully *killed* the formation pressure with the hydrostatic head of mud, drilling may be resumed.

Offshore Engineering

Should it be necessary to reduce the pressure within the well bore at any time during the well kill operation, this may be effected via the choke manifold connected to the drilling spool. The choke contains a variable orifice through which well pressure may be vented to atmosphere.

A blowout occurs when a rapid increase in reservoir pressure displaces the drilling mud from the hole, causing a release of oil, gas or solids. This situation is virtually unrecoverable and should obviously be avoided at all costs. When all else fails activation of the shear rams should cut and seal the drill pipe and provide valuable time in which corrective action, or abandonment of the installation may be effected.

Further information pertaining to blowout prevention equipment can be found in the following documents which are available from the American Petroleum Institute (API):

i) **API RP 53,** Recommended Practise for Blowout Prevention Equipment Systems for Drilling Wells (officially discontinued but still available).

ii) **API RP 64,** Recommended Practise for Diverter Systems, Equipment and Operations.

iii) **API RP 16E,** Recommended Practise for the Design of Control Systems for Drilling Well Control Equipment.

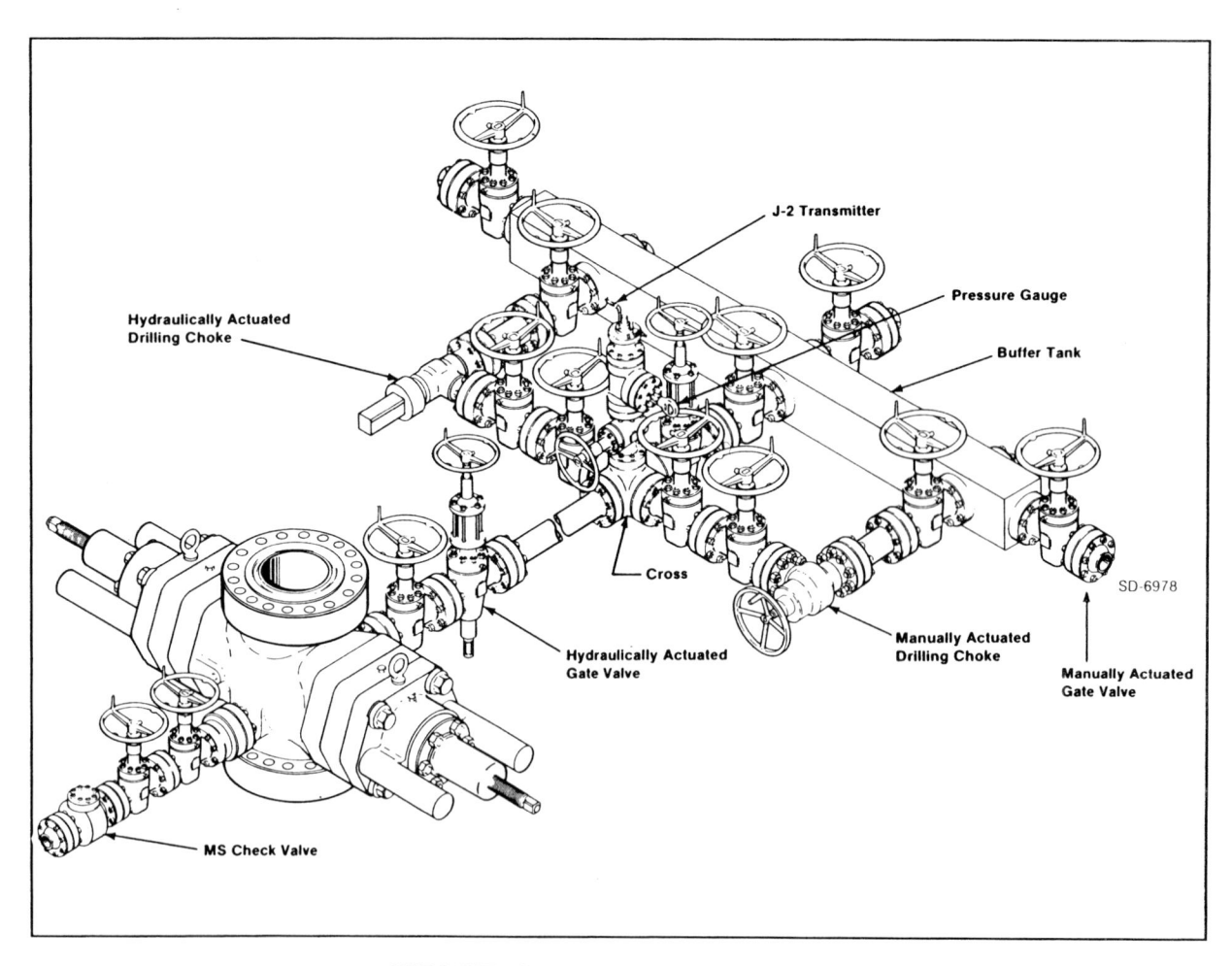

CHOKE AND KILL MANIFOLD
(Reproduced with permission of Cooper Oil Tools, Houston, Texas)

268

Part 6. DRILL SHIP EQUIPMENT

Having described the process and equipment used to drill a well from a bottom supported structure such as a *jack-up* or *fixed installation*, we must now consider the situation where water depths dictate that drilling operations are carried out from a drill ship or semi-submersible vessel (semi-sub).

The procedure for drilling a well is relatively unaffected by the type of vessel employed, be it fixed or floating. The major changes occur in the selection of equipment used to occupy the space between the sea bed and the drill floor. This equipment must include a motion, or *heave compensation system* that will accommodate vessel movement relative to the sea bed.

The sketch shows the various components required to carry out drilling operations from a floating vessel.

i) GUIDE BASE

The first operation entails installation of a guide base on the sea bed through which drilling operations may proceed. Wires extending from the drill ship to the guide base assist in the location of the drill string and wellhead equipment.

Prior to commencement of drilling the conductor pipe must be *spudded* (piled) through the guide base and into the sea bed. The conductor stabilises the soft sea bed, provides a passage for the disposal of drill cuttings and incorporates a suspension system for the support of the intermediate casing strings. The top of the suspension system is fitted with a machined collar designed for attachment to a wellhead coupling.

ii) BLOWOUT PREVENTER (BOP) STACK

When drilling operations are carried out from a floating vessel, the blowout preventer is mounted on the sea bed. This ensures that should an emergency situation necessitate suspension of the drilling programme and departure of the vessel, the well may be left in a secure condition.

The BOP stack consists of a combination of *annular* and *ram* type BOPs housed within a tubular framework. The framework provides protection during transportation and incorporates location devices to assist in the attachment of the stack to the guide base. Remotely operated hydraulic couplings connect the base of the BOP stack to the conductor, and the top of the stack to the riser. The uppermost connector is frequently referred to as the *emergency disconnect package* (EDP).

iii) MARINE RISER FLEXIBLE JOINT

A flexible ball joint located between the EDP and the marine riser can tolerate a certain amount of lateral movement of the drill ship relative to the wellhead. A second ball joint is attached to the top of the riser, frequently incorporated within the diverter.

iv) MARINE DRILLING RISER

The length of steel pipe extending from the BOP stack to the drill ship is referred to as the marine drilling riser and it provides a return passage for the drilling mud. The riser incorporates a telescopic joint, the top half of which is attached to a tensioning system.

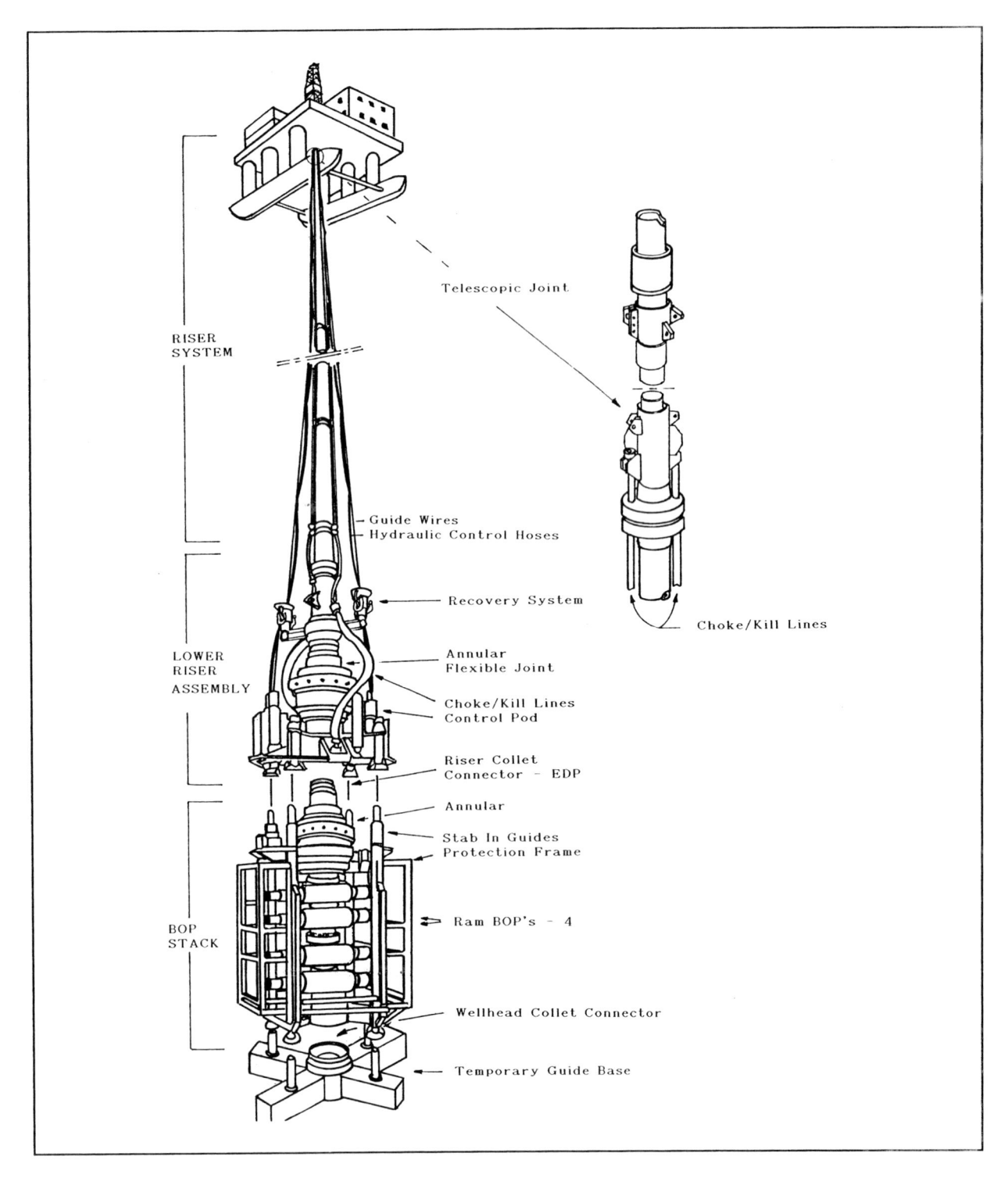

DRILL SHIP EQUIPMENT

v) TELESCOPIC JOINT

The telescopic joint permits relative vertical movement between the stationary lower section of drilling riser which is attached to the sea bed and the upper section of drilling riser which is connected to the *drilling derrick*. It consists of a hydraulic or pneumatically activated packing element which is located in the top section of drilling riser.

vi) THE RISER TENSIONING SYSTEM

The riser tensioning system, or *heave compensation system*, is designed to keep the drilling riser in tension and to minimise the effects of hull movement. It can accommodate up to 15 metres (50 feet) of vertical wave induced motion or heave. Wires emanating from the upper section of drilling riser are connected to hydraulic/pneumatic cylinders which act like shock absorbers, extending as the vessel rides up and compressing as the vessel moves down.

vii) DRILL STRING TENSIONING SYSTEM

The *travelling block* is fitted with a heave compensation system that is similar to, if somewhat less sophisticated than that employed to tension the riser. The hook is separated from the travelling block and mounted within a pneumatic/hydraulically activated piston assembly which ensures that a constant load is maintained on the *drill string* at all times.

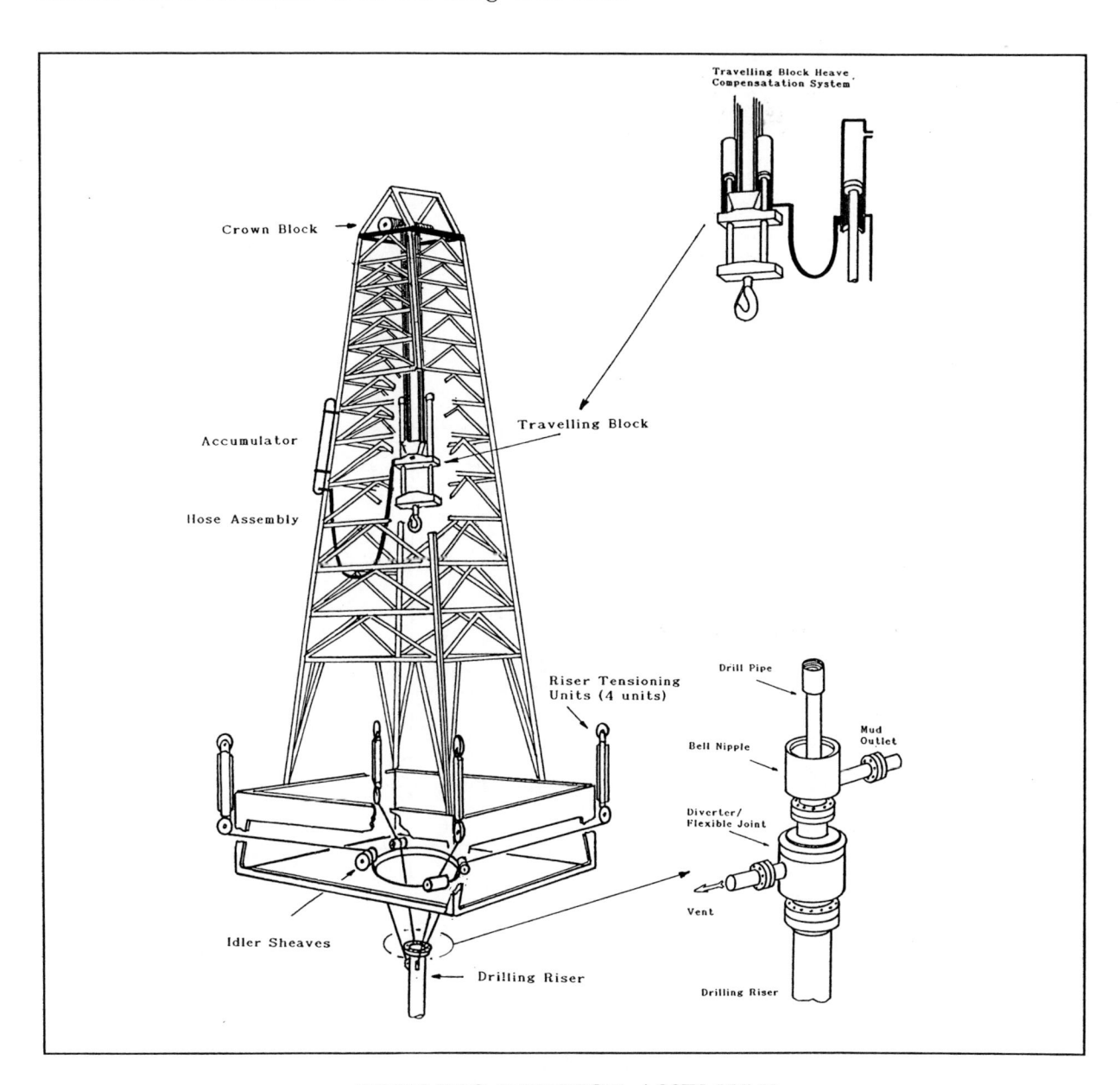

DRILLING DERRICK ASSEMBLY

Offshore Engineering

viii) DIVERTER

The diverter sits on top of the drilling riser and provides a means of controlling an influx of hydrocarbons into the well bore during the early stages of drilling, should one occur.

ix) CHOKE AND KILL LINES

Sub-sea choke and kill line connections are arranged slightly differently to those on a surface well in as much as they are manifolded to permit pressure release, or the pumping in of drilling mud through either connection.

They are run down the outside of the marine drilling riser in hard steel pipe with flexible couplings providing the means of attachment to the BOP stack. The choke and kill line connections to the vessel are by means of flexible hoses suitably dimensioned to absorb wave induced motion.

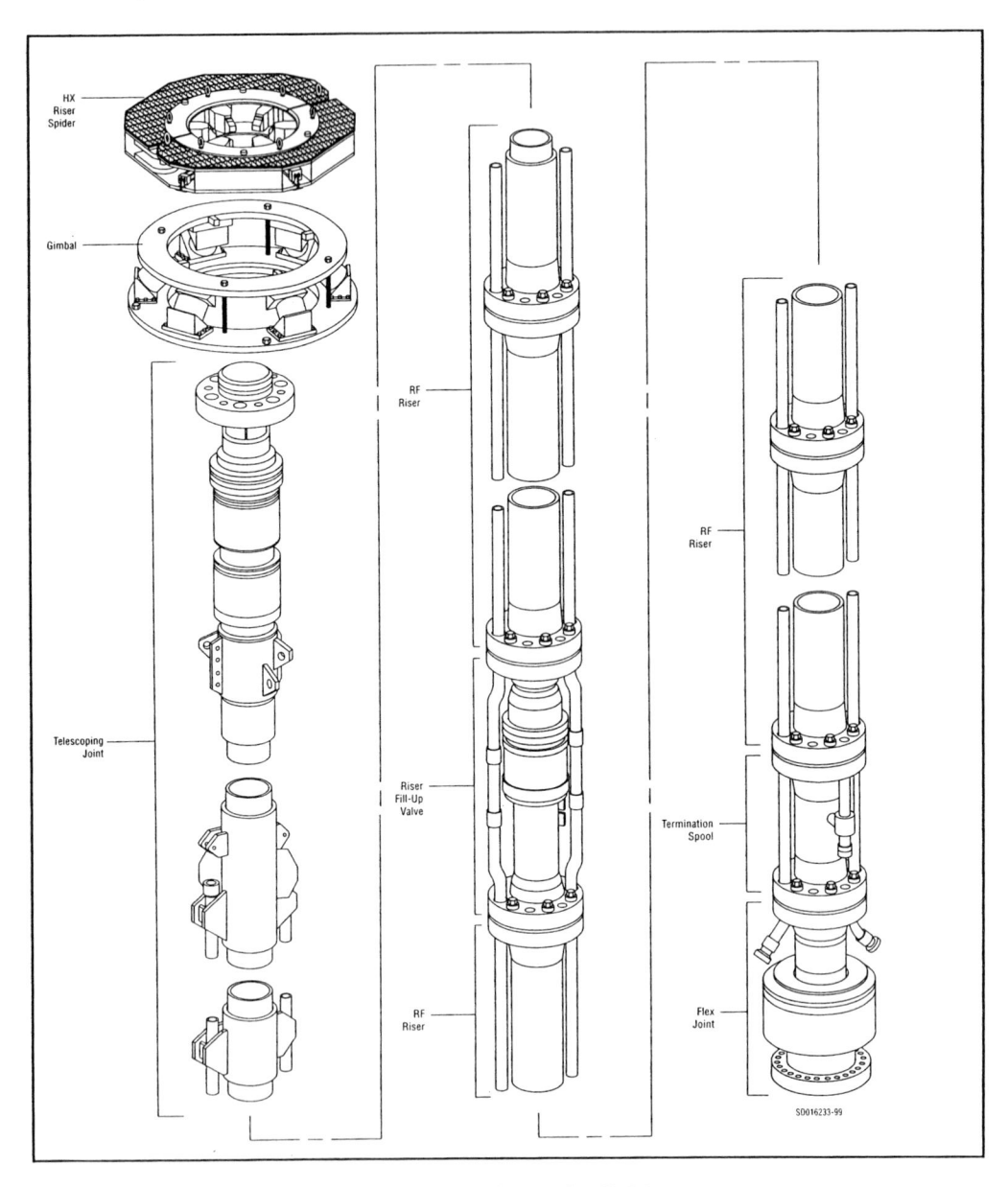

SUB-SEA RISER SYSTEM
(Reproduced with permission of Cooper Oil Tools, Houston, Texas)

Varco BJ
DRILLING SYSTEMS

Description:
The Type V Racking System was designed primarily to improve the efficiency and safety of drilling operations on Drillships and Semi-Submersibles where vessel motion makes handling tubulars dangerous and inefficient.

The Type V Racker System consists of three pipe racker assemblies normally located on the same side of the derrick as the setback area. Each assembly consists of a racker arm capable of moving in and out (towards and away from the rig centerline) and a carriage that moves laterally across the derrick. A racker head assembly on each arm is used to grip the pipe. In a typical tripping operation, the upper racker arm guides the upper end of the stand between well center and the fingerboard while the intermediate racker arm lifts and guides the lower end of the stand. The operation of the upper and intermediate racker arms are controlled from the derrickman's and assistant driller's consoles, respectively.

The lower racker arm is used to assist in the handling of tubulars at the rig floor. The operation of handling the kelly, casing, collars, and marine risers can now be easily and safely performed. The lifting force and vertical motion of the intermediate racker lifting head is provided by a stand lift cylinder assembly controlled from the assistant driller's console. In a variation on the basic three arm Type V Racker System, the lower racker arm is replaced with the hydraulically powered Telescoping Arm which is normally mounted on a pedestal on the rig floor. The substitution of the lower racker arm can provide more operating flexibility in some cases.

For maximum efficiency and reduction in trip times the system should be combined with a remotely operated Varco pneumatically powered fingerboard, air operated elevators and a retract system on the Top Drive/Traveling block.

Benefits:
- Greatly improves rig floor safety
- Simple Electro/Hydraulic System
- Provides remote control racking
- Lower Maintenance Costs
- Reduces trip time when used in conjunction with a retract system.

Specifications:
Pipe Size (inches): 3-1/2 to 9-1/2
Lifting Capacity: 20,000 lbs
Max Reach: Customer specified
Max Push: 4,000 lbs

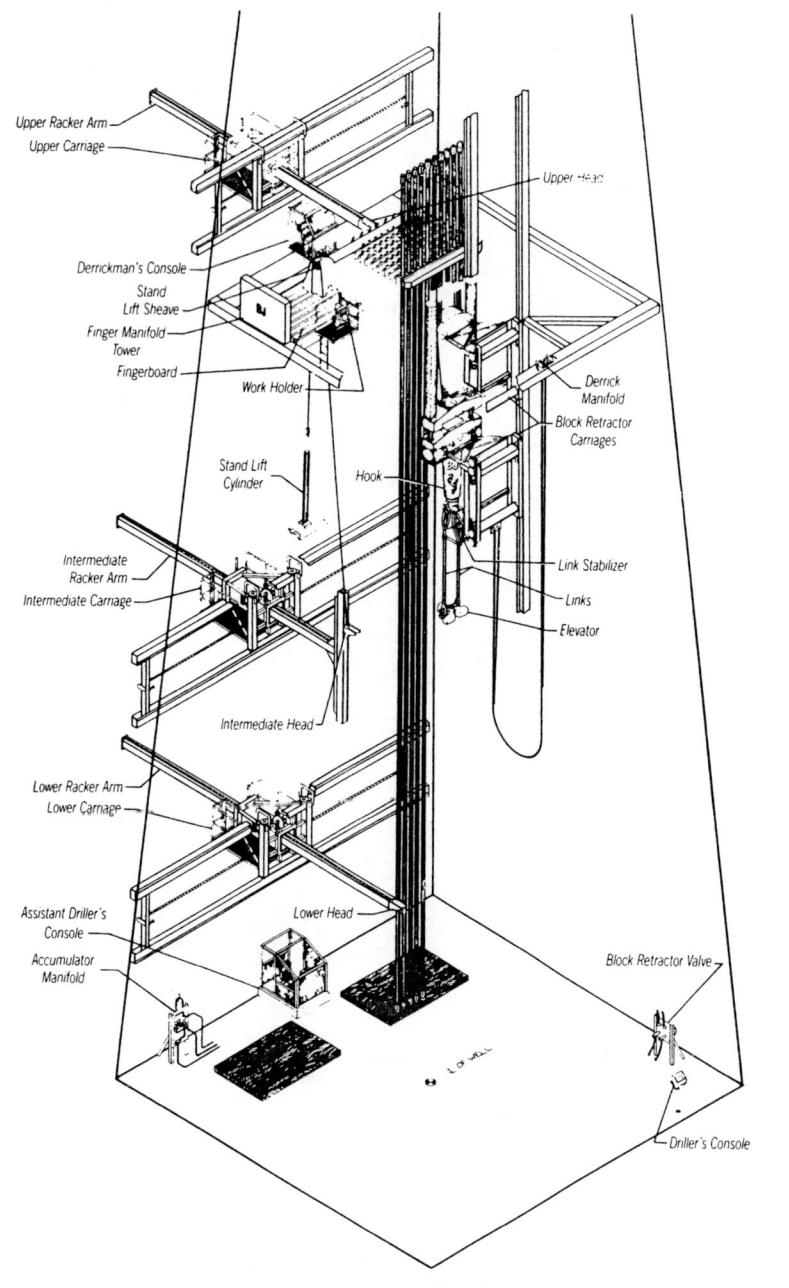

Varco BJ Type V Three-Arm Racker System

AUTOMATED DRILL FLOOR EQUIPMENT
(Reproduced with permission of Varco BJ, Orange, California)

Offshore Engineering

a) Jet Bit

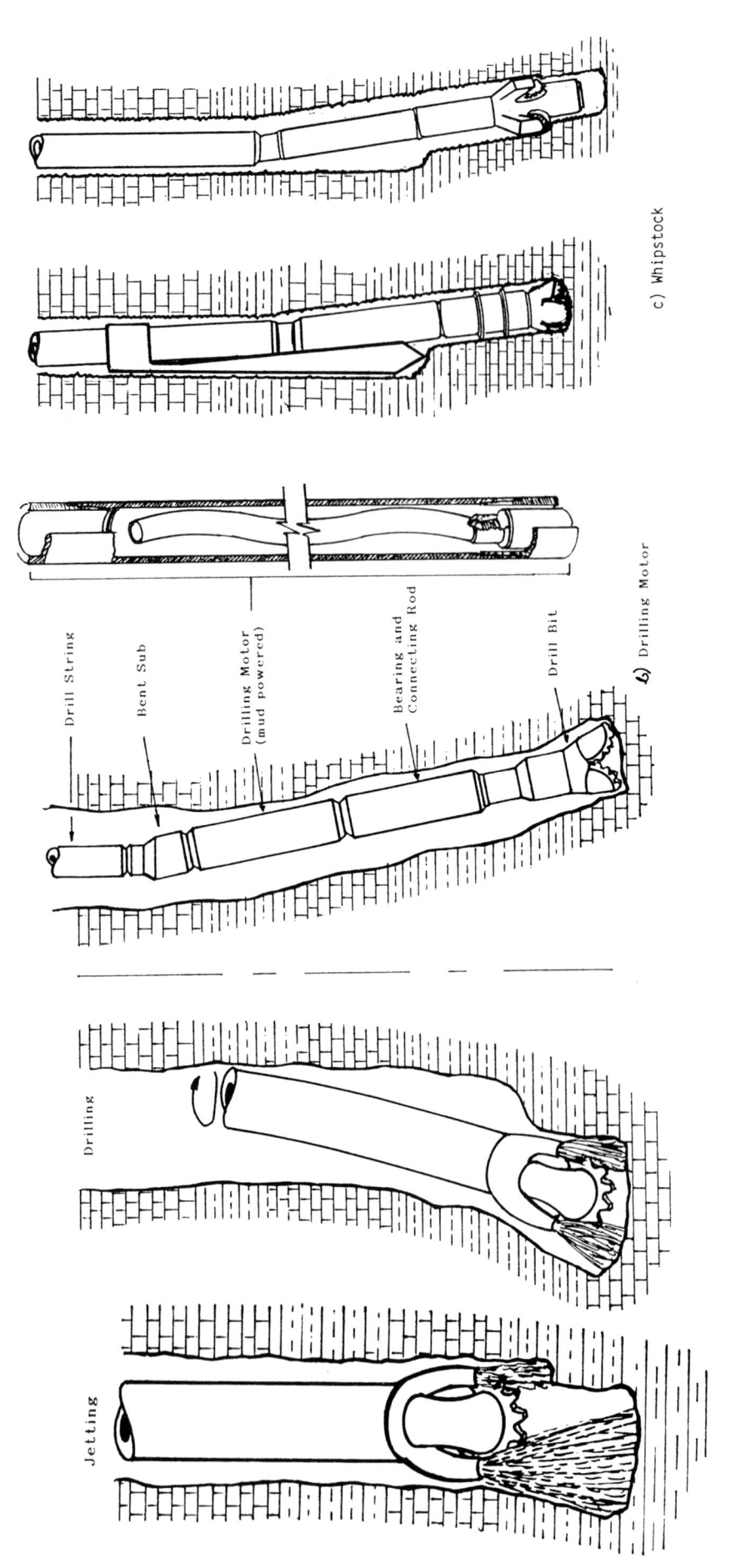

Jetting

Drilling

b) Drilling Motor

Drill String
Bent Sub
Drilling Motor (mud powered)
Bearing and Connecting Rod
Drill Bit

c) Whipstock

DIRECTIONAL DRILLING

Part 7. DEVIATED WELLS

1. DIRECTIONAL DRILLING TECHNIQUES

Directional drilling was developed primarily to permit the exploitation of hydrocarbon bearing formations at locations horizontally distant from fixed installations that would otherwise necessitate an additional platform or sub-sea well.

The most common techniques employed to drill a deviated hole are:

i) JET BIT

The jet deflection processes utilize a triple jet drill bit in which one of the drilling mud outlet connections or jets is considerably larger than the other two. The large jet is positioned in the desired direction of deviation whilst high pressure drilling mud is pumped through the stationary drill bit. The impact force provided by the mud erodes a hole into which the drill string will track on resumption of drilling. The process is simple and effective when the formation is soft but it has largely been superseded by more modern methods.

ii) WHIPSTOCK

The whipstock has been used since the early days of the oil industry as a means of changing the direction of the drill bit. Made of steel, it is shaped with a concave tapered groove set at an angle of 2–3° and is approximately 1.5–3 metres (6–12 feet) in length. One of the disadvantages of the whipstock is that it produces an under-gauge hole which must subsequently be reamed out. Largely superseded by more sophisticated directional drilling tools, a whipstock is still used to kickoff *multi-lateral* wells.

iii) DRILLING MOTOR

The vast majority of directional drilling operations are carried out using a drilling motor to rotate the drill bit and a bent-sub assembly to steer the drill string, although more recently adjustable motors have been developed which combine the functions of both pieces of equipment.

The drilling motor comprises a hydraulic turbine or a positive displacement hydraulic motor, drilling mud being used as the power fluid. The drill string does not rotate, it simply provides a conduit for the supply of drilling mud to the drilling motor. The well is steered to the target by means of the bent-sub, an offset attachment supplied in angles ranging from 1°-3° positioned between the drilling motor and the drill string. This technique is referred to as drilling in *sliding mode*, the drill string sliding after the drill bit as the hole is progressed.

2. HORIZONTAL WELLS

Directional drilling technology has developed radically since the early 1990s and wells can now be drilled which extend a distance of 8 kilometres (5 miles) from the wellhead, and which have horizontal sections which exceed 2,500 metres in distance. These wells are referred to as *horizontal wells*, the main advantage being that they expose a larger and often more favourable section of the reservoir to the well bore. This improves reservoir drainage and enhances production, particularly from thin hydrocarbon bearing formations, and can result in a considerable reduction in the number of wells required, often by up to 25%.

Offshore Engineering

DRILLING HIGHLY DEVIATED AND HORIZONTAL WELLS

The drilling of a highly deviated or a horizontal well is carefully planned by first calculating the geometric co-ordinates of the reservoir from seismic maps. A well bore trajectory is then prepared to take the drill bit to the target.

The drilling programme commences with the vertical section of the well being drilled to a predetermined depth with a conventional rotary table or top drive motor being used to rotate the drill string, the equipment being referred to as a rotary bottomhole assembly (BHA). When the desired vertical depth has been reached, a *bent-sub* and *drilling motor* are inserted between the drill string and the drill bit to provide the means by which the hole will be *kicked off* from the vertical plane and steered to the target.

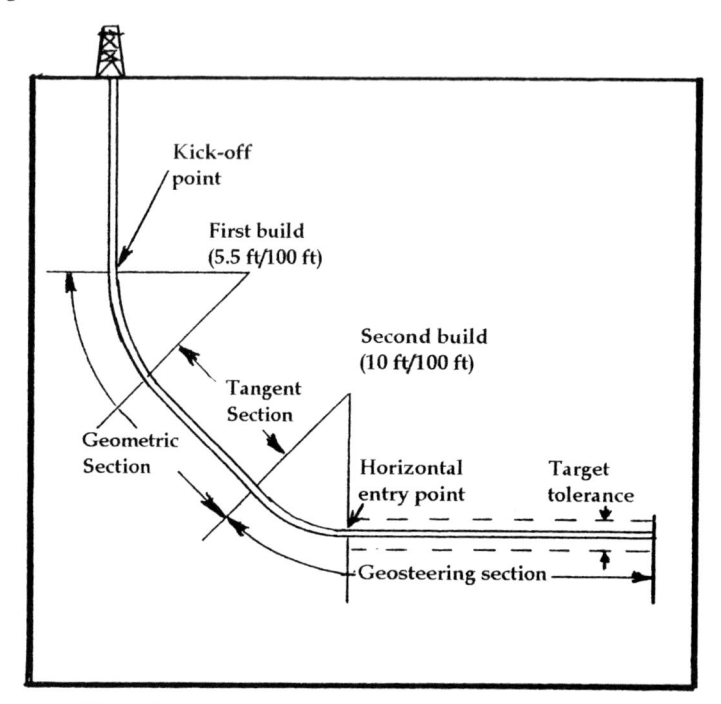

HORIZONTAL WELL PLAN

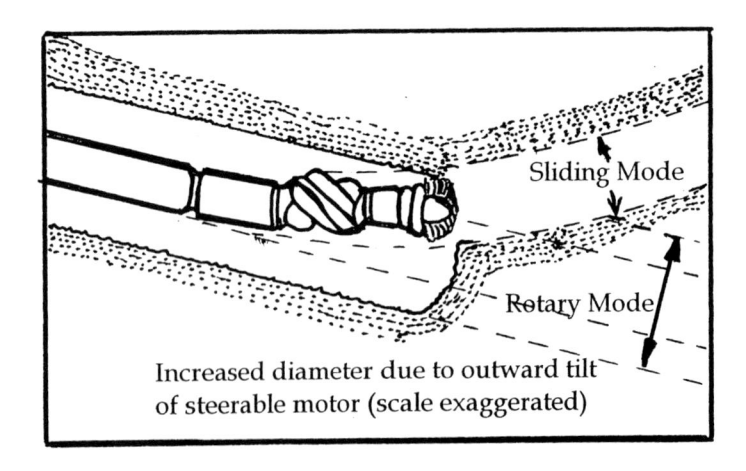

In rotary mode, both the bit and drill string rotate and the bit cuts a straight path parallel to the axis of the drill string above the bent sub. In sliding mode, only the bit rotates and the hole follows the axis of the bent housing below the bend, thus changing the trajectory.

As the hole is progressed, the position of the drill bit is plotted continuously to ensure that it follows the desired trajectory. The position of the drill bit at any particular time can be calculated geometrically from the compass heading, and the length and inclination of the drill string. The length of drill string can be established by counting the lengths of drill pipe in the hole whilst *measurement-while-drilling* (MWD) tools measure inclination (angle from the vertical) and azimuth (compass direction). Accelerometers measure inclination and magnetometers measure azimuth, responding to the earths magnetic field. The measurements are electronically converted into a binary code and are transmitted to the surface by mud-pulse telemetry, a system whereby pressure signals are sent through the drilling mud in what can best be described as a form of Morse code.

Numerous changes to the direction of the drill bit will be required during the course of the drilling programme and these are effected by means of the bent sub, the acuteness of the angle determining the severity of the trajectory of the hole. When large changes in direction are required, drilling is suspended and the drill bit is lifted from the bottom of the hole whilst the drill string is rotated slightly to re-orientate the bent sub, drilling is then resumed.

Once sufficient inclination has been achieved, the horizontal section of the hole is normally drilled with the steerable motor whilst rotating the drill string. When compared to operation in the sliding mode, rotational drilling is both quicker, and reduces the chance of the drill string sticking in the hole. When further changes in direction are required, rotation is suspended whilst the drilling motor is used to change direction, rotational drilling is then resumed.

Deviation of the hole from the vertical to the horizontal position is a very gradual process and once drilled, the hole may be completed in a similar manner to a vertical well. Conventional casing is run in the hole although it is not cemented in the horizontal section. The casing can be pre-perforated prior to installation and sand screens may be fitted to prevent the formation invading the well bore.

As previously stated, one of the main advantages of horizontal wells is that they permit the exploitation of small, thin reservoirs by effectively providing an extended drainage channel which far exceeds what could be achieved with a vertical well bore. However, the drilling of horizontal wells is fraught with problems, some of which we will now consider before providing a brief outline of some of the technology and equipment that has been developed to overcome these problems.

OVERCOMING LIMITATIONS IN HORIZONTAL DRILLING

In recent years, the information provided by enhanced 3-D seismic survey techniques has enabled the geologists to determine the location of potential oil and gas bearing formations with a far greater degree of confidence than was previously the case and small, intricate reservoirs are now being defined that would previously have gone undetected. Unfortunately, there are limitations in the accuracy of the geometric co-ordinates produced from seismic surveys and a variation of 5% in sound velocities can lead to errors in calculating the depth of the target by several hundred feet. These limitations are less of a problem when drilling into large deposits but small, complex reservoirs can be missed altogether if there are significant discrepancies in the depth and lateral co-ordinates.

The technique of *geological steering* or *geosteering* was developed to ensure higher rates of success in reaching these smaller and less clearly defined targets.

3. GEOSTEERING

Geosteering is defined as *real-time steering of horizontal and high-angle wells using while-drilling formation evaluation*. It enables the driller to guide the drill bit and thus the well to the optimum geological destination, rather than directionally steering the well to a pre-determined, possibly non-optimum, geometric location.

At the heart of the new equipment is a comprehensive array of *measurement-while-drilling* (MWD) and *logging-while-drilling* (LWD) tools linked to an interactive graphical interface. The logging and measurement tools are incorporated in the drill string so that the drilling team can determine with a very high degree of accuracy the location of the drill bit in the reservoir, whilst drilling.

PLANNING

Prior to commencement of geological drilling operations, a computer generated model of the formation is constructed based on a combination of seismic survey information and on wireline logs taken from similar wells located in the immediate vicinity of the area under investigation. From the reservoir model, a simulation of the predicted response of the formation in terms of resistivity is prepared and this is then compared with the actual resistivity measurements recorded whilst drilling.

Resistivity, which describes the response a material has to electromagnetic induction, is one of the most reliable techniques for determining the geology of the formation. Different formations have differing resistivities and the resistivity is further affected by the presence of liquids, either oil or water. By comparing the actual results, measured-whilst-drilling, with the predicted results, the position of the drill bit can be plotted accurately on the geological model of the reservoir. Should any discrepancies occur between the predicted values and the measured values, the model can be updated and the drilling co-ordinates revised. Any inaccuracies in the initial geometric co-ordinates are thus compensated for and the chances of the drill bit being lead astray are thus reduced significantly, or at least corrective action can be taken prior to the tool wandering off course by a significant distance. In this manner, the drill bit can literally be navigated in to the pay zone along geological boundaries.

A detailed description of these highly specialised tools is beyond the scope of this book. However, the basic principles of operation will now be described.

GEOSTEERING EQUIPMENT

There is nothing radical in the design of what may best be described as the drilling hardware, that is the drilling motor and the surface adjustable bent housing (or the adjustable motor). It is the sensing and measuring equipment, and the telemetry systems which relay the measurements back to the drill floor where the greatest advances have taken place in recent years.

In order to steer a well more efficiently with fewer unplanned lateral and vertical undulations, tools must be provided which can:

- survey the formation;
- determine the inclination of the drill string;
- determine the compass direction of the drill bit, and
- relay this information to the drill floor.

To achieve the desired speed of response, these sensing tools must be placed as near to the drill bit as possible if changes in direction are to made with the minimum of delay.

GEOSTEERING TOOL

The GeoSteering Tool (GST) provides the driller with the equivalent of *a window in the drill bit* and the facility to steer the drill bit either on a geometric path or to navigate it along a geological path through the formation.

The primary components of the Geosteering tool are the PowerPak steerable down-hole motor, an instrumented housing, and a wireless telemetry system.

The instrumented housing is located between the motor-bearing section and the surface-adjustable bent housing and contains the measuring sensors. It also acts as a housing for the driveshaft that transmits mechanical energy from the PowerPak motor section through to the drive motor bearing assembly. The near-bit sensors provide inclination, azimuthal resistivity and azimuthal gamma ray readings which report both on the drill bits position, and on the composition of the formation.

MEASUREMENT SYSTEMS

i) BIT RESISTIVITY

A 1500-Hz alternating current is forced into the formation via electrical coils mounted on the drill collar and bit, the drill collar, formation and coils effectively acting like a large transformer. The flow of current is governed by the resistivity of the formation and by comparing this with the voltage induced in a toriod coil positioned approximately 1.5 metres (5 feet) above the bit face, the resistivity of the formation can be computed. The magnitude of the induced current varies inversely with the resistivity of the formation.

ii) AZIMUTHAL RESISTIVITY AND GAMMA RAY

Azimuthal resistivity and gamma ray readings assist in determining where the formation boundaries are in relation to the drill bit. This can make the difference between drilling through the pay zone and missing it, or landing in the pay zone and drilling along it, as shown in the sketch. By being able to identify the shale-sand interface and changing the direction of the drill bit immediately that a variation from the predicted formation is measured, a more productive well can be achieved.

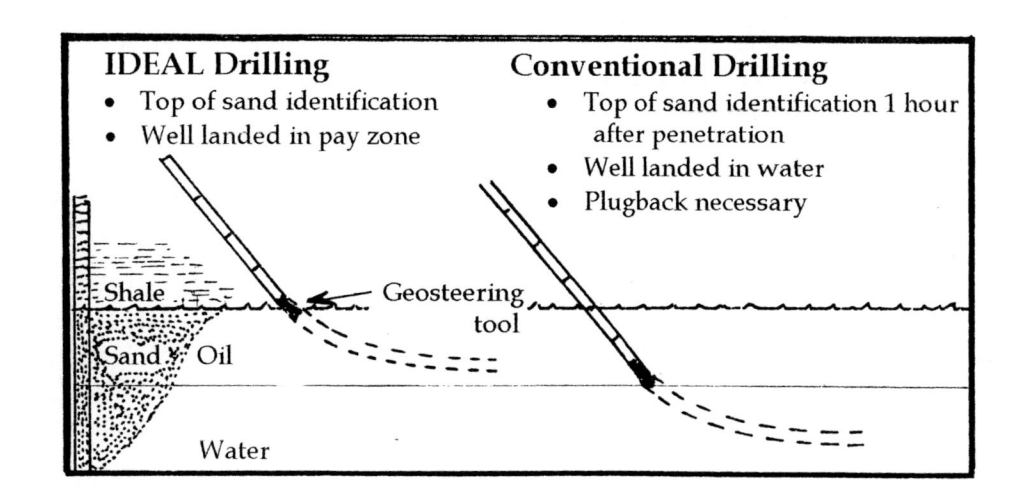

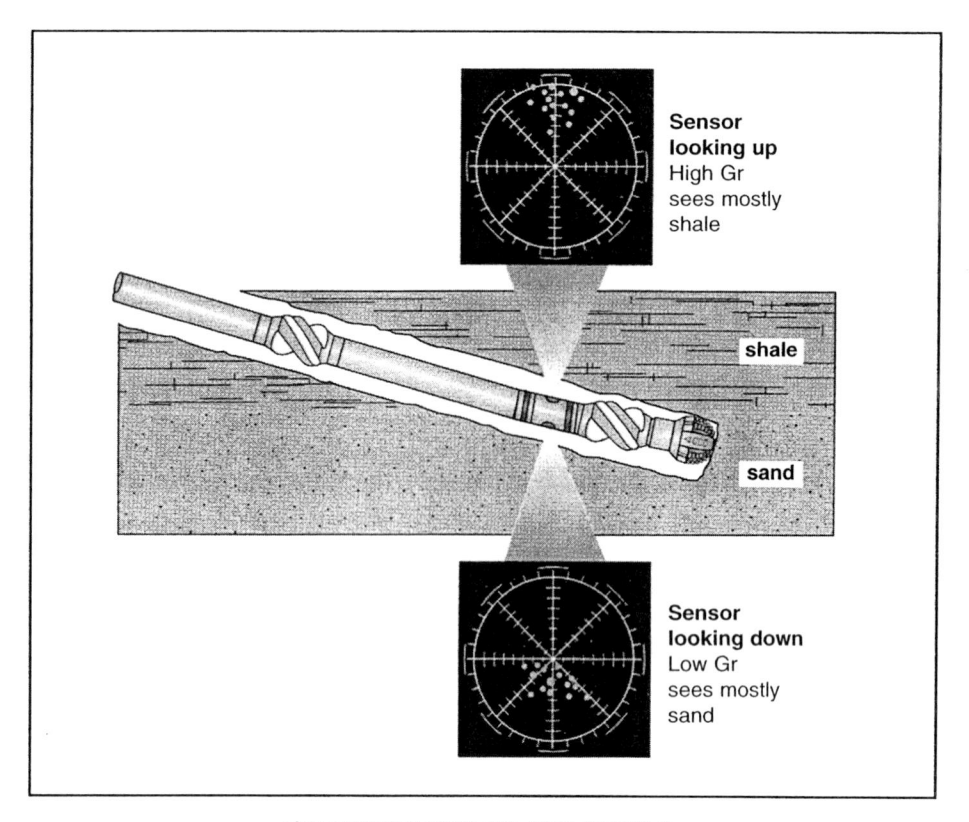

GEOSTEERING SENSORS
(Reproduced with permission of Schlumberger Oilfield Services)

iii) INCLINATION

Inclination measurements taken both at the drill bit and 18 metres (60 feet) behind the bit remove any uncertainties as to where the drill bit is pointing. They provide the driller with a *feel* for what the bottom hole assembly is doing and enable him to determine whether or not the drill bit is drifting in response to formation conditions.

iv) RWOB

In addition to the tools described above a *receiver, weight on bit* (RWOB) tool may be included which measures downhole weight and torque on the drill bit. This allows the driller to optimise the power curve which improves penetration, reduces the chance of the drill pipe sticking, and provides an early indication of problems with the drill bit.

TELEMETRY SYSTEMS

The technical leap that allows measurements to be made at the drill bit is the wireless telemetry system.

The measurement tools and sensors located near the drill bit are battery powered, self contained units. A wireless telemetry link sends data from the sensors to the MWD tool which may be located up to 200 metres behind the bit, a path that bypasses the intervention drilling tools such as the steering motor. The MWD system records and transposes the signals and then sends the data to the surface using mud-pulse telemetry.

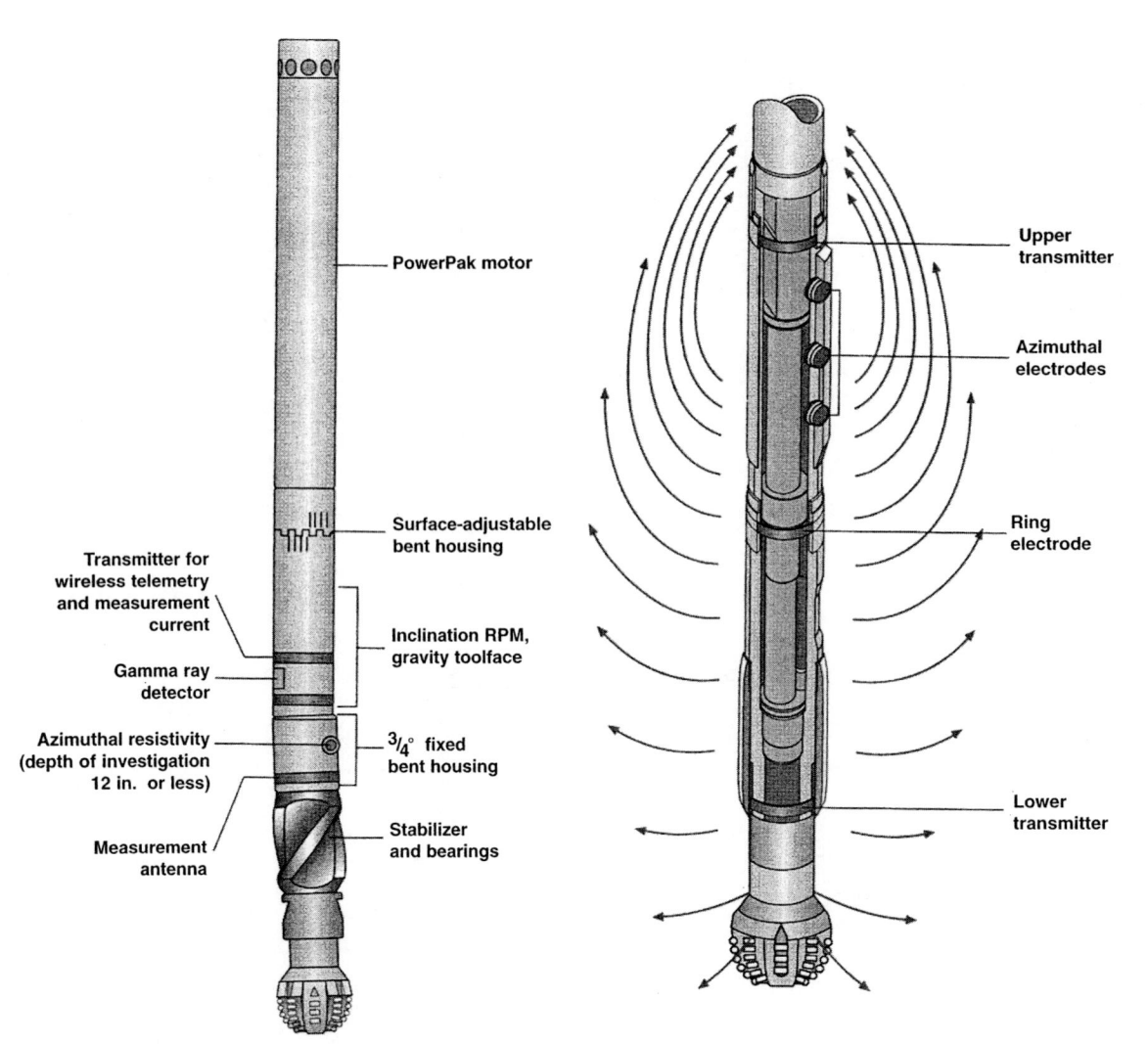

GEOSTEERING AND RAB TOOLS
(Reproduced with permission of Schlumberger Oilfield Services)

The Geosteering and RAB tools. The Geosteering tool is an instrumented steerable motor, meaning it can be used in rotary or sliding mode. The RAB tool is an instrumented stabilizer on a

4. MULTI-LATERAL WELLS

A new technology attracting considerable attention at the moment is the multi-lateral well, an arrangement whereby two or more wells will share the same well bore.

The development of multi-lateral wells has taken place over the same period of time as the development of *horizontal wells* but there is no direct association between the two, they are developments which stand on their own merits and one is not reliant on the other. Indeed multi-lateral are used extensively both in conventional, and horizontal wells.

There are two main reasons for opting to drill a multi-lateral well. The first is simply cost, because using an existing well bore saves on drilling time and materials and savings of up to 25% can be made compared to drilling a completely new well. The second reason multi-laterals are chosen is to improve reservoir drainage, a successfully drilled multi-lateral improving production by up to 500% in ideal cases.

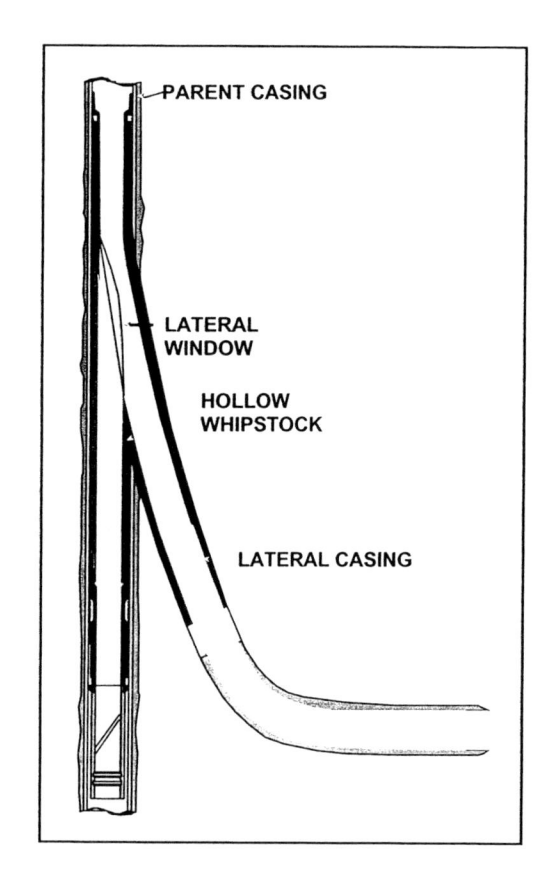

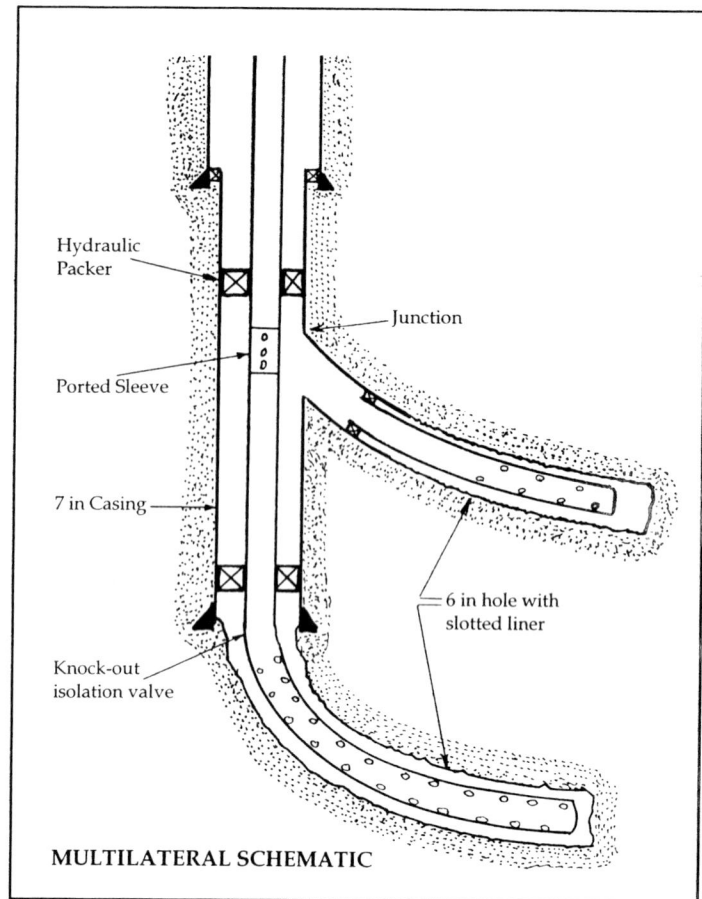

MULTI-LATERAL WELLS

A multi-lateral is defined as *multiple boreholes drilled from an existing single bore well*. The new bore hole may be horizontal or deviated to reach the desired bottomhole location, a *branch* being the term used to describe a lateral drilled from a horizontal lateral in the horizontal plane whilst a *splay* is defined as a lateral which has been drilled from a horizontal lateral in the vertical plane.

Multi-laterals are formed by cutting a hole through the side of the casing of an existing well with a milling cutter attached to the end of a drill string. A *whipstock*, a hardened steel wedge, is used to deflect the milling cutter at the desired location. The angle of the whipstock is between 1.5–3.0° so the deflection of the multi-lateral is extremely gentle and once the junction is completed, the well is conventional in all other respects and may be either cased or uncased.

There are basically six types of junction and a standard classification system has been prepared by the industry depending on the complexity of the junction. Cased junctions involve inserting a tubular casing string through the junction point and they are used to support a weak or unstable formation, to simplify re-entry of the well during workovers, and when pressure isolation is required.

DRILLING

Land Drilling System Components

Cameron Bop Stack
1. D.Annular BOP
2. Double U BOP
3. Drilling Spool
4. Manual FL Gate Valve
5. Hydraulic FL Gate Valve
6. R Check Valve
7. Single U BOP
8. Casing Head Spool
9. Casing Head Housing

Choke Manifold
10. J-2 Transmitter
11. Pressure Gauge
12. Manual FL Gate Valve
13. Hydraulic FL Gate Valve
14. Drilling Choke
15. Choke Control Console
16. Standpipe Pressure Gauges, FL Gate Valves, J-2 Transmitter

BOP Control System
17. Closing Unit
18. Pipe Rack
19. Remote Control Panel
20. Weight Indicator
21. Mud Pumps and Manifold, Reset and Shear Relief Valves, Pressure Gauges
22. Mud Tanks, Mud Valves, Degasser

LAND DRILLING RIG
(Reproduced with permission of Cooper Oil Tools, Houston, Texas)

Chapter Eight

THE WELL – COMPONENT PARTS

PART 1.   CHRISTMAS TREE

PART 2   SURFACE WELLHEAD

PART 3   MUDLINE SUSPENSION SYSTEM

PART 4   MUDLINE SAFETY VALVE – (SCSSV)

PART 5   PRODUCTION PACKER

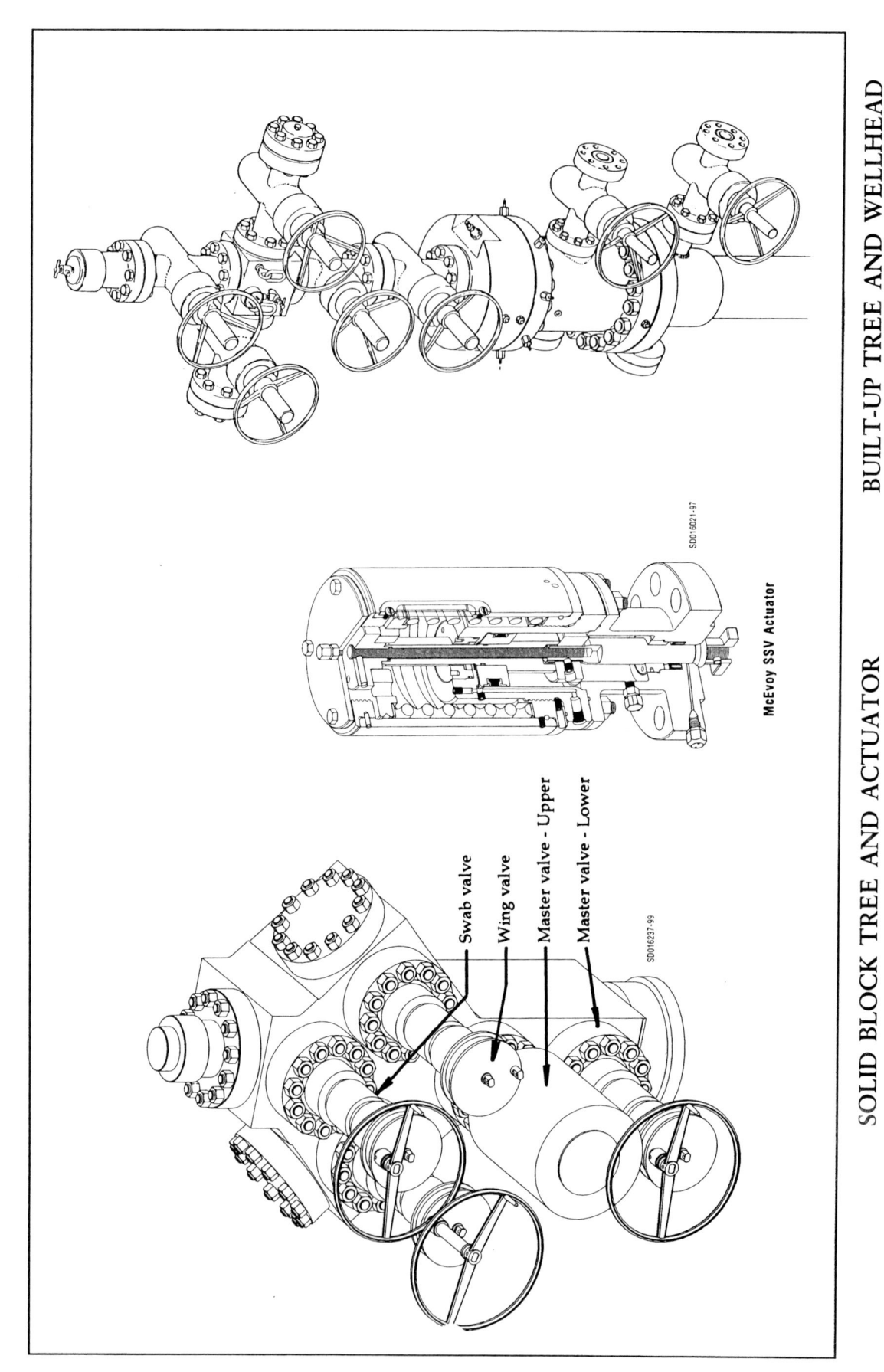

BUILT-UP TREE AND WELLHEAD
(ONSHORE TYPE)

McEvoy SSV Actuator

SOLID BLOCK TREE AND ACTUATOR
(Reproduced with permission of Cooper Oil Tools, Houston, Texas)

Swab valve
Wing valve
Master valve - Upper
Master valve - Lower

THE CHRISTMAS TREE

Part 1. THE CHRISTMAS TREE

The Christmas tree sits on top of the wellhead and provides the controlling mechanism for the isolation of individual wells, and for the regulation of the flow of hydrocarbons to the production and process facilities. The Christmas trees play an essential role in the emergency shutdown (ESD) system.

The Christmas tree is essentially an assembly of gate valves and the term *Christmas tree*, is thought to have been used to describe the onshore trees which comprise a number of single valves bolted together. Solid block trees are preferred for offshore service because they provide fewer potential leak paths being manufactured from a single alloy steel or stainless steel forging or casting into which the valve chests are machined. Depending on the composition of the reservoir products, the bores and sealing surfaces may be protected against corrosion and erosion by cladding (over welding) with a material such as Inconel.

Occasionally a combination of solid block and individual valves are used.

1. SOLID BLOCK CHRISTMAS TREE

A solid block Christmas tree will normally comprise the following valves:

1.1 MASTER VALVE(S)

Most offshore Christmas trees are provided with two master valves which are referred to as the *upper master* and *lower master* valves. The valves are mounted one above the other in the vertical plane and provide a *double block* arrangement for the isolation of the well.

The upper master valve is opened by a hydraulic or pneumatically powered actuator which is controlled from the wellhead control panel. The lower master valve is almost always a manual valve which is only closed during prolonged periods of shutdown such as during maintenance programmes. The lower master valve is opened first and closed last. This ensures minimal flow of hydrocarbons over the valve seats, thus providing protection from abrasive particles and ensuring a good seal is maintained.

Modern Christmas trees are fitted with a *wire cutting* master valve. Should an emergency situation arise when wireline operations are in progress the valve will close completely, cutting the wireline in the process. The wire and tool string will fall into the well and can be recovered by *fishing* at a later date.

1.2 WING VALVE

Christmas trees may be supplied with one or two wing valves, the valves being offset from the vertical plane so that a clear entry into the well is maintained through the *swab valve* for wireline work. One valve, the *production wing* is permanently connected to the hydrocarbon process system and is normally fitted with a hydraulic or pneumatically powered actuator. The other valve, the *service wing* is manual in operation and permits the injection of chemicals or gases into the well without disturbing production pipework.

The flow of gas from the well is normally regulated by a choke valve which is fitted either above the wing valve, or further downstream in the vicinity of the production header.

1.3 SWAB VALVE

The swab valve is the uppermost valve on the Christmas tree and is positioned directly above the master valves. It facilitates the entry of wireline equipment into the well.

The valve is manually operated and the top connection is fitted with a swab cap to protect the threads and prevent the accumulation of debris.

2. WELLHEAD CONTROL PANEL

A wellhead control panel or *wellhead skid* controls the operation of the Christmas trees and *mudline* or *surface contolled subsurface safety valves* (MLSV, SSSV or SCSSV). The skid permits valves to be operated locally, remotely or via the ESD system and timing mechanisms provide a means of controlling the speed and sequence of valve operation. For valves fitted with actuators the sequence would normally be:

TO CLOSE

i) Production wing valve.
ii) Upper master valve.
iii) Mudline safety valve.

During an emergency shutdown (ESD) situation, complete closure of the Christmas tree valves should be effected within approximately 45 seconds to comply with API RP 14H, the only document which provides guidance on this particular aspect.

TO OPEN

i) Mudline safety valve.
ii) Upper master valve.
iii) Wing valve.

3. GATE VALVES AND ACTUATORS

3.1 GATE VALVES

Gate valves are the preferred choice of valve type for Christmas trees because they are robust and the sealing surfaces can be easily replaced should damage occur. Gate valves are most effective when used in either the fully open or fully closed position and are not generally used to regulate flow, this being carried out by a choke valve located downstream of the production wing valve, if required.

3.2 VALVE GATES

Two types of gate valve may be encountered, the *slab gate* and the *split* or *segmented gate*. The slab gates rely solely on well pressure to push the gate against the seat to establish a seal and they are simpler and cheaper to produce than the split gates. Consequently, they are most effective in high pressure wells and may only seal securely in one direction, the direction of flow.

The segmented gate valve employs two gates held apart by a spring or expander and it is the expander that forces the gates against the seats as the valve closes. The resulting seal is effective in both directions and because it does not rely on well pressure, it works well at both high and low pressures.

3.3 VALVE SEATS

As previously stated, both the gates and the seats of a gate valve can be replaced with the minimum of problems. Once the valve bonnet has been removed, the gates are simply unscrewed from the valve stem whilst the seats can be *jacked* out of the valve chest. Fixed gates tend to be used with segmented gates, that is to say they are a metal to metal interference fit in the valve chest. Floating seats are generally used with slab gates, an elastomeric seal sealing the seat in the valve body, the seats being free to move slightly to assist in the forming of an effective seal against the gate.

3.4 VALVE ACTUATORS

All the valve actuators are *fail-safe* in operation. The valves are held open by oil or air pressure against a compressed coil spring. Venting of the power medium permits the spring to push the valve plate into the closed position. The actuators are normally designed so that they can be replaced should the need arise without disturbing the gate valve. Position indicators may also be fitted, either visual or electrical proximity switches to show when the valve is open or closed.

Illustrations of gate valves and actuators can be seen in Chapter 4.

4. HORIZONTAL TREES

Unlike conventional Christmas trees, the *horizontal* or *spool tree* is a through bore design with the valves mounted external to the vertical bore of the well. The well is produced from the side of the tubing hanger, the reservoir products flowing horizontally through the master valve.

The development of horizontal Christmas trees has taken place over the same period of time as the development of *horizontal wells* but there is no direct association between the two, they are developments which stand on their own merits and one is not reliant on the other.

Whilst horizontal trees are available for installation on conventional offshore platforms, the market has grown considerably because of the interest in deep water developments using sub-sea wells.

The main advantage of the horizontal production tree is that it permits complete access to the well bore without removing the tree because there are no valves in the vertical plane above the production tubing. Major workovers can therefore be carried out without the need to recover the sub-sea tree, the BOP stack being landed directly on top of the Christmas tree. They are thus ideal for use in wells where frequent tubing-downhole completion recoveries are anticipated such as when *electrical submersible pumps* (ESP) are used. There are also economic benefits because larger diameter production tubings can be installed which permit a greater throughput of reservoir products.

The only perceived disadvantage of the horizontal tree is that with the absence of valves in the well bore, *double block valve isolation* of the well is not provided. However, two barriers are provided in the form of plugs which are set in the top of the tree.

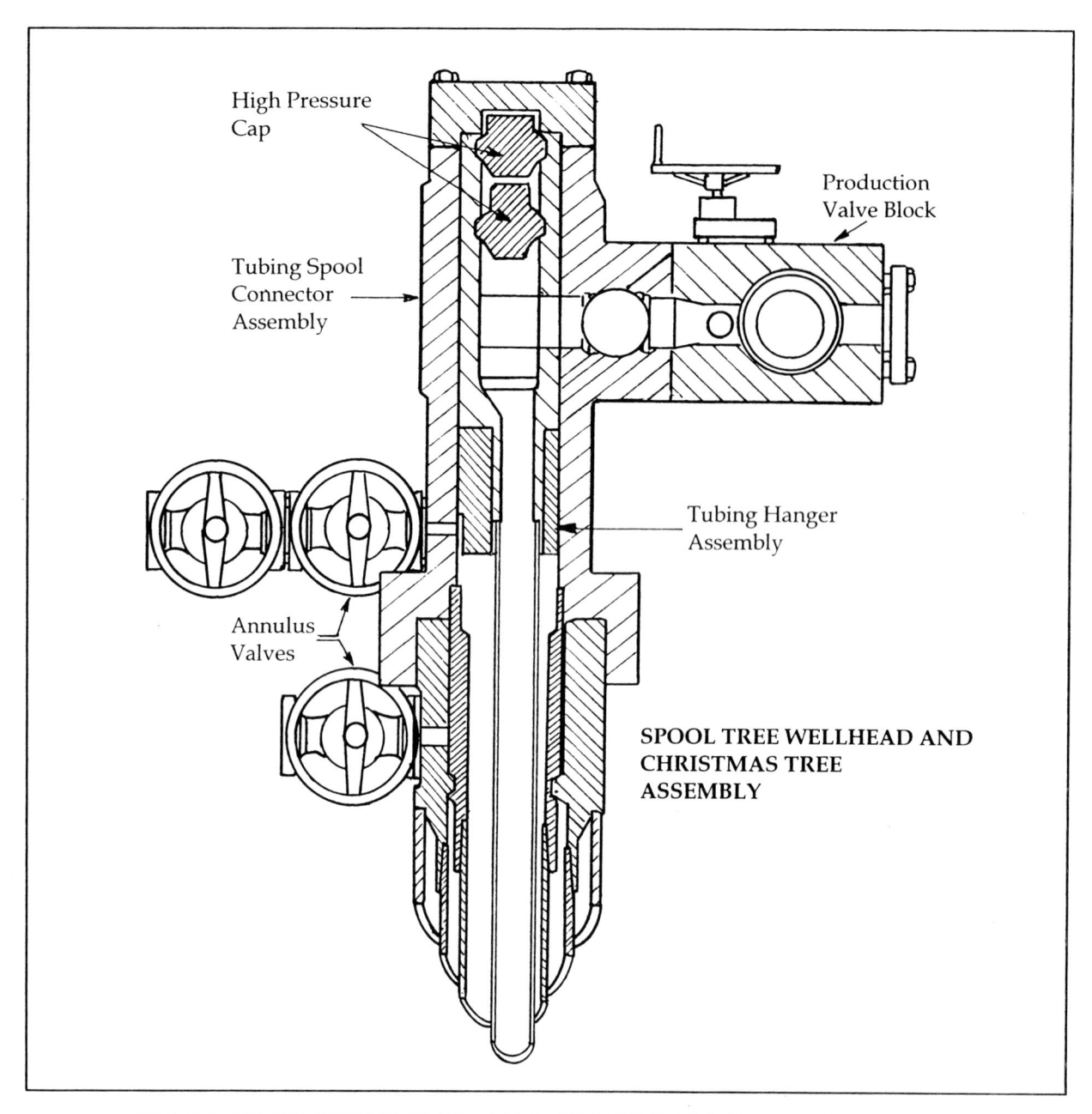

SPOOL TREE WELLHEAD AND CHRISTMAS TREE ASSEMBLY

GUIDANCE NOTES AND REFERENCE STANDARDS

(i) API Spec. 6A, Specification for Wellhead and Christmas Tree Equipment.

(ii) API RP 6AR, Repair and Remanufacture of Wellhead and Christmas Tree Equipment.

(iii) API RP 14D, Recommended Practise for Wellhead Surface Safety Valves and Underwater Safety Valves for Offshore Service.

(iv) API RP 14H, Recommended Practise for Installation, Maintenance, and Repair of Surface Safety Valves and Underwater Safety Valves Offshore.

(v) The Department of Energy Guidance Notes on Offshore Installations: Guidance on Design, Construction and Certification, Part II, Section 43.

Part 2. THE SURFACE WELLHEAD

The wellhead assembly on a fixed installation consists of a collection of components designed to support and seal the *intermediate casings, production tubing* and *Christmas tree*. It also provides a base for the location of the *blowout preventer* (BOP) during drilling operations.

Assembly of the wellhead commences with the installation of the conductor pipe and the various stages can best be described with the aid of a sketch.

1. BASE PLATE/LANDING RING

Once the conductor has been driven to the desired depth the base plate or landing ring can be positioned and welded. Landing rings are used to help distribute wellhead loads through the conductor and into the formation. It should be noted that these components are designed specifically to support the casing strings and are not required to be pressure retaining. At this stage in the drilling operation the formation should be of a type which cannot support the existence of hydrocarbons.

2. STARTING HEAD

Depending on the design the starting head may be screwed onto, or welded to the top of the first intermediate casing string and it may incorporate a landing ring. The function of the starting head is to transmit the loads created by subsequent casing strings through to the conductor, and to provide the initial location for the blowout preventer.

The blowout preventer will remain in place for the remainder of the drilling operation being moved onto each subsequent casing head in turn.

3. CASING HEAD SPOOLS

Following the installation of the starting head, all subsequent casing strings will be terminated within a casing head spool. The spool is designed to support and seal the upper section of casing against oil or gas pressure and provide connections through which the condition of the annulus spaces may be monitored. Generally speaking, the casing head spool is installed after the casing has been cemented, but before the cement has cured.

Examples of casing head support and seal arrangements are shown highlighted in the sketch. The *casing slips* or *hangers* are manufactured with a serrated steel internal face designed to bite into the casing string and thus provide support within the casing head, and a metal to metal seal. A secondary rubber or neoprene *pack off* seal may then be activated under the compression forces provided by external lock down screws or internal cap-head screws. This two fold sealing arrangement may vary slightly from one manufacturer to another but the basic principles of operation remain the same.

Offshore Engineering

Casing Spools (3 off)

Lock Down Screws

CASING HANGER
- upper slips
- elastomer seal
- lower slips

CASING HEAD SPOOL, CASING AND HANGER

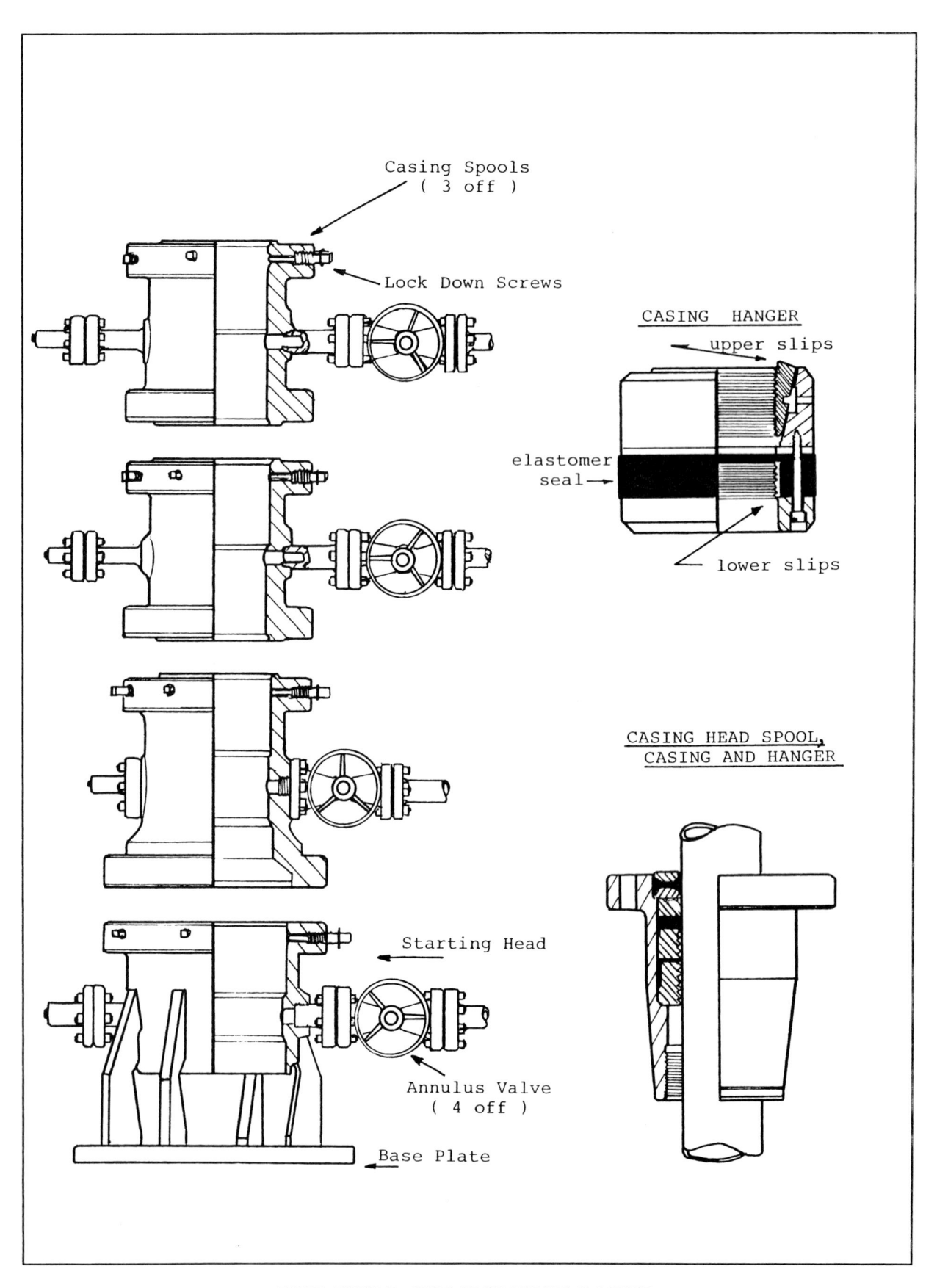

Starting Head

Annulus Valve (4 off)

Base Plate

THE WELL COMPONENT PARTS

THE WELL – COMPONENT PARTS

4. TUBING HEAD SPOOL

The tubing head spool is designed to support and seal the *production tubing* within the final intermediate casing. The production tubing sits in a bushing or *tubing hanger* and is the only item within the well which receives direct exposure to reservoir pressure. The intermediate casings and casing head spools are designed to resist well pressure but will only be subjected to pressure should failure of the production tubing packing seals occur, or a gas lift enhanced oil recovery system is fitted.

5. CHRISTMAS TREE

The final operation consists of installing the Christmas tree in preparation for bringing the well on stream. The Christmas tree permits isolation of the reservoir products from the process equipment.

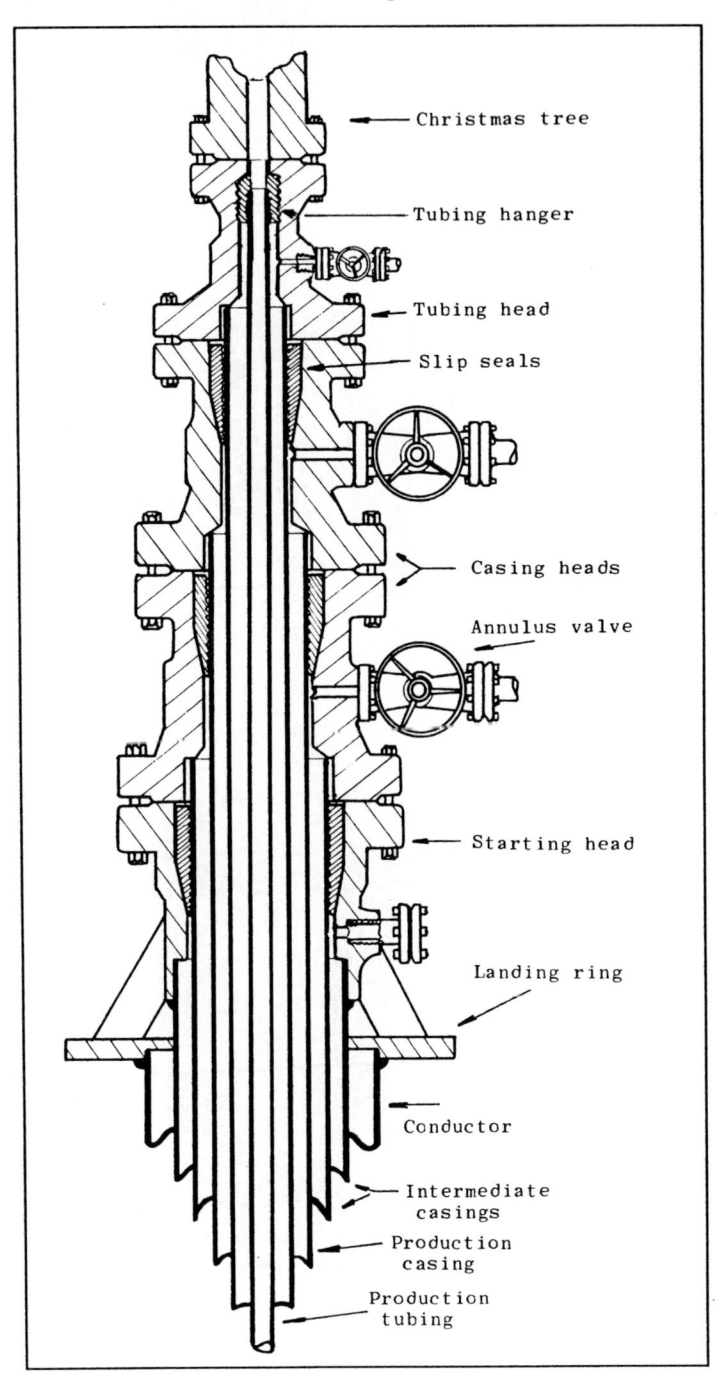

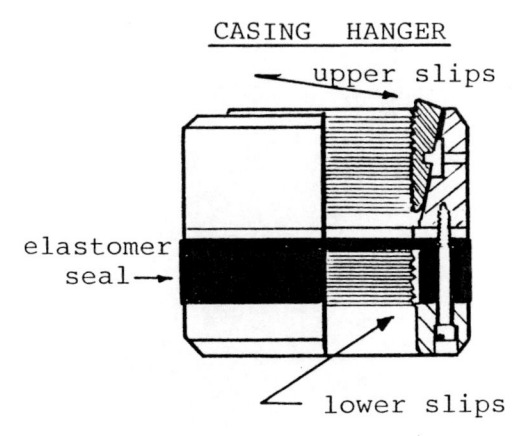

CASING HANGER

WELLHEAD SCHEMATIC

CASING HEAD SPOOL, CASING AND HANGER

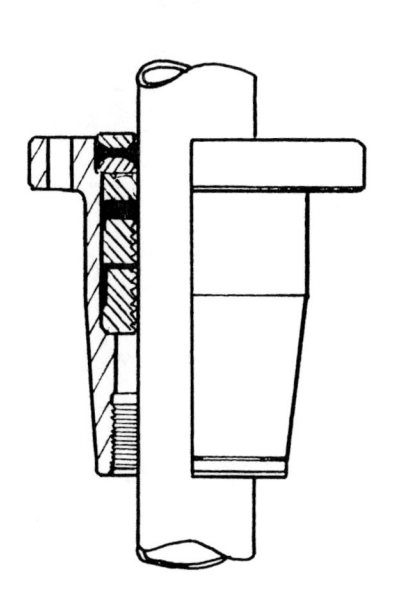

Offshore Engineering

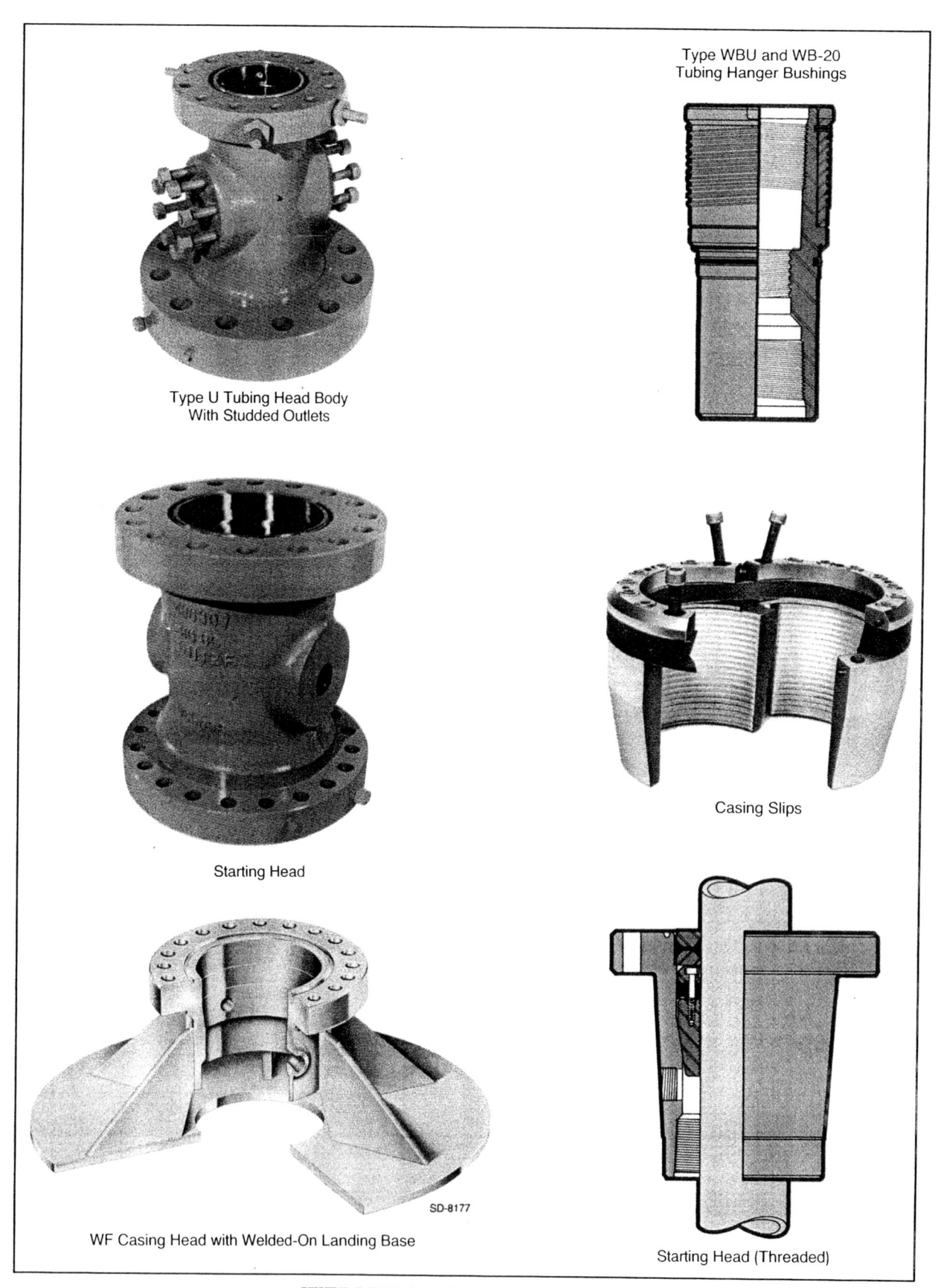

Type U Tubing Head Body
With Studded Outlets

Type WBU and WB-20
Tubing Hanger Bushings

Starting Head

Casing Slips

WF Casing Head with Welded-On Landing Base

Starting Head (Threaded)

WELLHEAD COMPONENTS
(Reproduced with permission of Cooper Oil Tools, Houston, Texas)

Part 3. *MUDLINE SUSPENSION SYSTEMS – (MLSS)*

A mudline suspension system provides an alternative to the surface wellhead for the support and sealing of *intermediate casing strings*. As the name suggests the suspension system is located at the *mudline* or sea bed and sits within the conductor pipe rather than on top of the conductor as is the case with the surface wellhead.

The sketch shows a simplified layout of a MLSS of the type installed by a bottom mounted drilling rig or jack-up. Before continuing with the description of the various components it would be beneficial to recap on how the conductor and intermediate casing strings are formed.

The conductor and casings are assembled from 9 metre (30 feet) lengths of pipe which have externally threaded ends. They are joined together and sealed against oil/gas pressure by internally threaded tubular couplings or *joints*. The MLSS employs special joints which incorporate a casing hanger that screws into the casing string as it is lowered into the hole.

The assembly process commences with the installation of the conductor. A joint containing a landing face is screwed into the conductor at a position that will coincide with the mud line. The first intermediate casing will locate or *hang off* this protrusion and each subsequent casing string will be supported in a similar fashion, that is *hung off* the preceding intermediate *casing hanger* joint. Some designs of mudline suspension system employ a spring loaded, expanding casing hanger which latches into the preceding casing to provide additional support.

The mudline suspension system has two main applications.

1. PRE-DRILLING OF WELLS

The drilling of wells for a production installation is a particularly time consuming operation. A considerable saving in time, and hence money, can be achieved by drilling the wells prior to the arrival of the permanent installation and the design of the MLSS is ideally suited to this application.

On completion of drilling, plugs are set in the well and the conductor and intermediate casing strings which extend from the sea bed to the drill floor are removed. They are simply unscrewed from the casing hanger and replaced with threaded *abandonment caps*. The process is reversed when the permanent installation arrives and the wells are required for production.

2. SUB-SEA WELLS

Sub-sea wells are drilled and completed in an identical manner to surface wells. However, the absence of a fixed structure to support a conventional wellhead necessitates the use of a either a MLSS or a sub-sea (marine) wellhead, both arrangements sharing the same basic installation concepts. As in the previous example, on completion of drilling the casing strings extending from the casing hanger to the drill floor are removed. They are then replaced with a *completion adaptor spool*, a location device which provides a suitable profile for the attachment of the sub-sea Christmas Tree.

Offshore Engineering

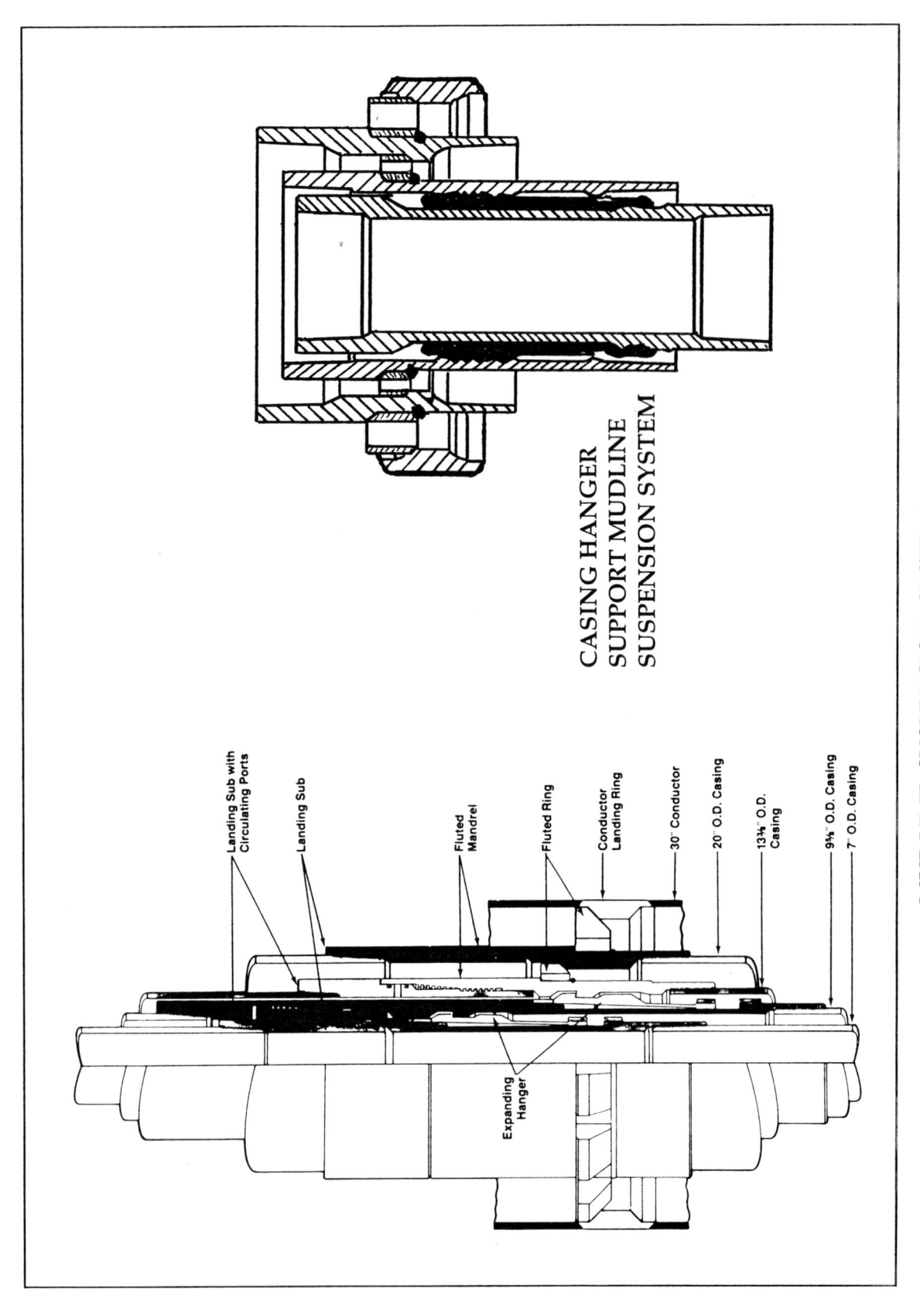

CASING HANGER SUPPORT MUDLINE SUSPENSION SYSTEM

- Landing Sub with Circulating Ports
- Landing Sub
- Fluted Mandrel
- Fluted Ring
- Conductor Landing Ring
- 30" Conductor
- 20" O.D. Casing
- 13 3/8" O.D. Casing
- 9 5/8" O.D. Casing
- 7" O.D. Casing
- Expanding Hanger

MUDLINE SUSPENSION SYSTEM
(Reproduced with permission of Cooper Oil Tools, Texas)

Part 4. MUDLINE SAFETY VALVE

The mudline safety valve (MLSV), or to give it its correct title the *surface controlled subsurface safety valve* (SSSV or SCSSV), is located in the *production tubing* at a depth of 60 to 90 metres (200 to 300 feet) below the surface of the sea bed. It provides a last ditch barrier to isolate the well should the installation be destroyed by an explosion or collision with a ship. The valve is *fail-safe* in operation.

The most graphic example of SCSSV operation was provided when the *Piper Alpha* installation was destroyed by fire and explosion in July 1988. The mudline safety valves on all 36 wells closed satisfactorily, but unfortunately two of the wells provided fuel for the fire via the annulus spaces which were being used for *gas lift* purposes, the isolation of which ultimately required the attention of *Red Adair*'s specialist fire fighting team.

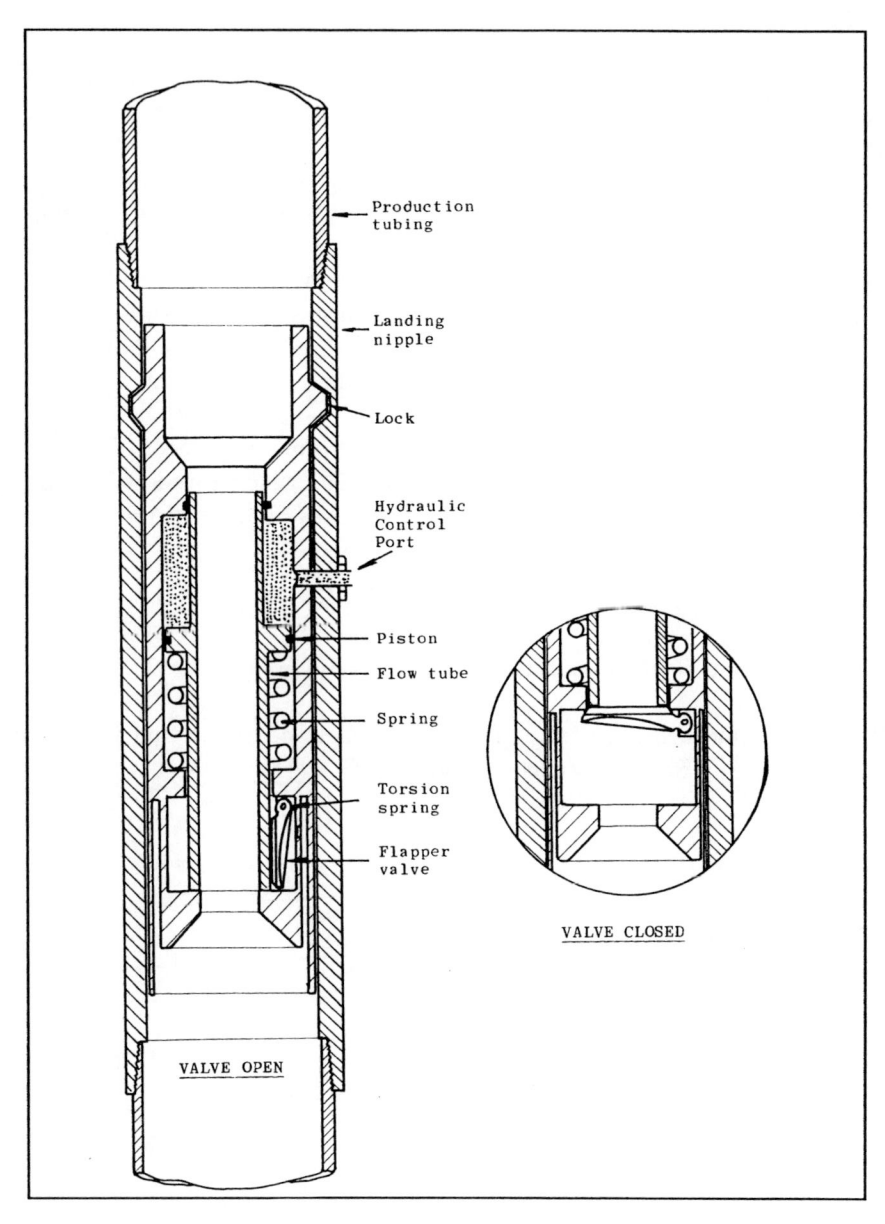

MUDLINE SAFETY VALVE

The valves should be checked for leakage on a regular basis, preferably monthly. This entails closing the valve and depressurising the production tubing up to the Christmas tree. The production tubing should remain free of any significant pressure build up for a period of 10 minutes. Should the valve fail or leak significantly it should be replaced at the first available opportunity.

The actual valve which may be of ball, but more probably a flapper type is located in a tubular housing. The housing design determines the method by which it is installed in the production tubing, that is it may be *tubing retrievable or wireline retrievable*.

(i) TUBING RETRIEVABLE

The tubing retrievable SCSSV is screwed into the production tubing as a joint when the tubing is run (installed). Replacement involves a workover and the pulling (removal) of the production tubing, a costly and time consuming exercise.

(ii) WIRELINE RETRIEVABLE

The wireline retrievable SCSSV can be recovered as the name suggests, by a wireline operation. These valves are often preferred for the ease in which replacement can be effected. The production tubing is fitted with a machined recess or landing nipple in which the valve is located and latched into position using a wireline tool string. The disadvantage of the wireline retrievable valve is that it restricts the throughput of the production tubing.

The sketches show the operation of a flapper type wireline retrievable subsurface safety valve. Hydraulic oil pressure is required to keep the valve open against spring pressure. The hydraulic oil is normally supplied via stainless steel instrument tubing routed through the wellhead although some of the older installations achieved the same result by pressurising the *annulus space*.

The hydraulic oil pressure is supplied too, or released from the SCSSVs by the *wellhead skid* under the influence of electronic control signals initiated by emergency shutdown (ESD) system. The valves can also be operated manually from the wellhead skid, or by rupture of the *fire loop*.

GUIDANCE NOTES AND REFERENCE STANDARDS

1. API Spec 14A   Specification for Subsurface Safety Valve Equipment

2. API RP 14B,   Recommended Practise for Design, Installation, Repair and Operation of Subsurface Safety Systems. (Contains ISO 10417:1993)

Part 5. THE PRODUCTION PACKER

The function of the production packer is to seal the base of the *production tubing* into the *intermediate casing* string. Both permanent and temporary packers are available but the permanent type are generally preferred for their more reliable operation over prolonged periods.

A sketch has been prepared to assist in the explanation of the mode of operation of a permanent packer. It should be noted that the packer is installed prior to *perforation* of the casing or liner so the well is full of *well preservation* fluid and devoid of all hydrocarbon pressure.

A packer can be run (inserted) as part of the production tubing or installed on a wireline tool string. Once in position the seals may be activated by a mechanical tool, hydraulic pressure, or an electrically activated gas charge, depending on design. The sketch shows a hydraulic packer and to activate the seals a steel blanking plug must first be set in the production tubing below the packer. The tubing is then pressurised in order to energise the setting piston which moves upwards snapping the shear pin and compressing the rubber sealing element. When the pressure is released the steel slips bite into the intermediate casing and prevent re-expansion of the sealing element. The packer is thus permanently set and should removal be required it must be milled out, a major operation.

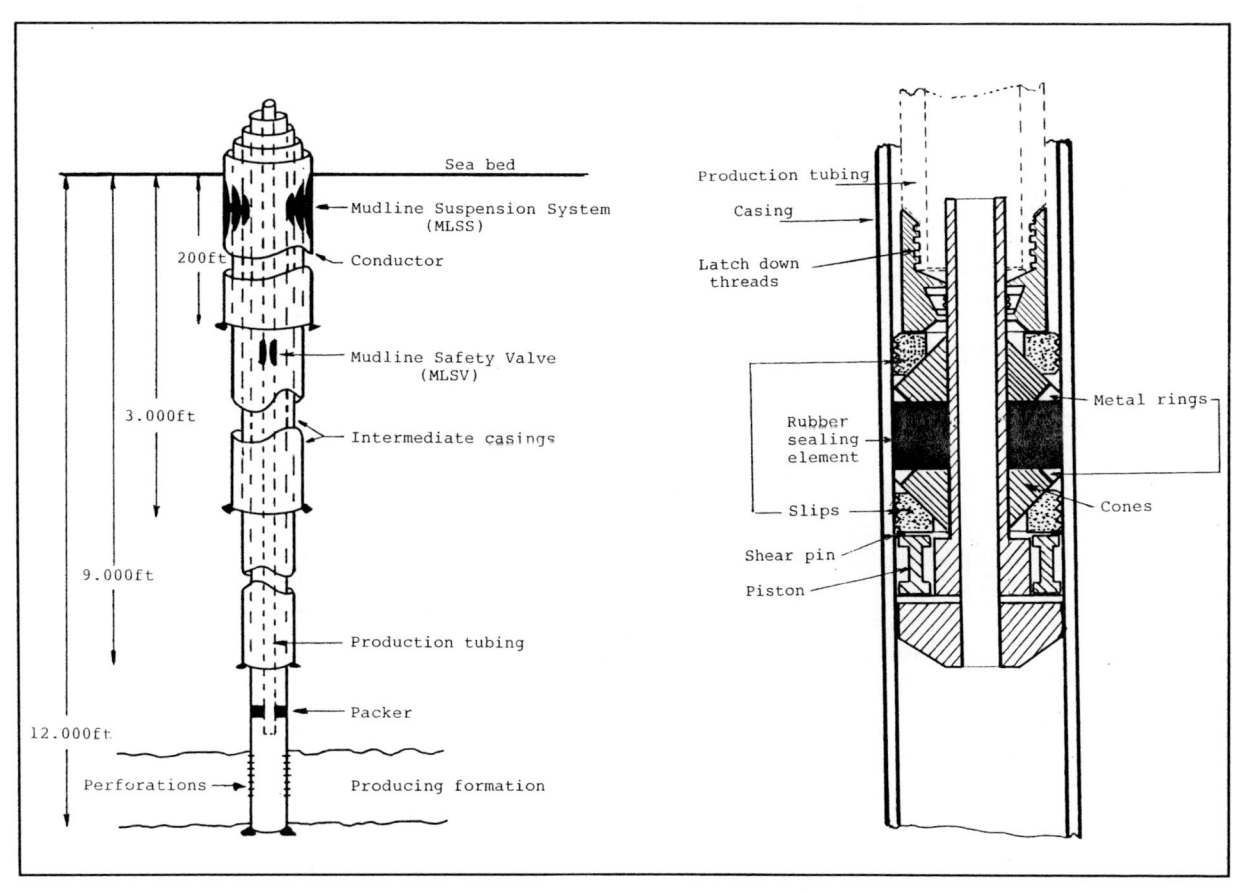

TYPICAL CASING ARRANGEMENT   **PRODUCTION PACKER**

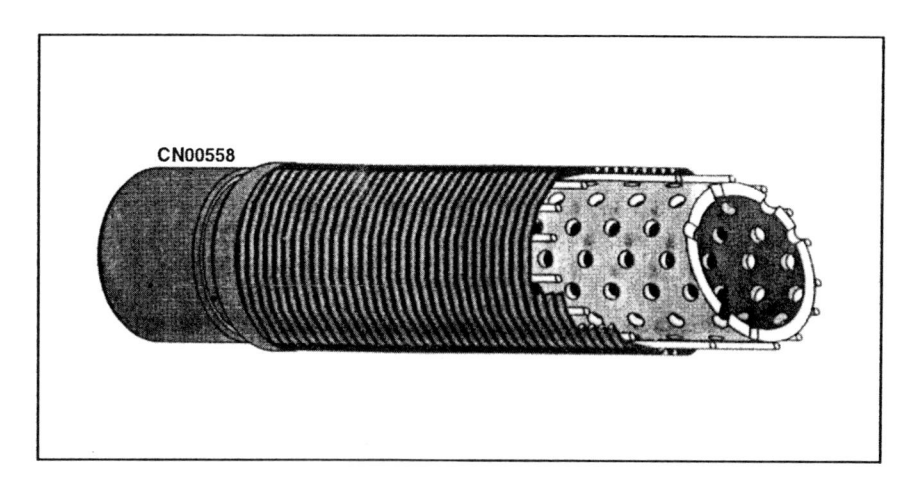

ALL-WELDED WIRE-WRAPPED SCREENS
(Reproduced with permission of Halliburton Energy Services)

Key shaped wire on ribs welded on to the base pipe decrease the chances of the screen from becoming blocked.

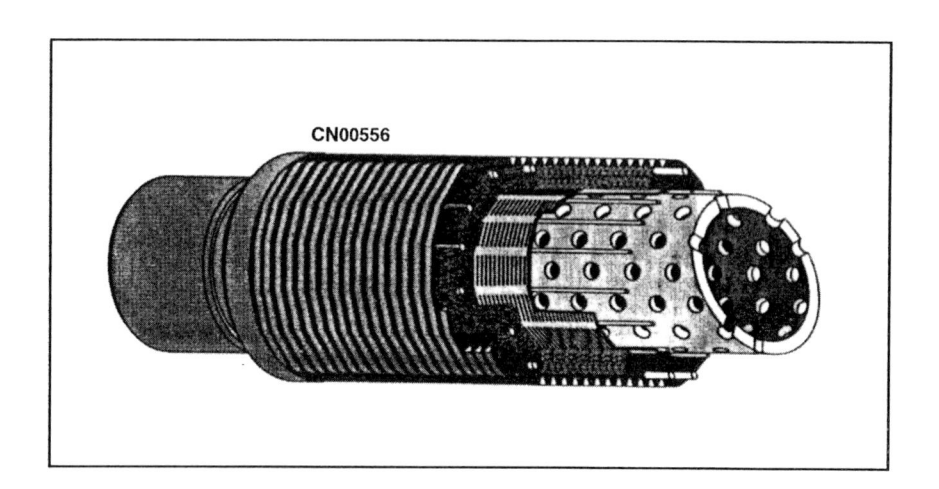

ENHANCED LOW-PROFILE PREPACK SCREENS
(Reproduced with permission of Halliburton Energy Services)

Suitable for installing in horizontal wells these screens can also be installed in gravel packed wells. They incorporate a strong inner screen and a prepack sand annulus to optimise filtering capabilities.

Chapter Nine

WELL MAINTENANCE

PART 1. WIRELINE OPERATIONS

    1. Wireline Equipment

    2. Wireline Tool String

    3. Wireline Operations

PART 2. WORKOVER OPERATIONS

    1. Well Stimulation

    2. Production Tubing Modifications

Offshore Engineering

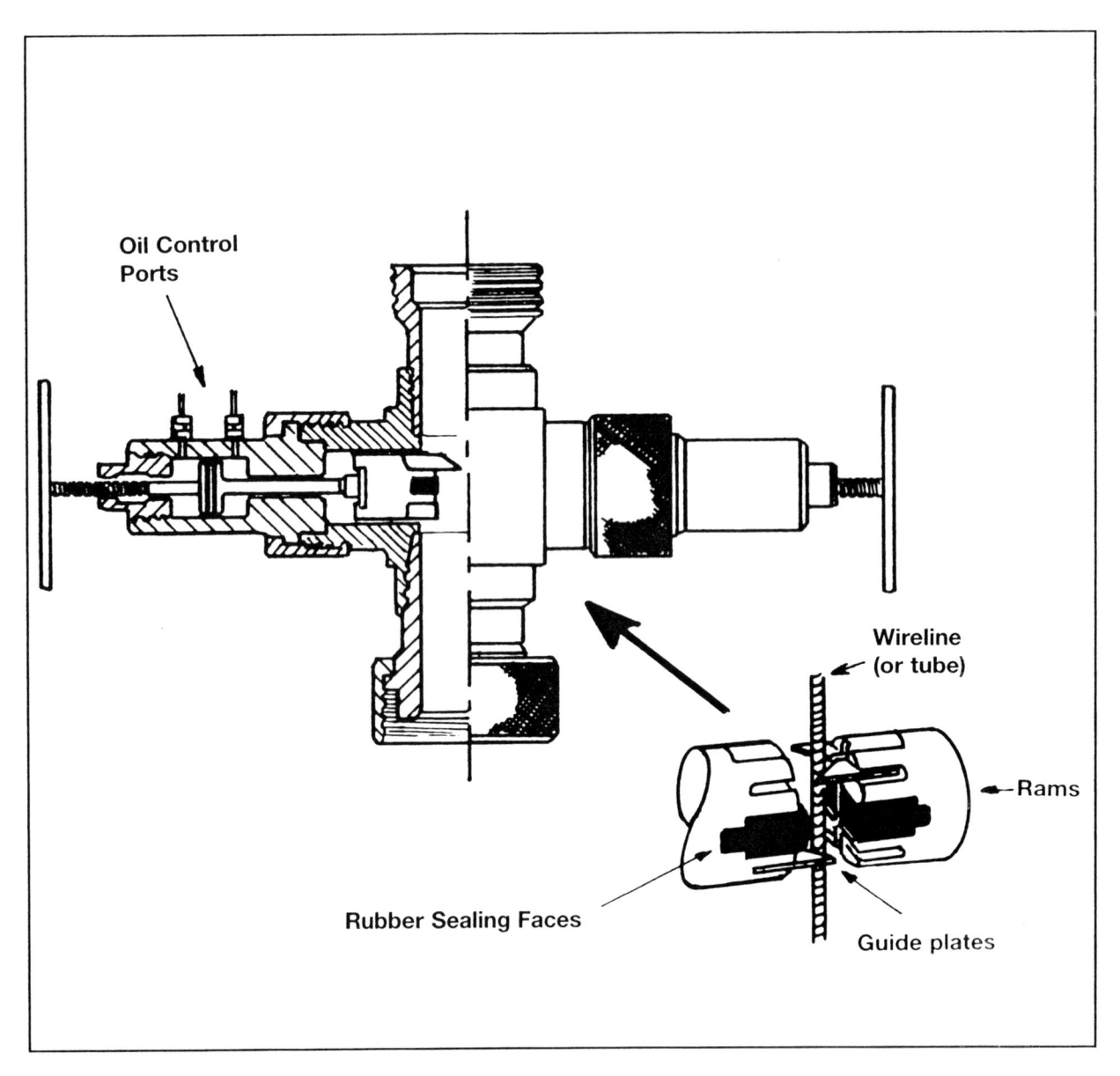

COIL TUBING/WIRELINE VALVE

WELL MAINTENANCE

Part 1. WIRELINE OPERATIONS

Wireline equipment has been in use since the early days of the oil and gas industry. It permits mechanical devices to be installed or removed in the *production tubing* with minimal interruption to the production process.

The equipment is attached (flanged or screwed) to the top of the *Christmas Tree* as shown in the sketch. Various wireline tools can then be attached to the wire and lowered through the Christmas tree swab valve and into the production tubing.

Most operations can be carried out using a single strand wire or *slick line* (also called *piano* or *music wire*) which is available in thicknesses ranging from 0.066 to 5/16 inch (1.65 mm to 8.0 mm) and supplied in lengths of 10,000 to 40,000 feet (3,030 to 12,120 metres). However, when a stronger line is required a braided wire of 3/16 to 5/16 inch (4.75 mm to 8.0 mm) diameter is normally used (often called torpedo or well shooters line).

An electric line may be used as an alternative to the simple wireline. The electric line facilitates the passage of electrical signals and permits the use of more sophisticated equipment. The electric line can even be used to operate a video camera which may be lowered into the well and used to visually identify leaks and blockages in the production tubing.

The wireline is stored on a winch drum which is powered by a diesel driven hydraulic pump. Hydraulic equipment is preferred on the grounds of safety for use in hazardous areas and a counter fitted to the winch drum provides the operator with an indication of the depth of the tool string in the well.

1. WIRELINE EQUIPMENT

The wireline assembly shown in the diagram contains the following items of equipment:

i) WIRELINE VALVE OR BLOWOUT PREVENTER (BOP)

The wireline valve lies at the heart of the wireline operation. It enables quick and safe isolation of well pressure from the lubricator sections mounted above it. The valve is often referred to as a BOP (blowout preventer), somewhat of a misnomer as the term *blowout* is only used in relation to drilling operations. However, the wireline valve is very similar in appearance to the drilling BOP if somewhat smaller, and they are both designed to contain reservoir pressure within the well bore so the confusion is understandable.

In its simplest form the wireline valve consists of a pair of rams fitted with rubber inserts which can be manually wound in to close and seal around the wire, thus sealing the production tubing.

A more sophisticated version is available which employs two pairs of hydraulically operated rams. The top rams are fitted with rubber inserts for sealing on the wire and the lower blind rams are designed to shut off well pressure should the wire become detached from the winch drum.

The wireline valve is attached to the Christmas tree at the Swab connection. A *crossover assembly* may be required between the two to compensate for diameter, flange or thread variations.

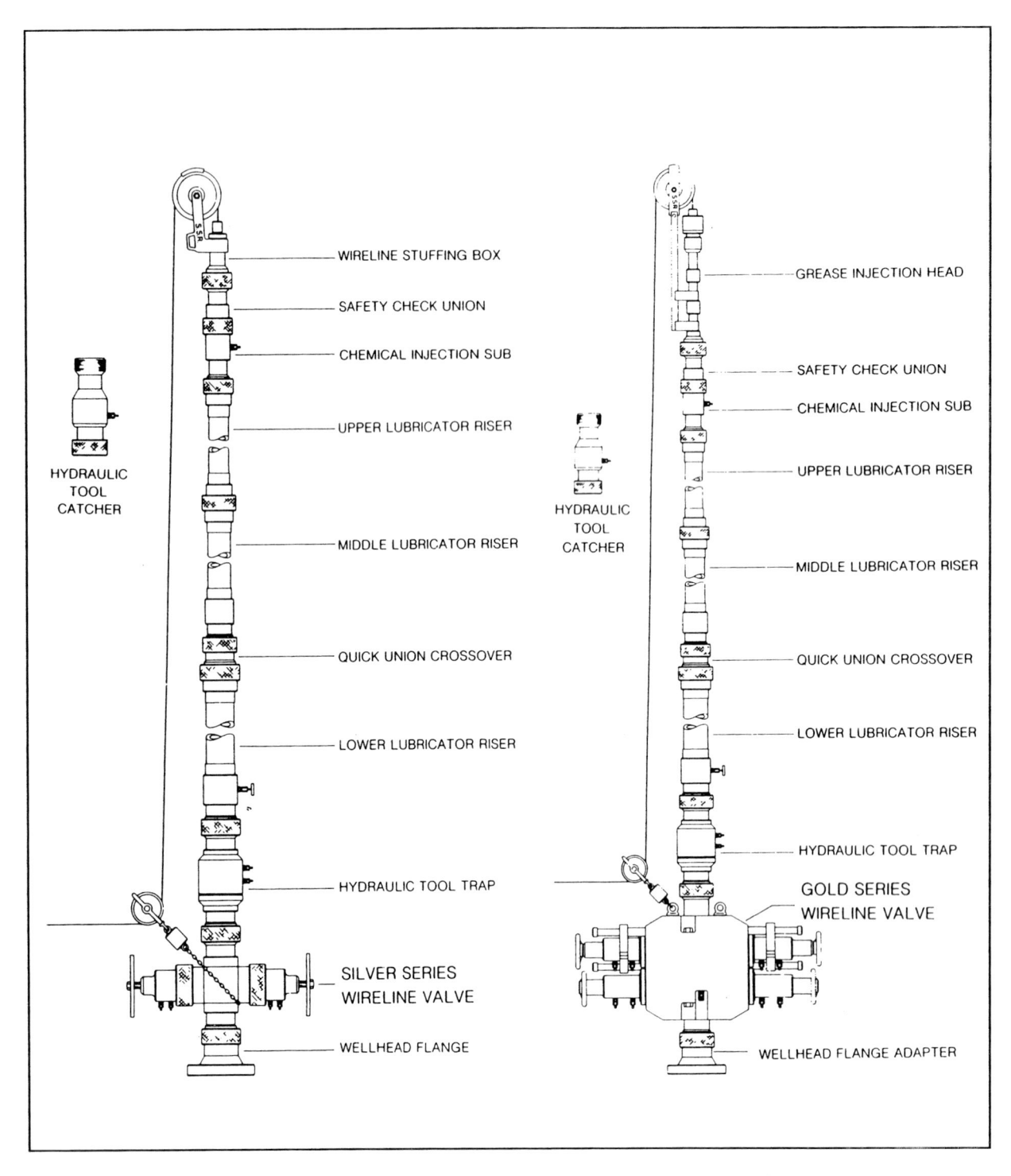

SLICK LINE LUBRICATOR STACK   **WIRE LUBRICATOR STACK**
(Courtesy of SSR (Int.) Ltd)

During wireline operations on a producing well the Christmas tree should remain connected to the emergency shutdown (ESD) system. This ensures that should an emergency situation develop, the Christmas tree master valve will close and sever the wire line. The wire, complete with tools will drop into the well thus allowing the mudline (sub-surface) safety valve to close and effect sealing of the well bore. The wire and tools can be obtained by *fishing* at a later date.

The master valve on older Christmas trees may not be designed to cut wires. Where this situation occurs a fail safe *wireline cutting* gate valve should be installed between the wireline valve and the Christmas tree and connected into the platform's ESD system.

ii) LUBRICATOR OR RISER

The name *lubricator* is misleading as no lubrication actually takes place within this component. It consists of a length of pipe, approximately 8 to 10 feet (2.4 to 3.0 metres) in length and of a diameter to suit a particular tool string, normally $3\frac{1}{2}$ to $5\frac{1}{2}$ inches (90 to 140 mm). It is fitted with male (*pin*) and female (*box*) unions (screwed or welded) at each end to permit ease of assembly. Several sections of lubricator may be joined together to give the necessary height above the wireline valve to facilitate the entry of the tool string.

iii) STUFFING BOX AND CHECK VALVE UNION

The stuffing box is a simple device attached to the top section of the lubricator whose function is to seal the wire against gas pressure. The tubular body is filled with layers of packing and may incorporate a plunger or ball valve designed to shut off gas pressure should the packing become damaged, or the wire break. In the absence of a plunger or ball valve a separate check valve union will be mounted below the stuffing box.

iv) HAY PULLEY

The hay pulley is attached to the top of the stuffing box, its function being to redirect the wire to the winch drum.

v) HIGH PRESSURE GREASE TUBE (HGT)

The high pressure grease tube or *grease injection head* is used in place of the stuffing box when braided wire is used in preference to a single strand wire. Grease is pumped through the injection head at a pressure approximately 500 psi (35 bar) above the gas pressure to seal and lubricate the wire.

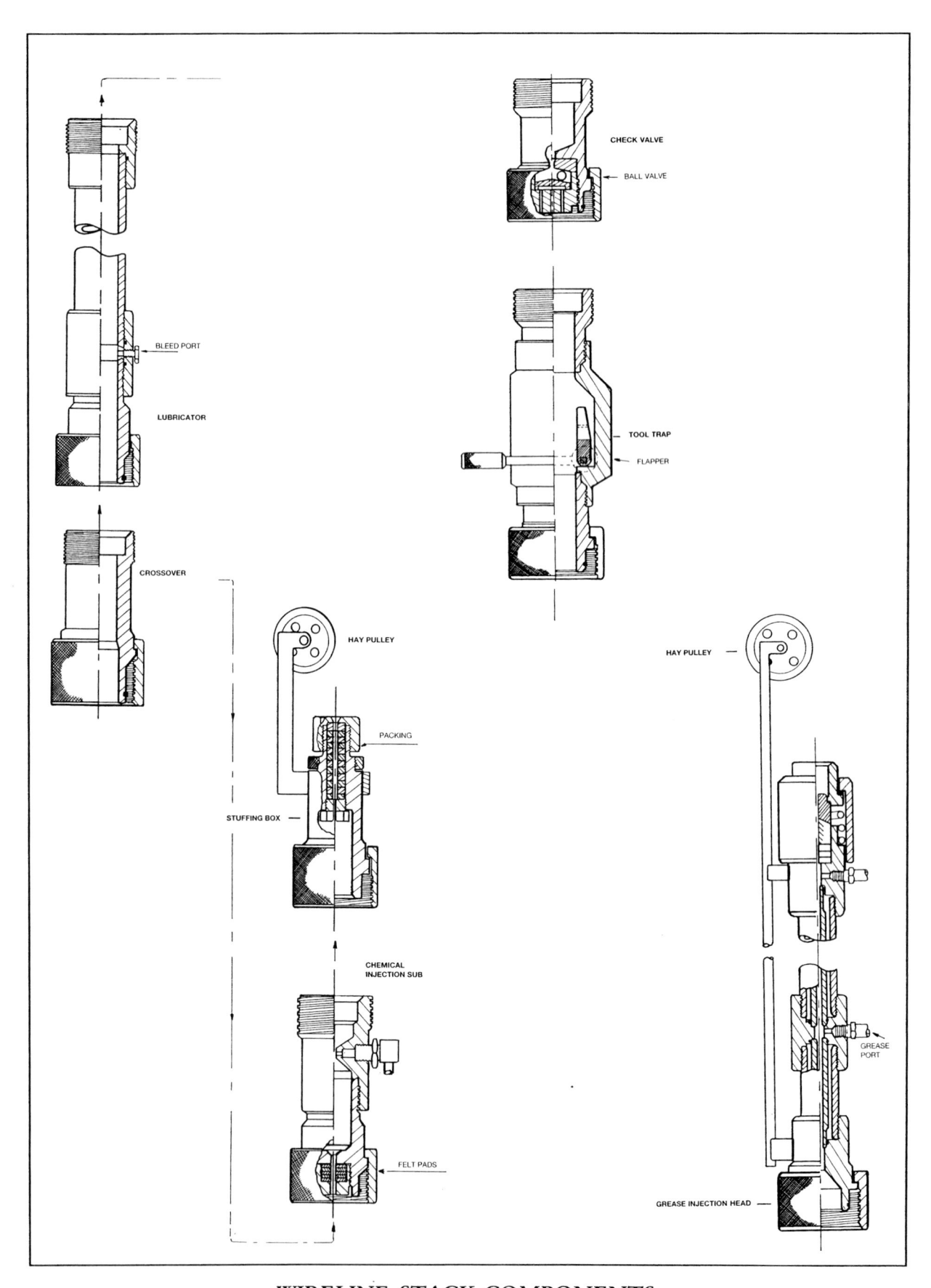

WIRELINE STACK COMPONENTS

WELL MAINTENANCE

vi) TOOL TRAP

The wireline winch drum is fitted with a counter to enable the operator to determine the depth of the tool string in the well. However, there is always the risk of stripping the wire from the rope socket as it is winched into the stuffing box and to prevent the tool string from dropping back into the well, a tool catcher may be fitted. Mechanical and hydraulic tool traps are available, some of which latch on to the rope socket under wireline tension and some which close under the toolstring as it is raised into the lubricator assembly. The tool trap will be located either at the top or the bottom of the lubricator stack depending on the design.

vii) CHEMICAL INJECTION SUB

The chemical injection sub provides the means by which chemicals may be applied to the wireline. The chemicals are introduced into the sub assembly at a side port and are absorbed by replaceable felt pads which wipe the wire as it enters and leaves the well. Methanol is frequently used to combat icing whilst corrosion inhibitors provide the wire with some protection against hydrocarbon products produced from sour reservoirs.

2. WIRELINE TOOL STRING

The sketch shows a typical wireline tool string arrangement. The wire rope socket, jars and knuckle joint are present during all wireline operations and special tools are added to the basic string depending on the task in hand.

The wire is fed through the stuffing box and lubricator prior to addition of the tool string and the complete assembly is then mounted on top of the wireline valve. The wireline toolstring is then lowered through the Christmas tree swab valve, master valve or valves and into the production tubing.

The components of a tool string may be described as follows:

i) WIRE ROPE SOCKET.

This device joins the tool string to the wire. The wire is threaded through the body and around a plug which is then pulled into a taper. Slip rope sockets are a variation on this basic theme and are designed to break at 50%, 60%, 70%, 80% or 90% of the wires tensile strength should the tool string become jammed. Failure of the wire simplifies tool string retrieval, the rope socket being designed to facilitate the attachment of a latching recovery or fishing tool.

ii) JARS

The simplest form of jar is a heavy length of tube and various weights are available. They are used to *jar* free tools stuck in the tubing (as in the previous case describing the rope socket) and to provide extra weight to assist latching operations.

Mechanical jars can only be used to jar in a downward direction. Hydraulic jars are required for jarring in an upward direction. Their operation is self-explanatory.

iii) KNUCKLE JOINT

The knuckle joint contains a ball and socket which provides flexibility within the tool string.

Offshore Engineering

iv) WEIGHT BARS

Weights are used to increase the effectiveness of jars and weigh from 5 to 33 pounds (2.25 to 15 kilograms).

v) PULLING/RUNNING TOOLS

Pulling refers to the removal of a device such as a mudline safety valve or pressure test plug from the production tubing. The installation process is known as running a tool and relies on a latching mechanism for emplacement.

3. WIRELINE OPERATIONS

A multitude of operations can be carried out using wireline equipment and four of the basic tasks will now be described.

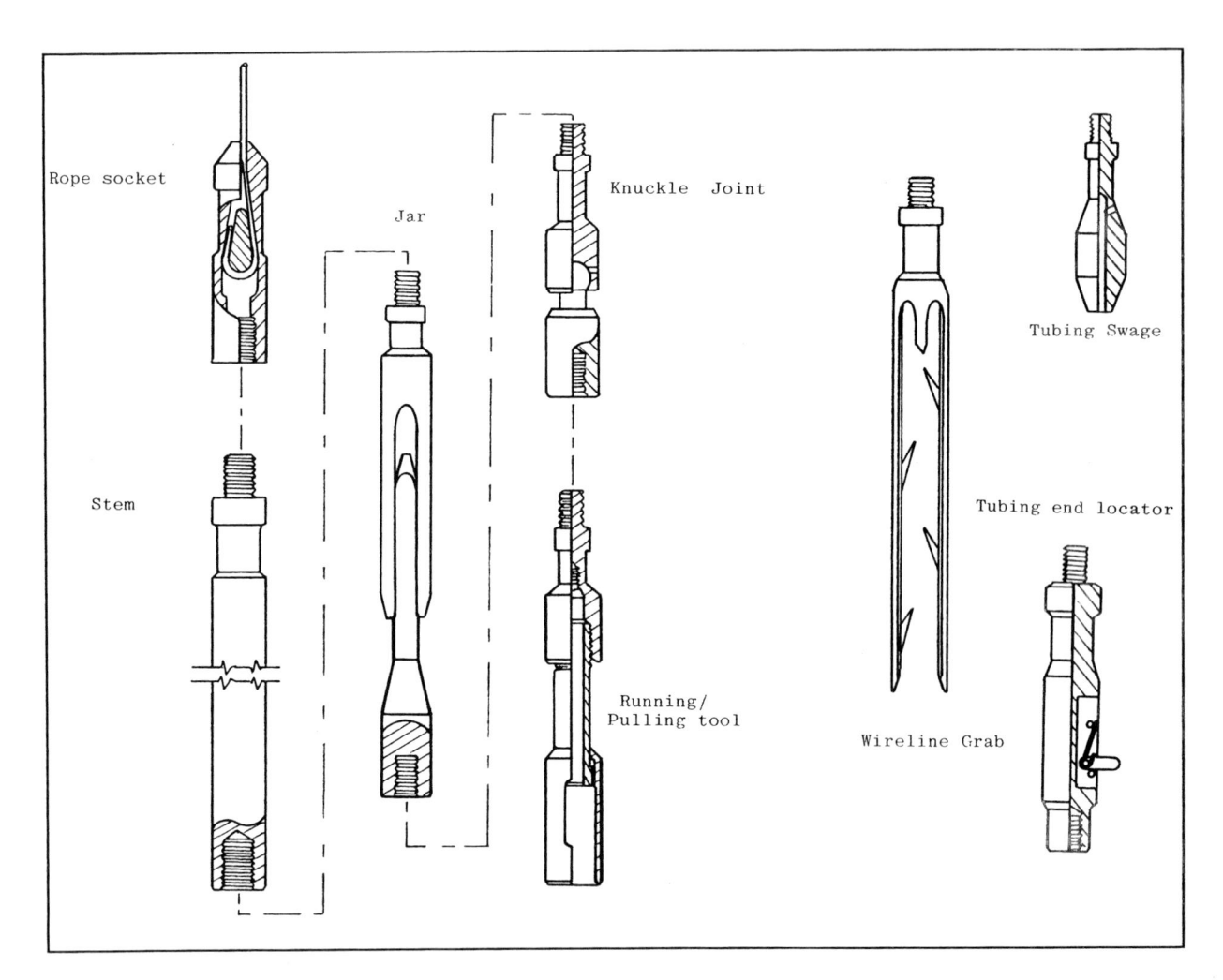

WIRELINE TOOL STRING **SELECTION OF TOOLS**

25 – The offshore industry works around the clock, 365 days a year. This night time shot shows the steam hammer of the Heerema Balder in action driving the foundation piles through the legs of the Shell/Esso 49/26 Foxtrot jacket. Inset – one of the Shell/Esso 48/19a Clipper jackets and foundation piles during transportation. Note the comparative size of the men on the barge, the unpainted lower steel structure and the depth graduations on the piles.

26 – Top   The roll-up of the outer legs of the Britannia jacket at the fabrication yard in Dragodos, Spain. For ease of fabrication, the jacket frames are welded in the horizontal position and when finished, are rolled up in to the vertical by using a combination of cranes and winches. The frames are then restrained by guide wires whilst the infill bracings are installed.

26 – Bottom   The massive 20,000 ton Britannia jacket being guided on to the transportation barge. (Pictures reproduced with kind permission of Conoco (U.K.) Limited and Chevron U.K. Limited.)

27 – Barge launch of deep water jacket.

27 – Barge transportation of shallow water jacket.

28 – Oil and gas deposits always seem to be found at locations where it is either very hot or cold. Here we see the construction crew in a particularly relaxed mood as they finish off the securing of the foundation piles inside the jacket legs, the location being the Arabian Gulf.

The Upper picture shows grout pouring from vent holes, the grout filling the space between the jacket leg and pile. The lower picture, of a different jacket, clearly shows the steel shim plates which locate the pile in the top of the jacket leg. The white paint provides a contrast whilst the welds are subjected to a magnetic particle inspection to determine the soundness of the welds.

*29 – The Lowering of one of the Shell/Esso 48/19a Clipper jackets
into the shallow waters of the southern North Sea by heavy lift crane vessel the Heeremac DB102.
Inset – installation of the jacket complete. Piles driven, welded, cut, dressed and awaiting the topside structure.*

*30 – The transportation to site of the Shell/Esso 49/19a Clipper topside structure.
The Heeremac DB 102 can be seen waiting in the background.*

31 – Absolutely ideal conditiions to lift 3,000 tonnes and demenstrate the capabilities of a semi-submersible heavy lift crane vessel (SSCV). The platform is the Amoco North West Hutton, physically one of the largest oil installations in the North Sea and the crane vessel is the Heerena Balder. Date 1981.

32 – At 11,000 tonnes, the single integrated deck of the Conoco/Chevron Britannia is the heaviest offshore lift ever recorded, the strain being taken by the Heeremac heavy lift crane barge the DB102. Date 1998

i) PERFORATING

Perforating is carried out on a new well prior to bringing it into service and on existing wells which have been subjected to modifications in the vicinity of the hydrocarbon bearing formation or pay zone. It is a wireline process by which a string of explosive charges is lowered through the production tubing to the depth at which the gas or oil producing formation exists. When the explosive charges are detonated the concentrated blast punches holes through the casing and into the formation. The numerous small holes produced permit hydrocarbons to flow into the production tubing and up to the rig. The casing may be perforated over a distance of several hundred feet.

ii) LOCATION PLUGS AND VALVES

Two of the most frequent operations carried out using wireline equipment are the change of wireline retrievable *mudline* or *sub-surface safety valves* (MLSV or SCSSV) and to a lesser extent the fitting of blanking plugs in the production tubing.

The *production tubing* is the innermost tubing string in the well through which hydrocarbon products are conveyed from the reservoir to the *Christmas tree*. As the tubing is installed it is fitted with a number of *landing nipples* (shouldered joints) set at different depths which are designed to accommodate the sub-surface safety valve, and on occasions, steel blanking plugs. The blanking plugs permit pressure tests to be carried out on the production tubing when searching for leaks, and provide the means by which the well can be isolated prior to changing a Christmas tree.

The installation of sub-surface safety valve is a straightforward wireline operation. They are attached to a suitable running tool and latched into the landing nipple, a pulling tool being used to separate the latch prior to removal.

iii) FORMATION SURVEYS

a) BOTTOM HOLE SURVEYS

An accurate assessment of the condition of a production well can be made by lowering self contained pressure and temperature sensing devices into the hydrocarbon bearing formation on the end of a wireline toolstring. The data obtained enables the reservoir engineers to determine whether or not the well is producing its maximum potential.

b) LOGGING

During drilling operations the hydrocarbon bearing potential of the formation can be determined by *logging* the hole using electric wireline operated tools. A log is a record of readings generated by a logging tool which transmits electric, sonic or radioactive signals into the formation. The return signals are computer analysed and a profile of the logged zone developed. Subsurface formations possess certain measurable characteristics which help to identify the type of rock structure, and the hydrocarbon content.

iv) WELL MAINTENANCE

Typical maintenance operations carried out with wireline equipment are the cleaning of the production tubing of wax, clearing blockages with a swage and removing sand. Special tools have been developed for each of these operations as shown in the sketches.

There is also a fishing tool for recovering broken wires. The term *fishing* may also be used to describe the process of latching on to the sub-surface safety valves or plugs prior to their removal.

v) **WIRELINE TRACTOR**

Deploying wireline equipment in highly deviated or horizontal wells presents certain problems because the equipment relies on gravity to take the tools to their destination. Wireline tools can be deployed with coiled tubing equipment which can push the tools downhole but more recently a down hole tractor has been developed for pulling wireline into horizontal wells.

The tractor system is controlled from a communications and motor control system located at the surface. An electrical AC motor integrated into the wireline tractor drives a hydraulic pump which forces the tractor wheels onto the side of the production tubing or casing. Once a pre-set pressure is reached the wheels start to rotate and drive the tractor forward. When the oil pressure is released the wheels retract into the tractor body and an electrical relay shifts the unit between the tractor and logging mode. Conventional wireline tools are used.

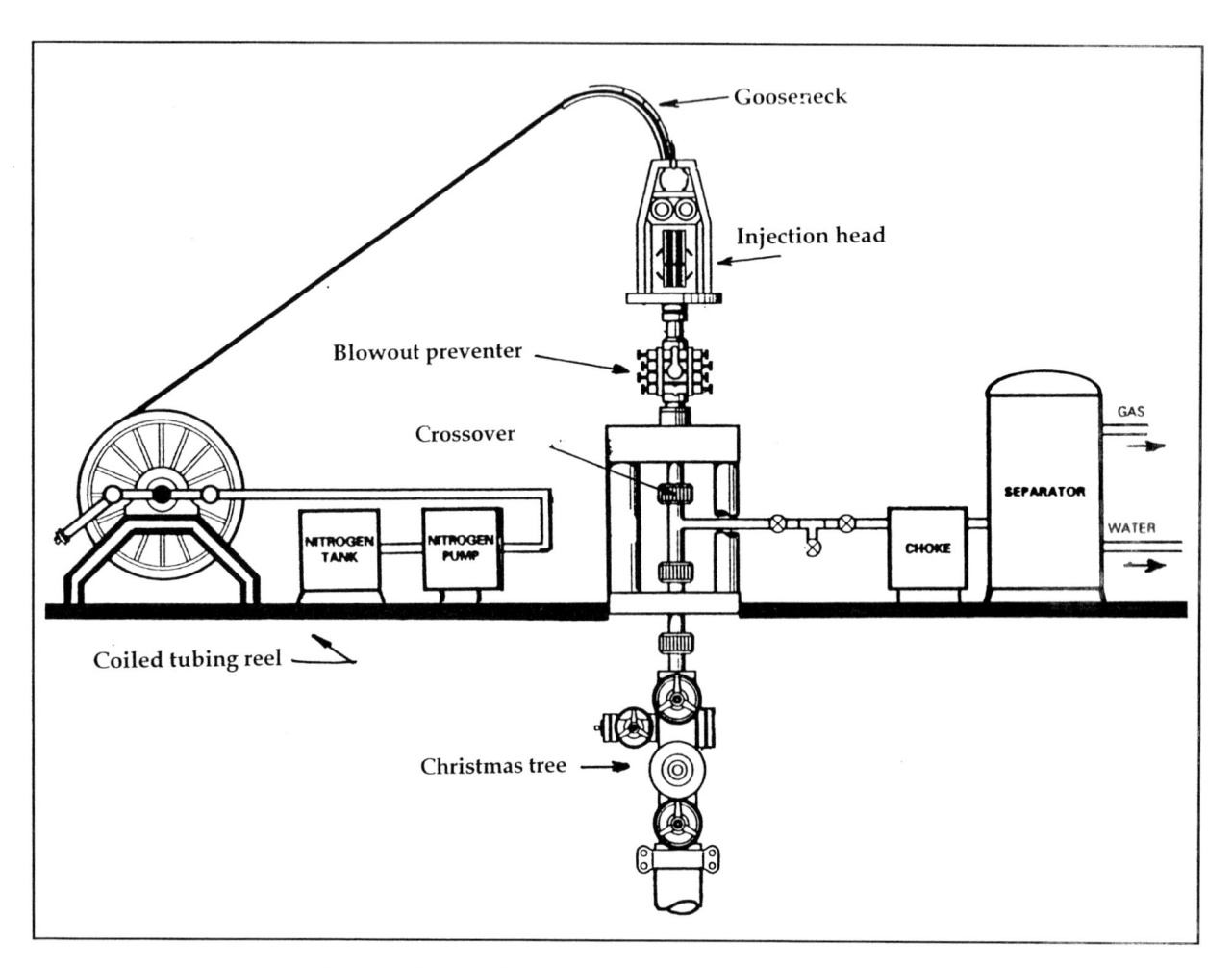

COIL TUBING EQUIPMENT SCHEMATIC

Part 2. *WORKOVER OPERATIONS*

During the life of a well which maybe 25 years or more, situations inevitably arise which call for the modification or repair of equipment associated with the *production tubing*. Mechanical problems such as leaks in *packers*, tubing joints and cement seals develop and the depletion of hydrocarbons may result in changes in reservoir conditions which can restrict output, typically caused by the build up of sand. These problems can be corrected by suspending production on the affected well and carrying out a *workover job*. Workovers involve the use of highly sophisticated equipment normally supplied by contractors who specialise in a particular field of operations. They may be loosely divided into two categories, that is workovers which deal with the rectification of defects in the production tubing and workovers designed to improve or stimulate the hydrocarbon bearing formation in the vicinity of the well bore.

TYPICAL WORKOVER OPERATIONS

1. WELL STIMULATION

Three of the more common well stimulation processes will now be discussed. They may be carried out using containerised equipment temporarily located on the platform, or from purpose built well stimulation ships, depending on the size of the task in hand, the operations involving the injection of gases (nitrogen) or liquids (acids) into the well. Coil tubing equipment may be used for some of the smaller tubing cleaning operations, the equipment outwardly resembling the wireline valve and lubricator stack used for wireline operations. However, it is a small diameter steel tube ($\frac{3}{4}$–$3\frac{1}{2}$ inch diameter) that passes through a coil tubing BOP into the Christmas tree and on into the well and not a wireline tool string. The inherent flexibility of the tubing enables it to be coiled around a large drum in continuous lengths of up to 20,000 feet when not in use, hence the derivation of the expression, *coil tubing*.

i) FRACTURING (FRACING)

Fracturing is used to improve the flow of hydrocarbons into the well bore. It involves the injection of oil or water based fluids at very high pressures into the rock formation. This induces cracks and opens fissures thus enlarging and extending the drainage area through the rocks.

Sand, or artificially manufactured proppants may be injected with the fracture fluid so that when the hydraulic pressure is released, the solid particles will remain in place holding the fissures apart as the liquid drains away.

ii) ACIDIZING

Acid can be used to remove damage near the well bore in sandstone and carbonate formations and to produce long linear flow channels away from the well.

Matrix acidizing involves injecting acid below fracture pressure to dissolve deposits restricting the perforations.

Fracture acidizing involves injecting acid at a pressure greater than fracture pressure. The acid forces the fracture line apart and dissolves a flow path which remains when the fluid pressure subsides and the fracture closes once more.

Additives are mixed with the acid solution to prevent the corrosion of steel, and the formation of sludges which could block the tubing perforations.

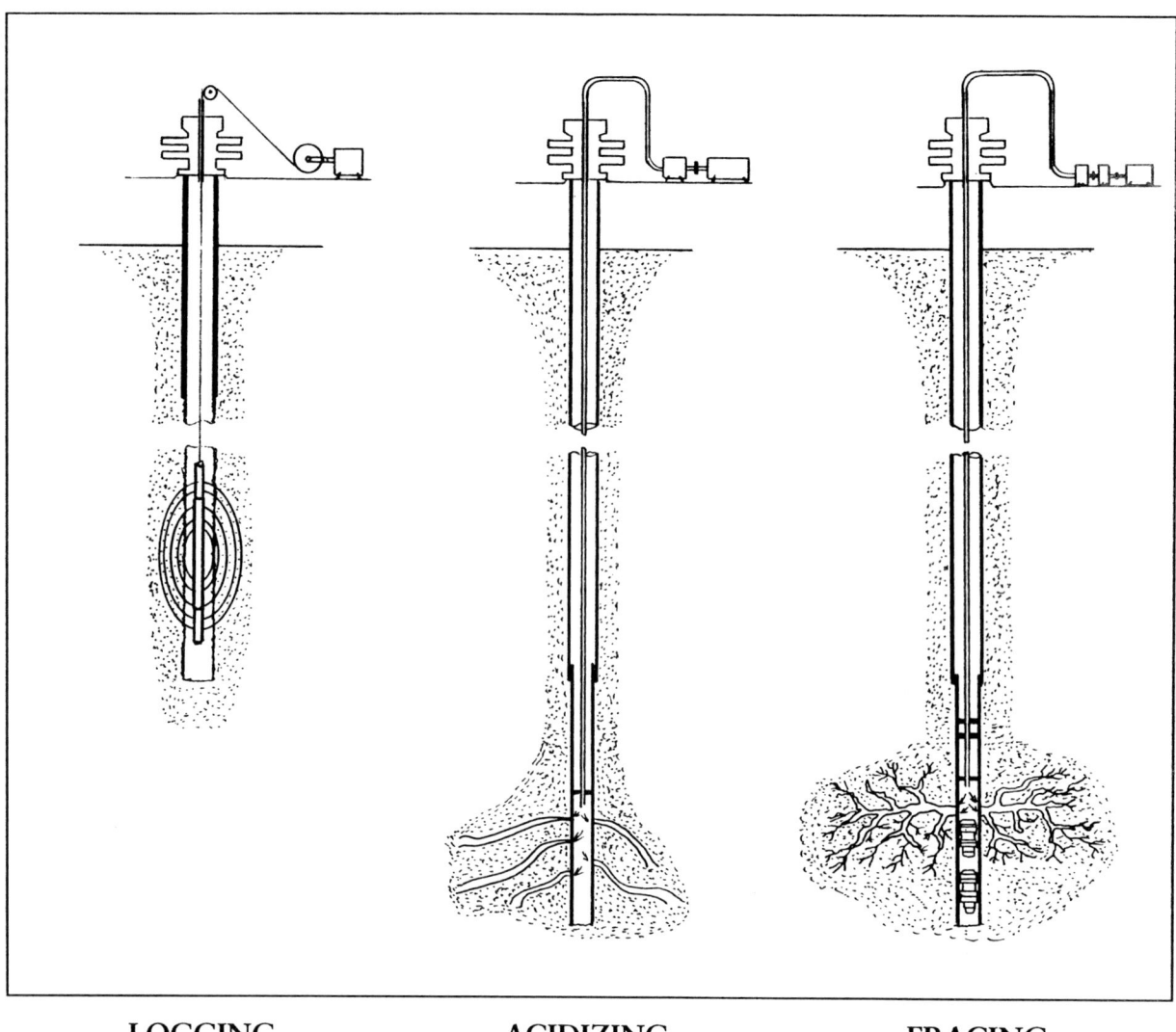

| LOGGING | ACIDIZING | FRACING |

iii) SAND CONTROL

Sand can congregate at the well base and block the casing perforations. Small quantities can be scooped out using wireline equipment but where large quantities exist, more drastic action is required.

A popular solution is to install a slotted screen in the open hole, or in the casing. Gravel packing the outside of the screen will permit the passage of hydrocarbons whilst hindering the passage of sand.

Another solution involves the precipitation of resin between the sand grains in order to bond them together and prevent the flow and further accumulation of sand.

2. WELL MODIFICATIONS

The removal of the production tubing is a major operation requiring the use of the drilling derrick, or a mobile workover rig. Problems which necessitate the removal of the production tubing tend to be fairly serious in nature and for safety reasons the well must be killed before work commences. This can be done most effectively by displacing the contents of the production tubing with brine in order to overbalance the hydrostatic head of the formation. With the well thus suitably *killed* the Christmas tree can be removed and replaced with a BOP stack prior to work commencing.

Some of the more basic well workover operations will now be described.

i) PRODUCTION TUBING DAMAGE

The production tubing is exposed to the full force of reservoir pressure and all the associated contaminants which accompany the hydrocarbons such as formation acids and abrasive sands. Consequently, it is not unusual for problems to develop as the well ages and leaks occasionally occur which require the replacement of all, or at least part of the production tubing.

ii) SUB-SURFACE (MUDLINE) SAFETY VALVE REPLACEMENT

Mudline safety valves may be of the wireline retrievable, or tubing retrievable type. As the name suggest the tubing retrievable valves require the removal of the production tubing in order to facilitate their replacement. Whilst the mudline safety valves tend to be very reliable in operation it is unlikely that they will operate reliably throughout the life of a well which could be 25 to 30 years. As the valve is located only 30 metres (100 feet) below the sea bed replacement does not necessitate the removal of the entire production tubing but it is still quite a considerable task.

iii) PRODUCTION TUBING REPLACEMENT

As the reservoir ages it is common practise to replace the production tubing with one of a smaller diameter in order to increase the velocity of well fluids. This ensures that impurities such as water and sand are brought to the surface where they can be separated and disposed of by the process plant rather than risk their accumulation in the production tubing which will eventually kill the well.

iv) PACKER REPLACEMENT

Quite a common problem that afflicts older wells is leakage of the packer which seals the production tubing into the innermost casing string. The remedy involves the removal of the production tubing so that the defective packer can be machined out. This is achieved using a milling cutter on the end of a drill string. With the casing suitably cleaned and all drilling debris removed the production tubing can be replaced and a new packer activated.

v) CEMENT SQUEEZE

Occasionally leaks develop in the intermediate casing strings at tubing joints, or where corrosion has taken place. This may manifest itself as a reduction in well output as reservoir products are lost from the well into the formation, or it may be indicated by increased water production when water seeps from the formation into the well. Again the solution necessitates the removal of the production tubing in order to isolate the damaged casing with blanking plugs. Cement is then pumped under pressure between the plugs from whence it is squeezed out through the leak zone where it will set to form a permanent repair. A similar procedure is used for blanking off redundant perforations.

vi) EOR EQUIPMENT

Some enhanced oil recovery aids such as deepwell pumps and *gas lift* equipment are installed or run with the production tubing. Therefore any repairs or modifications will involve a workover in order to remove the production tubing.

Appendices

APPENDIX I – STRUCTURAL STEEL

INTRODUCTION

It is beyond the scope of this book to enter into a discussion on the design of an offshore installation. Historically the only readily available publication on the subject is produced by the American Petroleum Institute, API RP 2A — *Recommended Practice for Planning, Design and Construction of Fixed Offshore Platforms*. The Code, which is now in its 20th Edition dates back to 1969 and was developed largely as a response to the destruction of 28 platforms by hurricanes in the Gulf of Mexico in 1964 and 1965.

Having established a suitable design, the structure must then be constructed in accordance with the requirements of a recognised construction specification. Traditionally structures have been constructed in accordance with the requirements of the American Welding Society's AWS D1.1 — *Structural Welding Code*. In common with the majority of engineering reference books which originate in the USA, AWS D.1.1 provides a wealth of information on all aspects of welding (including weld procedures and welder qualification tests) and includes guidance on the selection of materials, workmanship, inspection and non-destructive examinations. More recently in Europe, the *Engineering Equipment and Material Users Association* have developed a standard aimed specifically at the North Sea, namely EEMUA 158, *Construction of Fixed Offshore Installations in the North Sea*. The EEMUA 158 defines the basic requirements for fabrication work on primary structures whilst recommending that steel work other than primary be fabricated in accordance with AWS D1.1.

If not referenced specifically, EEMUA 158 and AWS D1.1 will almost certainly provide the basis from which Company specific structural construction specifications are prepared and as such will be encountered repeatedly on new building, repair and modification projects.

1. STRUCTURAL STEELWORK

The component parts of the jacket and topside support framework of an offshore installation are categorised as either *primary* or *secondary structure* and the steels used for construction are similarly identified as *primary* and *secondary steels*.

i) PRIMARY STRUCTURE

Primary structural members are those which contribute to the overall integrity of the installation and include such items as jacket legs, piles, braces, columns and beams. Where these components are subjected to significant tensile loads they must be constructed from steels with guaranteed through thickness properties and are often classed as Special Structural primary steels, examples of which are nodes and pad eyes/ears.

ii) SECONDARY STRUCTURE

Secondary structural members can be described as *those components whose contribution to the integrity of the structure as a whole is not significant* and includes items such as deck stringers, crane pedestal supports and stairways. Failure of a secondary structural member is *unlikely to affect the overall integrity of the installation.*

Some of the more readily identifiable structural components are:

Can — Large cylinder formed from rolled plate and longitudinally welded. Welded end to end they are used in the construction of jacket legs.

Bracing — Tubular support member arranged to provide stiffening to the jacket legs. Maybe manufactured from cans, seamless or welded pipe depending on the diameter required.

Node — Intersection of bracings. May be welded or cast with cast steel components preferred for modern deep water jackets.

Plate girder — "I" section beam manufactured from steel plate and used as a main deck support, normally referred to as a PG.

Pad eye/ear — Fabricated or cast steel lifting attachment through which (eye) or around which (ear) wire strops are attached for the lifting of the module during installation.

2. MATERIAL SPECIFICATIONS

Structural steels used for the construction of offshore installations will normally comply with one of four specifications.

i) BRITISH STANDARD 4360 – WELDABLE STRUCTURAL STEELS

For many years BS 4360 was used extensively both in the UK and abroad for the manufacture of structural steels with properties suitable for use offshore. Before being withdrawn it was used to provide the basis of a European Standard with applications specific to offshore structures, EEMUA No. 150.

ii) EEMUA No. 150 – STEEL SPECIFICATION FOR OFFSHORE STRUCTURES

The major European purchasers of engineering products formed an organisation known as EEMUA, Engineering Equipment and Material Users Association. The EEMUA No. 150 specifies parameters for the selection of structural steels and provides comparison tables for British, European and American material grades. In Europe it is now the preferred specification for primary grade structural steels.

iii) ASTM – AMERICAN SOCIETY FOR TESTING AND MATERIALS

The ASTM Organisation provide a range of specifications covering low strength carbon steels through to high strength low alloy steels, most of which are suitable for offshore structures.

iv) **API SPECIFICATION 5L – LINE PIPE**

Whilst Spec. 5L pipes are manufactured primarily for pipeline service, they are also used extensively as jacket bracings. The project fabrication specification must be studied to confirm the suitability of the materials and establish if any additional mechanical tests are required prior to use.

3. MATERIAL PROPERTIES

The specifications listed above contain numerous grades of steel which may be compared firstly by tensile strength, secondly by yield strength and thirdly by impact (charpy) properties. The materials are generally referenced using a number to indicate tensile or yield strength and a letter to give an indication as to the resistance to brittle fracture. For example EEMUA 355D has a minimum *yield strength* of 355 N/mm^2 with impact values recorded at $-20°$C, this material is equivalent to the old BS 4360 50D material where the 50D designation refers to the minimum *tensile strength* of 500 N/mm^2 (actually 490 N/mm^2) and the D again refers to the achievement of acceptable impact values at $-20°$

In order to simplify the selection of steels for a particular project the fabricator or client may tabulate materials into groups or types depending on their application. The numbering system employed has not been standardised so the previously mentioned EEMUA 355D may be listed as a type 2 material by one fabricator whilst a competitor lists it as a type 4. To avoid confusion, all materials should be selected in accordance with the requirements listed by the latest revision of the project specific fabrication specification.

The massive 20,000 ton *Britannia* jacket being guided on to the transportation barge. Ballast water is discharged throughout the loading operation to maintain the barge on an even keel, level with the quay. Note the mud mats and pile clusters at the base of each leg and the temporary flotation tanks (unpainted) located on the top half of the structure which will be used to assist in the positioning of the jacket.
(Picture reproduced with kind permission of Conoco (U.K.) Limited and Chevron U.K. Limited.)

Offshore Engineering

The massive 20,000 ton Britannia jacket being guided on to the transportation barge. Ballast water is discharged throughout the loading operation to maintain the barge on an even keel, level with the quay. Note the mud mats and pile clusters at the base of each leg and the temporary floatation tanks (unpainted) located on the top half of the structure which will be used to assist in the positioning of the jacket.

(Picture Reproduced with kind permission of Conoco (U.K.) Limited and Chevron U.K. Limited)

APPENDIX II – WELDING

Welding is an extremely complex subject and the preserve of the Welding Engineer. However, an appreciation of the more significant specifications and terms used by the industry will greatly enhance the readers understanding of construction activities.

1. TERMINOLOGY

QA/QC – Quality assurance and quality control are not welding terms in themselves but wherever fabrication and welding are carried out there will almost certainly be a Quality System in evidence. A Quality Assurance System in accordance with the British Standard 5750, (or an equivalent such as ISO 9000) is required to ensure that the end product is manufactured in accordance with the specification.

WP – The purpose of the Weld Procedure test is to prove the compatibility of the welding consumables with the construction materials, to confirm the suitability of the welding process and equipment and to establish the parameters for items such as the geometry of the weld preparation, preheat/interpass temperatures and heat inputs. On completion of welding the test piece is subjected to both non-destructive and destructive examinations in accordance with the specification to which approval is sought. Whilst some specifications such as AWS D1.1 *Structural Welding Code* permit the use of pre-qualified procedures this practice tends to be frowned on by the offshore industry. The weld procedures are invariably qualified prior to commencement of construction using samples of materials and consumables that will eventually be used in production.

The American Petroleum Institute publication API RP 1104 – *Standard for Welding Pipelines and Related Facilities*, is a self-contained reference document which deals specifically with welding procedures. It provides a step by step guide on how the tests should be conducted, tested and documented and is highly recommended reading. The AWS D1.1 – *Structural Welding Code* is also very informative.

WPQR – The Welding Procedure Qualification Record exists as a permanent record of the welding procedure test and contains all the relevant information categorised as either *essential* or *non-essential variables*, and includes the non-destructive and destructive test records.

WPS – The Weld Procedure Specification is prepared from the WPQR and provides the welder with specific guidance on how to approach a particular weld. It includes information pertaining to the size and type of electrode, the number of runs, the voltage, amperage and polarity to be used. A number of WPSs may be prepared within the limits of the essential variables permitted by the WPQR to cover such items as variations in component size, thickness and material type.

SWIS – The Site Welding Instruction Sheet is associated with EEMUA 158 *Construction Specification for Fixed Offshore Structures in the North Sea* and fulfils an identical function to the WPS.

Welder Qualification Test – To ensure that a welder possesses the necessary skills to produce a sound weld he must complete a test piece which is representative of conditions that will be encountered during production welding. Whilst the test weld must comply with a particular WPS it will normally permit the welder to weld a considerable number of similar WPSs.

Preheat — Preheat is used primarily to slow the cooling rate of a welded component in order to reduce the shrinkage stresses, prevent excess hardening and loss in ductility. It also assists in the diffusion of hydrogen and thus reduces the likelihood of underbead cracking. Preheat may be applied by propane gas torch, or by the attachment of electric resistance mats. The degree of preheat is governed largely by the thickness of the material and can be calculated from the Carbon Equivalent formula.

$$\text{Carbon Equivalent, Ceq} = \%C + \frac{\%Mn}{6} + \frac{\%Ni}{15} + \frac{\%Mo}{4} + \frac{\%Cr}{4} + \frac{\%Cu}{13}$$

The approximate preheat temperatures based on carbon equivalent values are listed below but in all cases the fabrication specification should be consulted to establish the exact requirements.

up to 0.45%	preheat optional
0.45% to 0.60%	95°C to 205°C (200°F to 400°F)
0.60% and above	205°C to 370°C (400°F to 700°F)

Interpass temperature — In a multi-pass weld, the temperature of the weld metal prior to commencement of the next pass is referred to as the interpass temperature and it must be controlled within pre-determined limits if weld defects are to be avoided.

Post-Weld Heat Treatment (PWHT) — Post-weld heat treatment or stress relieving is carried out to reduce the stresses caused by welding. For general fabrication work it is only required when material thickness exceed 30–40 mm but for some process pipework it may be required for material thickness as thin as $\frac{3}{4}$ inch (19 mm).

Typically it involves heating the component to a temperature of 590°C to 650°C (1100°F to 1200°F), holding or soaking for a period of 1 hour per inch of material thickness, followed by controlled cooling to 400°C before cooling slowly under insulation to ambient temperature. PWHT may be effected by the attachment of electric resistance heating mats by insertion into a furnace.

As with preheat requirements, the fabrication specification should be consulted to establish the material thickness which require stress relieving and the means by which the duration of heat application and the temperature are to be calculated.

SMAW — Submerged Manual Arc Welding using conventional stick electrodes is the most widely used of the various arc welding processes. The stick or welding rod consists of a steel wire covered with a flux designed to decompose under the heat of the arc and generate a shielding gas which will prevent the formation of oxides and nitrides in the molten weld pool.

SAW — Submerged Arc Welding or *sub-arc* is the term used to describe an automatic or semi-automatic welding process. The welding wire is fed from a reel whilst powdered flux is poured over the weld preparation to ensure that the arc remains submerged. Metal deposition may be up to 10 times that achieved with SMAW welding. The process is only suitable for downhand (flat) welding and whilst circular components can be welded, they must be rotated with the initial pass or *root run* being manually welded.

FCAW — Flux Cored Welding is a development of SMAW welding, can be used in all positions and exists in both automatic and semi-automatic forms. The flux is contained within the core of the welding wire which is in effect a stick electrode turned inside out. The welding wire is supplied in reels and is automatically fed through a gun so that the operator can weld continuously without the inconvenience of the frequent interruptions required to change electrodes.

GMAW — Gas Metal Arc Welding is a semi-automatic welding process normally referred to as MIG (metal inert gas) and utilises either carbon dioxide, or an argon and helium gas mixture as a shielding agent in preference to a traditional flux. The gas is circulated around the welding wire which is fed from a reel. MIG welding permits high deposition rates and is suitable for use on steels, alloy steels, aluminium, copper and magnesium.

GTAW — Gas Tungsten Arc Welding is normally referred to as TIG (tungsten inert gas) welding and employs a non combustible tungsten electrode to create an arc against the work piece whilst the filler wire is fed either manually or automatically into the molten weld pool. Protection against oxidation is provided by an argon, or argon/helium shielding gas mixture. Steels, aluminium and copper alloys can be TIG welded and whilst a relatively slow process it is particularly well suited to the welding of small diameter pipes where root penetration must be kept to a minimum.

Carbon Air Arc Gouging — Air arc gouging is frequently employed in preference to grinding for the removal of large quantities of weld metal prior to repairs or modifications to the weld preparation. The gouging process employs a carbon electrode to create an arc against the work piece and the subsequent pool of molten metal is removed by a jet of compressed air.

Plasma Arc Cutting — This is a relatively new process which is used primarily for the cutting out of steel components from plate in the workshop. An electric arc is generated against the work piece by a tungsten electrode, the arc subsequently being fuelled by a high velocity, super heated plasma gas stream which may be a mixture of argon and hydrogen, or nitrogen and hydrogen. The process is both quicker and more accurate than oxy-acetylene cutting and may be used on materials such as aluminium alloys in addition to ferrous materials. The finished article can normally be welded without further surface preparation.

Consumable — The term *consumable* refers to the items consumed during the welding process namely electrodes, filler wires, fluxes and shielding gases. With the exception of the gases, consumables are generally manufactured in accordance with AWS (American Welding Society) specifications.

2. WELDING SPECIFICATIONS

The standards listed below represent those which will be most frequently encountered during the construction of an offshore installation and its equipment.

i) BRITISH STANDARDS

B.S. EN 287 — Approval testing of welders for fusion welding (formerly BS 4871).

B.S. EN 288 — Specification and approval of welding procedures for metallic materials (formerly BS 4870).

B.S. 5135 — Metal arc welding of carbon and manganese steels.

EEMUA 158 — Construction specification for fixed offshore structures in the North Sea.

ii) AMERICAN STANDARDS

AWS D1.1 — Structural welding code.

ANSI/ASME B31.3 — Chemical plant and petroleum refinery piping.

ASME IX — Boiler and pressure vessel code.

3.0 SUPERVISION

The welding of all critical components such as pressure vessels, pressure piping and structural steelwork should be supervised by suitably qualified welding inspectors and subsequently examined by qualified NDT (non-destructive testing) technicians. The subject of supervision is dealt with in more detail in Appendix III.

For those who wish to expand on their knowledge of welding *The Procedure Book of Arc Welding* should be obtained. This moderately priced publication is produced by the Lincoln Electric Company, Cleveland Ohio, USA and is readily available in Europe. Over the years it has become a standard work of reference for welding both on and offshore and will be found sharing shelf space with the more formal specification wherever welding is carried out.

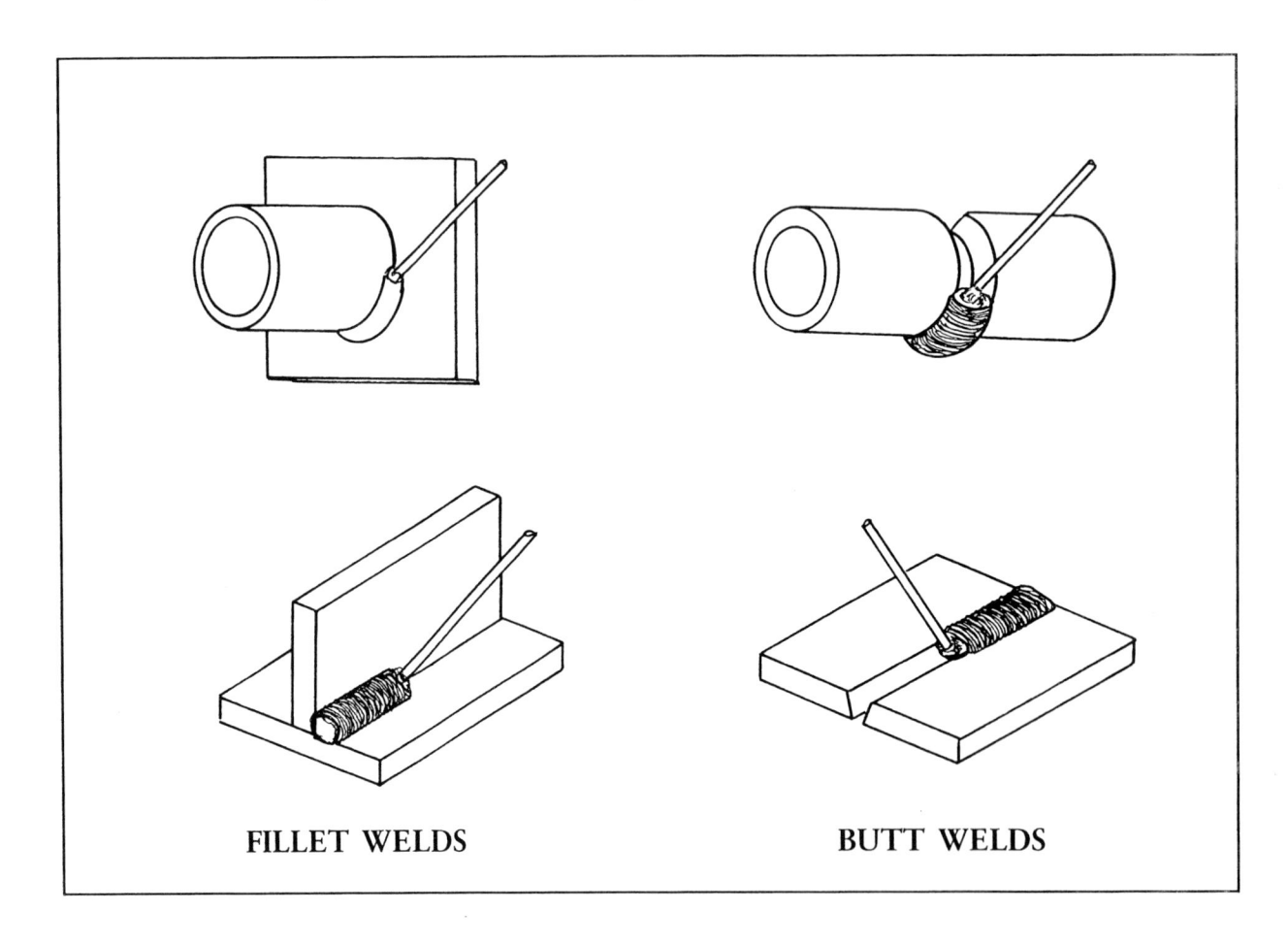

FILLET WELDS **BUTT WELDS**

APPENDICES

**MANUAL FLUX CORED WELDING.
SUB-SEA TEMPLATE PILE GUIDES**

Offshore Engineering

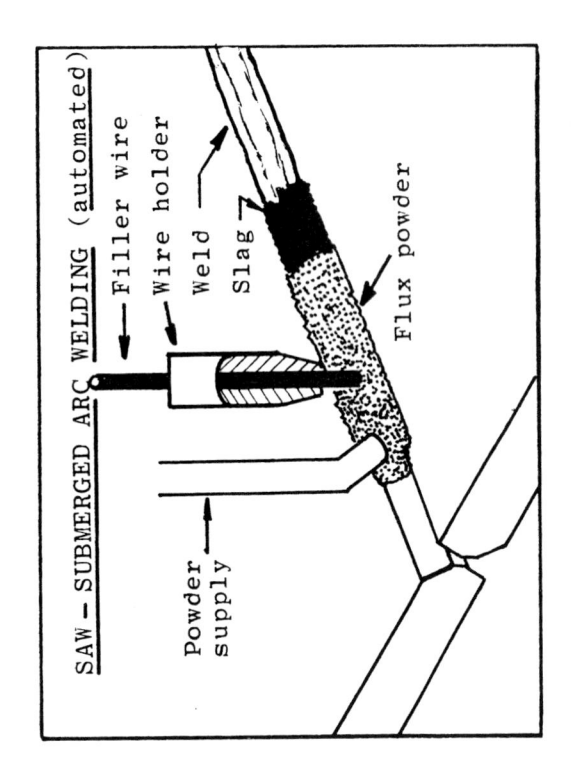

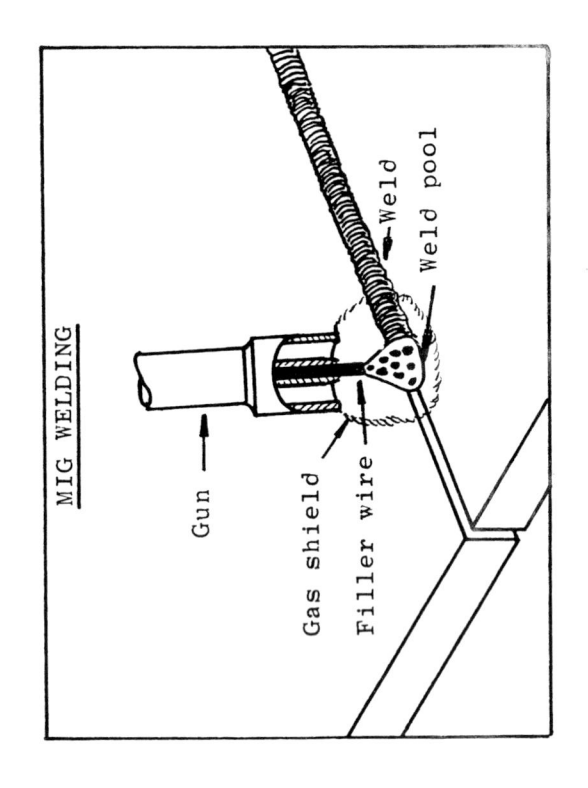

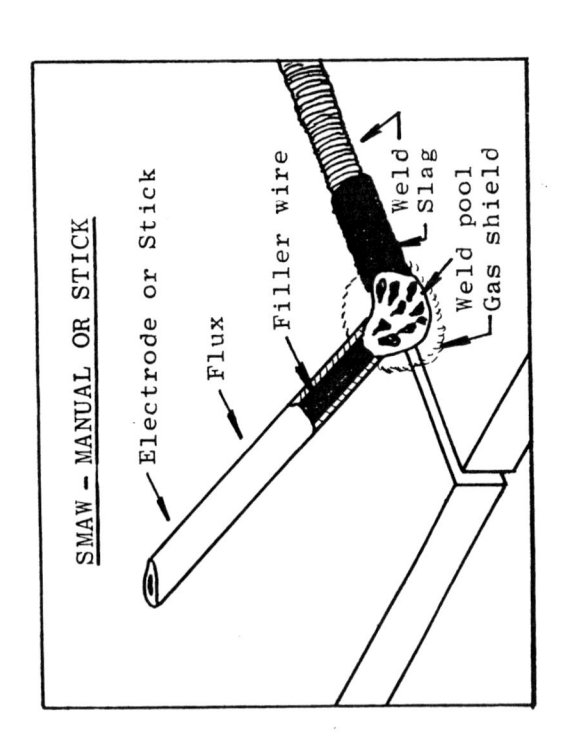

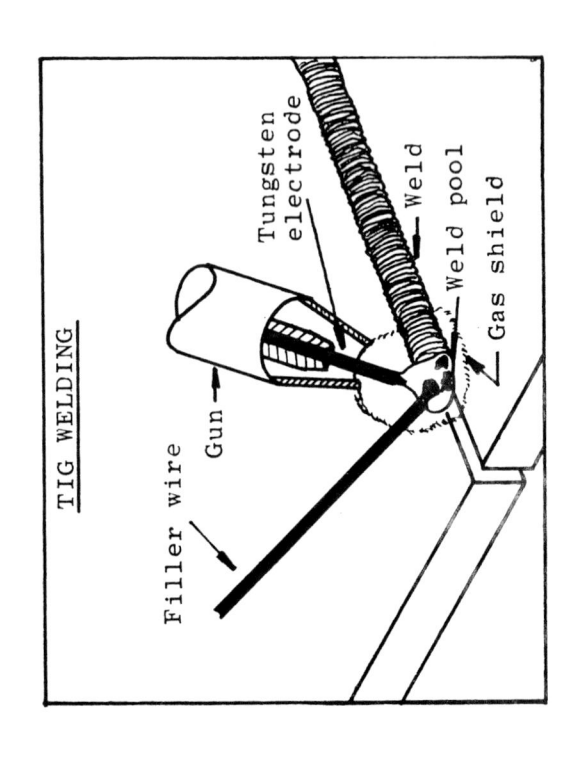

SAW – SUBMERGED ARC WELDING (automated)
- Filler wire
- Wire holder
- Weld
- Slag
- Flux powder
- Powder supply

MIG WELDING
- Weld
- Weld pool
- Gun
- Gas shield
- Filler wire

SMAW – MANUAL OR STICK
- Electrode or Stick
- Flux
- Filler wire
- Weld
- Slag
- Weld pool
- Gas shield

TIG WELDING
- Tungsten electrode
- Weld
- Weld pool
- Gas shield
- Filler wire
- Gun

WELDING PROCESSES

APPENDIX III — NON-DESTRUCTIVE EXAMINATION

Non-destructive examination (NDE), or non-destructive testing (NDT) provides the means by which components may be examined for defects during the construction of an offshore installation. The type and extent of NDE required will be determined by the fabrication specification and EEMUA 158 *Construction Specification for Fixed Offshore Structures in the North Sea* provides particularly informative reference tables on the subject.

There are a number of NDE processes, the more important of which will now be described. They can be divided into two categories, the first being used for the detection of defects on the surface of the component whilst the processes which fall into the second category are used for the detection of internal defects.

1. SURFACE INSPECTION TECHNIQUES

i) CLOSE VISUAL INSPECTION — CVI

Close, or thorough visual inspection with the naked eye is the most basic and still the most important of all the non-destructive examination techniques and one which should never be overlooked.

ii) MAGNETIC PARTICLE INSPECTIONS — MPI

MPI is the most sensitive and reliable of techniques for the detection of surface and near surface defects in ferro-magnetic materials (iron and steel). To enhance defect definition the object under examination is sprayed with white background paint prior to magnetisation with either a permanent magnet, or an electromagnet. The component is then liberally coated with particles of iron oxide in either liquid or dry powder form. The particles align themselves with any discontinuities in the magnetic field and the defects appear as a clear black line which is emphasised by the white background paint.

NOTE:– Permanent magnets should be restricted to the examination of materials of thickness less than 6 mm ($\frac{1}{4}$ inch). An AC electromagnetic yoke is preferred for thicker sections due to the more intense nature of the field it produces.

iii) EDDY CURRENT INSPECTION

Like MPI, eddy current test equipment relies on the disturbing effect that defects have on a magnetic field. A test probe containing an electromagnetic coil is traversed over the article under inspection and the response signal is displayed on an oscilloscope. One of the main advantages of eddy current testing is that it can be used on painted components, although some of the offshore paint systems are particularly thick and may require removal prior to testing.

iv) DYE PENETRANT EXAMINATION

Whilst not as effective as MPI or eddy current inspections *dye pen* is used extensively on both ferrous and non-ferrous components. It involves the application of a red dye which is absorbed by capillary action into surface breaking defects. Surplus dye is removed prior to the application of a developer which highlights the dye retained within the defect.

2. VOLUMETRIC INSPECTION TECHNIQUES

i) ULTRASONIC EXAMINATION — U/T

An ultrasonic examination can be carried out on most homogeneous materials and when performed by a skilled operator it represents the most sensitive of volumetric NDT techniques. It employs an ultrasound signal to search for defects, the results generating a signal which can be interpreted on an oscilloscope. Ultrasonic testing is particularly well suited to the location of cracks and laminations in fusion welded components.

For many years radiography has been preferred to ultrasonic testing primarily because a permanent record of the examination is produced in the form of a radiograph. However, with the advances in computer power, automated ultrasonic testing equipment is now available which can produce both a pictorial report and an electronic record. Ultrasonic testing is therefore challenging the domain of radiography being both quicker, and requiring fewer safety precautions due to the absence of radiation.

ii) RADIOGRAPHY — R/T

Industrial radiography involves the passing of ionising radiation through an object in order to record the results on radiographic film. It is most frequently used for the examination of circumferential welds and often referred to as *bombing*. The radiation may be generated by an X-ray machine or by a radioactive isotope such as Iridium or Cobalt. Whilst X-ray equipment generally produces superior results particularly on thin components, isotopes are preferred for site work because they are more compact and do not require an external power source or a supply of coolant.

3. NDE OPERATOR QUALIFICATIONS

All welding should be examined by suitably qualified welding inspectors and when required, further examined by suitably qualified NDE technicians. All inspection work should be carried out in accordance with written procedures which are generally based on specifications prepared by National Standards Organisations (NSO) such as the British Standards Institute, ASME, AWS and EEMUA.

There are two universally recognised organisations which operate schemes for the assessment and certification of welding inspectors and NDT technicians, the Welding Institute in the UK and the American Society of Non-Destructive Testing (ASNT) in the USA.

The Welding Institute examine inspectors and technicians on behalf of an independent management board and issue CSWIP/PCN certificates to successful candidates. The certificates are valid for five years and are transferable from one company to another.

CSWIP — Certification scheme for weldment inspection personnel.

PCN — Personnel certification in non-destructive testing.

The ASNT scheme differs from the Welding Institute's because the qualification certificates are issued by the company that employs the technician. The qualifications are specific to the type of work carried by the issuing company and are not transferable to other companies.

SNT T 1A — ASNT certificate categorised into three levels of qualifications.

4. NON-DESTRUCTIVE EXAMINATION SPECIFICATIONS

The British Standards (BS) listed below have formed the basis of non-destructive examinations in the UK since the offshore industry started in the 1960's. Whilst the majority of the standards have been withdrawn and replaced by the BS EN standards in recent years, the old standards are likely to be referenced for many years to come, particularly where installations are being modified and upgraded. They are also still referenced in other documents which have yet to be updated such as the British Standards relating to pressure vessels and pipeline welding.

i) **BRITISH STANDARDS**

BS 2910	Radiographic examination of fusion welded circumferential butt joints in steel pipes.
BS 2600	Radiographic examination of fusion welded butt joints in steel.
BS 3923	Ultrasonic examination of welds.
BS 5289	Code of Practise. Visual inspection of fusion welded joints.
BS 6443	Penetrant flaw detection.
BS 6072	Methods for magnetic particle flaw detection.

BS EN 1435:1997   Non-destructive examination of welds. Radiographic examination of welded joints.

BS EN 1714:1998   Non-destructive examination of welded joints. Ultrasonic examination of welded joints.

BS EN 970:1997   Non-destructive examination of fusion welds. Visual examination.

BS EN 571-1:1997   Non-destructive testing. Penetrant testing.

iii) **AMERICAN STANDARDS**

ASME V — Boiler and Pressure Vessel Code

Article 2 — Radiography
Article 4 — Ultrasonic inspection
Article 6 — Liquid penetrant flaw detection
Article 7 — Magnetic particle inspection
Article 8 — Eddy current testing
Article 9 — Visual inspection

AWS D1.1 — Structural Welding Code
(See Chapter 6 — Radiography/Ultrasonic.)

APPENDIX IV – UNITS OF MEASUREMENTS

1 tonne	= 7.5 barrels (crude oil)
1 long tonne	= 1.0165 tonnes
1 barrel	= 42 US gallons (35 imperial gallons)
	= 0.1589 cu metres
1 cubic metre	= 35.31 cubic feet.
1 billion cubic metres	= 0.83 million tonnes of oil equivalent.
1 cubic metre of gas	= 0.36 Therms
1 tonne of fuel oil	= 406 Therms
1 Therm	= 100 cubic feet = 1000,000 British Thermal Units

METRIC CONVERSIONS

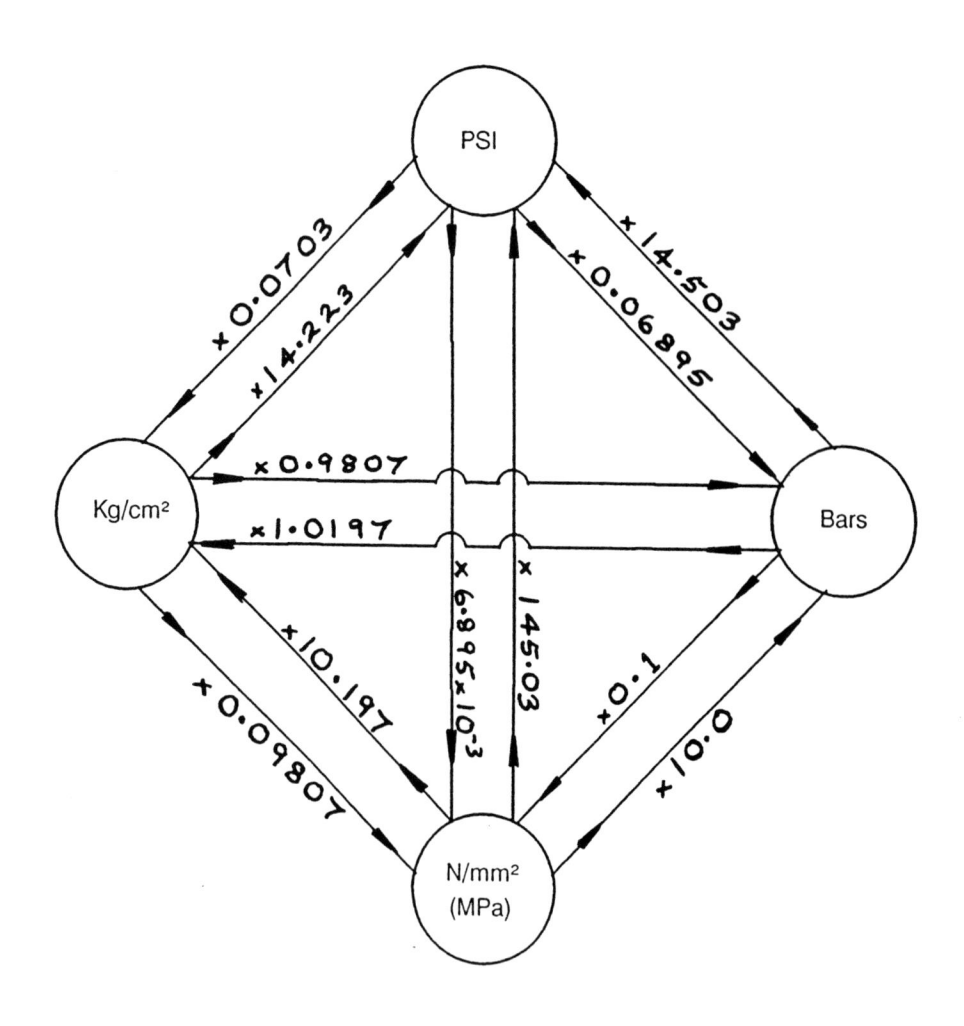

Conversion diagram between PSI, Kg/cm², Bars, and N/mm² (MPa):

- Kg/cm² ↔ PSI: ×0.0703 / ×14.223
- PSI ↔ Bars: ×0.06895 / ×14.503
- Kg/cm² ↔ Bars: ×0.9807 / ×1.0197
- Kg/cm² ↔ N/mm² (MPa): ×0.09807 / ×10.197
- PSI ↔ N/mm² (MPa): ×6.895×10⁻³ / ×145.03
- Bars ↔ N/mm² (MPa): ×0.1 / ×10.0

APPENDIX V — TABLE OF LINE PIPE DIMENSIONS

PLATFORM PIPING — DIMENSIONS AND MAXIMUM WORKING PRESSURES
ASTM A106 GRADE B PIPE
[Design temp. −29/204° C(−20/400° F)]

Nominal Size Inches	Outside Diameter Ins/mm	Nominal Wall Thickness Inches	Nominal Wall Thickness mm	Weight Class	Schedule No.	Working Pressure psig
2	2.375	0.218	5.34	XS	80	2489
	60.3	0.344	8.74	—	160	4618
		0.436	11.70	XXS	—	6285
2.5	2.875	0.276	7.01	XS	80	2814
	73.0	0.375	9.52	—	160	4194
		0.552	14.02	XXS	—	6850
3	3.500	0.300	7.62	XS	80	2552
	88.9	0.438	11.13	—	160	4123
		0.600	15.24	XXS	—	6090
4	4.500	0.237	6.72	STD	40	1140
	114.3	0.337	8.56	XS	80	2276
		0.438	11.13	—	120	3149
		0.531	13.49	—	160	3979
		0.674	17.12	XXS	—	5307
6	6.625	0.280	7.11	STD	40	1206
	168.3	0.432	10.97	XS	80	2062
		0.562	14.28	—	120	2817
		0.719	18.26	—	160	3760
		0.864	21.95	XXS	—	4660
8	8.625	0.322	8.18	STD	—	1098
	219.1	0.406	10.21	—	60	1457
		0.500	12.70	XS	80	1864
		0.594	15.08	—	100	2278
		0.719	18.26	—	120	2838
		0.812	20.63	—	140	3263
		0.875	22.22	XXS	—	3555
		0.906	23.00	—	160	3700
10	10.750	0.365	9.27	STD	40	1023
	273.0	0.500	12.70	XS	60	1485
		0.594	15.08	—	80	1811
		0.719	18.20	—	100	2252
		0.844	21.43	—	120	2700
		1.000	25.41	XXS	140	3271
		1.125	28.58	—	160	3737

Nominal Size Inches	Outside Diameter Ins/mm	Nominal Wall Thickness Inches	Nominal Wall Thickness mm	Weight Class	Schedule No.	Working Pressure psig
12	12.750	0.375	9.52	STD	—	888
	323.9	0.406	10.31	—	40	976
		0.500	12.70	XS	—	1246
		0.562	14.28	—	60	1425
		0.688	17.48	—	80	1794
		0.844	21.43	—	100	2258
		1.000	25.40	XXS	120	2730
		1.125	28.58	—	140	3114
		1.312	32.00	—	160	3700
14	14.000	0.375	9.52	STD	30	807
	355.6	0.438	11.07	—	40	971
		0.500	12.70	XS	—	1132
		0.594	15.08	—	60	1379
		0.750	19.10	—	80	1794
		0.938	23.80	—	100	2304
		1.094	27.00	—	120	2734
		1.250	31.75	—	140	3171
		1.406	35.71	—	160	3616
16	16.00	0.500	12.70	XS	40	988
	406.4	0.656	16.66	—	60	1345
		0.843	21.41	—	80	1780
		1.031	26.19	—	100	2225
		1.218	30.95	—	120	2675
		1.437	36.51	—	140	3212
18	18.000	0.500	12.70	XS	—	876
	457.2	0.562	14.25	—	40	1001
		0.718	19.10	—	60	1319
		0.937	23.80	—	80	1771
		1.156	29.30	—	100	2232
		1.343	34.92	—	120	2632
20	20.000	0.325	9.52	STD	—	499
	508.8	0.500	12.27	XS	—	669
		0.812	20.62	—	60	1098
		1.031	26.19	—	80	1404
		1.280	32.52	—	100	1760
		1.500	38.10	—	120	2079
		1.750	44.50	—	140	2446
		1.968	50.00	—	160	2776
24	24.000	0.375	9.52	STD	—	415
	609.6	0.500	12.70	XS	—	556
		0.562	14.25	—	30	625
		0.687	17.48	—	40	768
		0.968	24.60	—	60	1091
		1.218	30.95	—	80	1383
		1.531	38.89	—	100	1753
		1.812	46.03	—	120	2093
		2.062	52.38	—	140	2399
		2.343	59.38	—	160	2752
30	30.000	0.375	9.52	STD		
	762.0	0.500	12.70	XS		
36	36.000	0.375	9.52	STD		
	914.4	0.500	12.70	XS		

INDEX

A
Abandonment 31
Abandonment cap 295
Accommodation 7, 23, 114, 130
Acidizing 311–312
ADS 215
Air arc gouging 321
Air gap 102–220
Air lock 132
ALARP 57
Ammonium Bisulphate 202
Anodes 219
Annulus 204, 292, 298
American National Standards Institute (ANSI) 161
American Petroleum Institute (API) 161
    API RP 2A 315
    API RP 2G 161
    API RP 14E 161
    API RP 14C 141, 145, 161
    API RP 16E 268
    API RP 14H 290
    API RP 53 268
    API RP 64 268
    API RP 500 140
    API RP 520 161
    API RP 521 161
    API RP 1104 319
    API RP 1111 161
    API Spec. 5L 167, 317
    API Spec. 6A 153, 168, 189, 290
    API Spec. 6AF 173
    API Spec 14A 161
    API Spec 14D 290
American Society of Mechanical Engineers (ASME) 161, 319–321
American Society for Testing and Materials (ASTM) 316
American Society for Non-Destructive Testing (ASNT) 319–320
American Welding Society (AWS) 315–327
ANSI standards for fittings 168–169
ANSI standards piping 168–169, 321
Appraisal 244
Assessment 71
ASME Boiler and Pressure Vessel Code 162–163
ASTM standards for materials 167–169

B
BASEEFA 133–138, 140
Bad oil 159, 184, 185
Barytes 19, 255
Bathy corrometer 221
Battery room 114
Bell housing 257
Black water 153
Blowout 263, 267–269

Bombing 326
BOP, 247, 263–268, 292, 303, 312
    annular 263–268
    ram 263–268
    stack 267, 269–272
Box 255
Bracing 316
Brine 248, 249
British Standards Institute 161
British Standards – Electrical 139
British Standards – NDE 164, 327
British Standards – Welding 162–163, 164, 321
BS and W 184
Bubble point 199
Bubble valve 191
Bunker 148
Butane 190
Butt weld 165

C
Caisson 24
Calorific value 43
Can 316
Carbon Dioxide 110, 181, 197
Carbon Equivalent 320
Casing, 24, 254, 259, 272, 295
    hanger 292, 295
    head 267, 291
    intermediate 263, 269, 291
    slips 291
Cat line 253
Catenary anchor leg mooring (CALM) 15
Cathodic protection 219
CE Mark 90
Cellar deck 24
Cement, 244, 249
    squeeze 313
CENELEC 90
Certificate of Fitness (C of F) 48
Certifying Authority (CA) 48, 105, 121, 217
Charpy 317
Check valve union 307
Chemical injection 307
Choke 178, 288
Choke and kill 272, 267
Christmas tree 177, 286, 293–295, 309, 312
    horizontal 289
Class 165
Coalescer 88, 184
Coast Protection Act 46
Coil tubing 244, 248, 311
Communications 73, 114
Compressed air, 152
    instrument 152
    utility 152

Compression 179, 190
Compressor,
  air 152
  gas 190
Concrete structures 5
Condensate 147, 155, 156, 178, 179, 184, 185
Conditioning 248, 249
Conductor, 24, 244, 263–264, 269, 292, 295
  guide frame 24
Connection 261–263
Contactor 178, 191
Continental Shelf, 26
  Act 46
Control room 73, 105, 112, 116
Corrosion,
  allowance 167
  inhibitor 148
  pipeline 189
  sweet 181
Cronox 148
Crossover 303
Crown block 252
Crude oil 189, 190
Cullen 20, 49, 51
Cunifer 167

D
DCR 51, 67–69, 76–80
Dead crude 189
Deaerator 202
Decompression chamber (DDC) 213
Deepwell pump 204
Degassing 158, 185, 202
Dehydration 191–192
Department of Energy 49, 104
Depletion drive 199
Derrick,
  drilling 23, 26, 201, 202, 249–253, 312
  man 259
Desalination 148
Dew point 189, 191
Diesel fuel 149
Differential settlement 220
Distillation 196
Diverter 247, 263–264, 268, 272
Diving, 133, 214
  safety information sheet 94
  spider 217
  umbilical 133, 213, 215
Diving support vessel 213, 214
Doghouse 253
Drains, 88
  caisson 185
  closed 88
  drains 88
  hazardous 153
  non-hazardous 152–153
  open 152
  sump 158
Draw works 253

Drill,
  bit 254, 257, 261, 263, 267, 275
  collars 254
  floor 261
  pipe 19, 253, 261, 263–264
  ship 7, 263, 269
  stands 254, 261, 263
  string 253, 311
  string tension 271
Driller 259
Drilling, 7, 14, 263–268, 269
  deviation 275
  directional 275
  horizontal 275
  motor 275
  mud 245, 247, 255–258, 267
  pre-drilling 295
  underbalanced 258
Dry powder 112
Duplex 167
Dye-penetrant 165
Dynamic positioning 7

E
Eddy current 325–327
EEMUA 315–325
Electrical equipment 133–138–140
Emergency disconnect package 269
Emergency shutdown (ESD) 73, 141–145
Emergency shutdown valve (ESDV) 87, 145
Enclosed compartments 132
Enhanced oil recovery (EOR) 149, 187, 190, 199–212, 229, 293, 313
Escape routes 51, 75
ESD 73, 141–145
ESDV 87, 144, 145
Ethane 192, 196, 198
European Directive 90
Evacuation, escape and rescue (EERA) 21, 74, 119–124
Expander 195
Exploration 7, 242
Export 179, 189, 190

F
Fast rescue craft (FRC) 21
Fatigue 218
Filtration 202
Finger board 253, 261, 263
Fire and gas 105
Fire, explosion, emergency response (FEA) 123
Fire-fighting, 103–113
  A-60 113–119
  active 105
  B-15 114, 117
  deluge 107
  detectors 105
  equipment 105
  extinguishers 110
  foam 108
  guidance notes 103–104

INDEX

F
Fire-fighting *(contd.)*
  H-120 115
  halon 110
  hoses 107
  loop 141, 298
  machinery spaces 114
  main 106
  monitors 108
  passive 113–119
  pump 106, 151
  remote stops 112
  sprinkler 108
  test, hydrocarbon 113–117
    standard 113–117
  water 151
Fishing 287
Fixed steel structure 2–4, 26–30
Flanges, 168–169
  ANSI 168
  AISI 168
  raised face 168
  ring joint 168
Flare, 148, 185, 187, 190, 191, 197
  stack 23
Flexible joint 269
Flexible riser and mooring system (FRAMS) 15
Float collar 249
Flooded member 219
Flotation unit 88, 185
Floating production system (FPS) 8–15, 221
Floating production, storage and off-loading (FPSO) 8–15
Fog horn 128–129
Fracking 311
Fracturing 312
Frangible bulb 108
Fuel gas 147, 148, 181, 190, 191

G
Galvanic 219
Gas,
  associated 43, 189, 190
  blanket 249
  distribution 43
  fuel 147, 148, 181, 190, 191
  lift 190, 202, 204, 293, 313
  power 147
  re-injection 190, 202
  sale 43
  stripping 196, 198, 202
  wet 148, 185–186
Gaskets, 169
  ring joint 169
  spiral wound 169
Gate valve 288
General alarm 119
Generators, gas engines 23
Geneva Convention 46
Geosteering 278–279

Glycol, (see also MEG and TEG) 147, 148, 178, 191–192, 197
  lean 191
  regeneration 191–192
  rich 191
  sweetening 197
Glycolic acid 192
Goal setting 52
Goose neck 255, 257
Gravel pack 312
Gravity base structure (GBS) 5
Grease injection head (HGT) 305
Grey water 153
Grout 27–30
Guidance Notes,
  construction 314
  fire 104
  life saving appliances 116
Guide base 269
Guide shoe 249

H
Halon 110
Hay pulley 305
Hazan 57
Hazardous,
  areas 130–133, 138, 147
  drains 153
  equipment 133, 138–139
Hazop 57
Health and Safety at Work Act (HSWA) 47, 50
Health and Safety Committee (HSC) 47
Health and Safety Executive (HSE) 47, 93–94, 99, 103, 119, 163
Heat exchanger 178, 189, 192, 195
Heating and ventilating 132
Heave compensation 269
Heavy lift 7
Helideck 23, 108, 128–129
Helifuel 152
Heliox 213
Holding tanks 159
Horizontal Christmas tree 207, 289
HVAC 132
Hydrate 148, 181
Hydrocarbon 240, 267
Hydro-cyclones 159–160, 185
Hydrogen sulphide 163, 184, 197, 200
Hydrotest 172–173

I
Identification panels 128–129
Independent. competent person (ICP) 61
Ingress protection 140
Isotopes 326
IMO 123
Institute of Petroleum (IP) 161
Interpass temperature 319
Intrinsically safe 130

333

J

J tubes 24
Jacket 4, 5, 24, 26–30, 218–220, 315
Jack-up 14, 252, 263, 269
Jar 306
Joints 244, 295
Joule-Thompson 195

K

Kelly 253, 255, 261,
Kick 262
Kick out (ko) drum 158
Kill 313
Knock out 147
Knuckle joint 307

L

Lamination 326
Landing ring 291
Leak test 172–173
Licence, exploration 240
Lifeboats 123
Life saving,
  appliances LSA 119–124
  buoys 125
  jackets 125
  plan 122
  rafts 124
Limitation 48
Lincoln, Arc Welding 322
Liner 247
Live crude 189–190
Logging 309
Lubricator 305
LWD 278

M

MAE 63
Magnetic particle inspection (MPI) 165, 218, 219
Major accident event 63
Major survey 49
Mantis 217
MAPD 87, 144
MAR 51, 81–84
Maintenance 89, 92
Marginal fields 4, 221, 244
Marine growth 151, 216, 219
Materials 167, 315
MEG 148, 181
Metering 177, 179, 189
Methane 130, 192, 196, 198
Methanol 148, 181
Microsub 217
Mineral Workings Act (MWA) 46
Moonpool 214
Monkey board 253
Montreal Protocol 110
Mouse hole 253, 261
MSS 169

Mud mats 24, 30
Mud pulse telemetry 280
Mudline, 219
  safety valve (MLSV) (SCSSV) 141, 247, 288, 297–298, 309, 313
  suspension system (MLSS) 295
Multi-lateral 281–282
Multiphase,
  meter 233–234
  pump 208, 232
MWD 278

N

National grid 44
National Association of Corrosion Engineers (NACE) 161, 163
Navigation aids, 126–127
NGL 147, 184, 189, 191, 195–196
Nippled 247
Node 217, 315
Non-destructive examination (NDE) 163, 325–327
Non-destructive testing (NDT),
  dye penetrant 325–327
  eddy current 325–327
  isotopes, cobalt 325
       iridium 325
  radiography 325
  ultrasonic 325
  visual 325
  X-ray 325

O

Offshore Installation Manager (OIM) 82, 98
Offshore Safety Division (OSD) 49, 93
Oil,
  bad 184, 185
  crude 189, 190
  dead crude 189
  distribution 43
  live crude 189
  slop 184, 185
Oily water separator 88, 153
Open deck 132
Operations manual 99, 130
Oxalic acid 192
Oxygen, dissolved 202

P

Pack off 291
Packer 244, 299
Pad ears 315–316
Pad eyes 315–316
Partnering 39
Perforation 248, 309
Performance standards 60, 75
Permit to Work 54, 72, 83
Petroleum Engineering Division (PED) 93
PFEER 51, 70–76
Pig 181

P *(contd.)*
Piggy-back 148
Pile, 26–30, 220
   cluster 24
Pin 255
Pipe,
   draw 253
   lay 7
   rack 253
Pipeline 8, 43–44, 51, 84–88, 150, 189, 195, 219
   limits 85
Piper Alpha 20, 46, 181, 297
Piping, 165
   systems 147, 161
   design 163–165
Plate girder 316
POB 123–124
Polishing unit 88, 185
Post weld heat treatment (PWHT) 320
Potable water 148
Power gas 147
PPE 125
Pre-heat 320
Preservation 248
Pressure,
   test 172–173
   vessels 155–159, 163–165
Pressurisation 130
Process,
   area 23
Produced water 88, 153, 177, 178, 184, 185, 201
Production header 177
Production tubing 201, 202, 204, 247, 272, 293, 294, 297, 299, 303, 304, 311, 313
Propane 192
PSR 51, 84–88
Public address 119
Pumps,
   deepwell 204
   oil 189
   submergible 151
   submersible 151
PUWER 51, 88–91

Q
QRA 54, 56
Qualification 48
Quality,
   assurance 319
   control 319

R
Radio mast 23
Radiography 165, 326
Raised face 165
Rat hole 253
Reboiler, 192, 196
   glycol 191–192
Recompressor 195

Refrigeration 197
Reflux coil 192
Relief valves 147
Remotely operated vehicle (ROV) 215–216
Reservoir 241
Reverse osmosis 148
Ring joint 168
Riser, 24, 218, 263–264
   clamp 24
   drilling, marine 269, 272
   tensioning 271
Risk assessment 55
Rock dumping 219
Rotary hoses 255, 257
Rotary table 249–253, 259
Rough field 43
Roughneck 259
Roustabout 259

S
Safe area 130
Safety,
   Case 49, 51–53, 59, 63–67, 91, 99–102
   Critical element 59
   Management 54, 87
Sand,
   control 312
   removal 309
   sea 219
Satellite navigation 7
Saver sub 255
Scour 219
Seawater 149
Sea Gem 46
Seismic 242
Selexol 197–198
Semi sub 7, 9, 11, 263–264
Separator,
   condensate 156
   flash 156, 192
   production 155, 156, 177
   test 156, 177, 189
   vacuum 156
Sewage system 153
Scour 220
Scrubber 157, 178, 190
Shale shaker 257
Shuttle tanker 13
Side pocket mandrel 204
Single buoy mooring (SBM) 15
Single point mooring (SPM) 15
Single anchor leg mooring (SALM) 15
Skimmer 88, 181
Slick line 303
Slips 261–263
Slop oil 184, 185
Slug catcher 157, 179
SMS 54–57, 87

S *(contd.)*
Snubbing 264
Socket fitting 163
Sodium Hypochlorite 151, 202
SOLAS 104, 119, 120
Solution drive 199
Sour 163, 181, 197
Sphere launcher 181, 189, 191
Spider deck 24, 217
Spiking 189
Splash zone 217
Spud can 14
Spudding 14, 269
Stabilisation 196–197
Standby boat 20
Starting head 291
Statutory Instruments
  SI 289 48
  SI 743 PFEER 51, 70–76
  SI 738 MAR 51, 81–84
  SI 825 PSR 51, 84–88
  SI 913 DCR 51, 76–80
  SI 1029 144
  SI 2885 SCR 50, 63–69
  SI 2932 PUWER 51, 88–92
Stinger 8
Stripping 263–264
Stripping gas 193, 202
  column 193–195
Structural steelwork, 315–316
  primary 315
  secondary 315
  special 315
Stuffing box 306
Sub-sea,
  pressure boosting 229, 232
  processing 229–231
  template 26, 223–224, 228
  wells 221, 272, 295
  wellhead 15, 216, 229
SUBSIS 230–231
Suction boot 88
Suction drum 157
Supply boat 19, 148, 151
Surface Process Shutdown (SPS) 141–144
Surveys
  bottom hole 309
  formation 309
  swimaround 217
  underwater 217–219
Swab valve 303, 304
Swage 309
Sweet 163
Sweetening 184, 197–198
Swivel 255

T
TEG 148, 180, 192
Telemetry 23
Telescopic joint 270
Temporary refuge 123
TEMPSC 74, 96, 123
Tension leg platform (TLP) 4–6
Test,
  hydrostatic 172–173
  leak 172–173
  pressure 172–173
Thompson, William 195
Threaded fitting 165
Throttling 195
Tongs 261
Toolpusher 259
Tool trap 307
Topsides 26–30
Total Platform Shutdown (TPS) 144
Travelling block 252, 261–263, 271
Trip 252, 261
Tubing,
  hanger 247, 293
  head 293

U
UKCS 26
UKOOA 95
Ultrasonic examination 165, 219–220, 326
Umbilical 133, 213, 215
Underwriters Laboratory 140
Utility systems 147, 151–154, 165

V
Vacuum flash drum 198
Valve,
  ball 170
  butterfly 173
  check 173
  choke 171
  clack 170
  diaphragm 173
  gate 171, 288
  master 288
  non-return 170
  plug 171
  swab 288
  wing 288
Vent, 180
  LP 148
  stack 23
  systems 147
Ventilation 113
Verification 52, 58–59, 69
Visual inspection 325

W
Water,
  black 153
  drive 199
  grey 153
  injection 201, 202

INDEX

W

Water, *(contd.)*
  potable 151
Weight jars 306
Welding, 163, 319–326
  consumable 319–326
  flux cored 320
  inspector 326
  Institute 326
  MIG 321
  plasma 321
  procedures 163, 165, 319–326
  qualifications 165, 326
  specification 321
  stick 320
  submerged arc 320
  test 320
  TIG 321
Well, 245
  control 263
  fluids 263–264, 267
  horizontal 275, 277
  legislation 78–80
  maintenance 309
  multi-lateral 281–282
  stimulation 311
  test 242

Well, *(contd.)*
  verification 78
Wellhead, 177, 204, 215, 292–293
  area 23
  diverless 225
  guidelineless 225
  skid, control 288
  subsea 225–227
Wet gas 185–186
Whipstock 275, 282
Wildcat 15, 244
Wireline, 287
  braided 303
  electric 303
  equipment 312
  slick 303
  toolstring 299
  tools 307–309
  valve 305
Wire rope socket 307
Workover, 211, 315
  rig 315
Written scheme of examination 61, 75–76
WSE 75–76

Z

Zone 0/Zone 1/Zone 2 130–133, 138